Health Benefits of FRUITS AND VEGETABLES

The Author

Dr. Kanaya Lal Bhat, Professor (Retired) Division of Vegetable Science and Floriculture, Sher-e-Kashmir University Jammu (J&K), received his Ph.D. in Vegetable Science from HPKV, Palampur in 1994. Before his superannuation, he was teaching to M.Sc and Ph.D students at SKUAST –Srinagar and SKUAST –Jammu for about twenty years and has guided a number of students in the discipline of Vegetable production. He has numerous research papers and has written extensively in scientific papers and for the general public. Dr. Bhat has participared in several national seminars and workshops on vegetable production. The author has to his credit three books *viz.*, Vegetables: Untapped Potential (2007), Physiological Disorders of Vegetable Crops (2009) and Brinjal (2011).

Health Benefits of FRUITS AND VEGETABLES

— Author —
Dr. K.L. Bhat

2014
Daya Publishing House®
A Division of
Astral International Pvt. Ltd.
New Delhi – 110 002

ISBN 9789351301660

Published by : **Daya Publishing House®**
A Division of
Astral International Pvt. Ltd.
– ISO 9001:2008 Certified Company –
4760-61/23, Ansari Road, Darya Ganj
New Delhi-110 002
Ph. 011-43549197, 23278134
E-mail: info@astralint.com
Website: www.astralint.com

Laser Typesetting : **Classic Computer Services**, Delhi - 110 035

Printed at : **Replika Press Pvt. Ltd.**

PRINTED IN INDIA

Preface

India is the 2nd largest producer of fruits and vegetables in the world. Today, horticulture alone accounts for 30 per cent of India's agricultural GDP from 8.5 per cent of the cropped area.It is playing a pivotal role in strengthening the country's nutritional security besides generating employment avenues. Nutritional knowledge of whatever we take, especially of fruits and vegetables is in high demand and result is a greater need for nutritional information, conveniently available and easily understood. Fruits and vegetables have been recognized as a good source of vitamins and minerals. They have been especially valuable for their ability to prevent vitamin C and vitamin A deficiencies. *Fruits and vegetables give you high nutrition and low calories. They have enormous therapeutic value with their rich nutritional composition, high fibre and water content and boost immunity, improve stamina and protect us from several health problems.* Fresh fruit and vegetables that are in season are the best, as if they are eaten raw and as fresh as possible, valuable nutrients are less likely to be lost. Fruits and vegetables should be included in your diet to maintain a healthy and balanced body, which then functions optionally thus leading a longer life span. They are low in fat and high in fibre- may be associated with a reduced risk of some types of cancer and cholesterol.The unnecessory pressure on our hospitals as many of today's illnesses are related to diet and could be avoided if people looked after themselves and ate more fruits and vegetables. Admittedly, a huge amount of money could be saved and thus spent where it is most needed. Therefore, make sure to include green, orange, and yellow fruits and vegetables - such as broccoli, carrots, cantaloupe, berries and citrus fruits. These fruits are rich in vitamin C and vitamin A, which may help protect against various types of cancer and other diseases. A diet high in fruits and vegetables will ensure that your body gets the small amounts of sugar it needs, while providing you with far more health benefits than refined sugar. Fruit, veggies and nuts contain powerful anti-ageing chemicals to keep you youthful.

This book is intended to serve a nutritional dictionary, a quick reference book for layman as well as for professionals. Its purpose is to provide basic information on nutritional status of various fruits and vegetables available round the year and can be used at home or while shopping form the market.Therefore, an attempt has been made in this book, *Health Benefits of Fruiits and Vegetables*, to give as much information as is possible. I hope that new book will serve to increase public awareness of the exciting connection between nutrition, well being and longevity.

'Let us make fruits and vegetables our medicine'

K.L. Bhat

Contents

Introduction

Fruits and vegetables give you high nutrition and low calories. They have enormous therapeutic value with their rich nutritional composition, high fibre and water content and boost immunity, improve stamina and protect us from several health problems. Research shows diets containing substantial amounts of varied vegetables and fruits reduce chances of cancer by 20 per cent and of stroke and cardio-vascular disease by 60 per cent. Consumption of natural fruits rich in flavonoids helps body to protect from lung and oral cavity cancers. Diets low in fat and high in fibre-containing grain products, fruits and vegetables may be associated with a reduced risk of some types of cancer.

Different coloured fruits and vegetables contain different minerals, nutrients and antioxidants and therefore it is recommended that we consume a wide variety of fruit and vegetables in order to receive the benefits from the various types. For example, dark green leafy vegetables such as watercress, cabbage or spinach contain certain carotenoids that protect, delay and may prevent the onset of degenerative age-related eye diseases such as cataracts or macular degeneration. They are also rich in vitamins C and E, which are both very powerful antioxidants. This means that eating dark green vegetables daily could help to protect the body from developing cancerous cells and from suffering heart disease. Fruits and vegetables are a natural source of energy and give the body many nutrients that may help protect you from chronic diseases. Red, orange and yellow coloured fruits and vegetables such as melon, tomatoes, carrots and apricots contain lots of vitamins A, C and E, which all help to fight certain types of cancer and act by neutralising free radicals in the body.

As well as containing large amounts of vitamins A, C and E, fruits and vegetables are also rich in vitamins B and K plus minerals such as potassium, calcium, phosphorous, manganese and iron. As long as you are consuming lots of fruit and vegetables, it doesn't really matter whether they are raw, cooked, tinned, dried, fresh or frozen. But fresh fruit and vegetables that are in season are the best, as if they are eaten raw and as fresh as possible, valuable nutrients are less likely to be lost.

Nutritionists say that foods with high content of vitamin A (carrots), vitamin C (oranges, strawberries and broccoli) and vitamin E (peanuts, spinach, almonds, hazelnuts) optimise sperm count, whereas vegetables (selenium), tomato and watermelon (lycopene) help prevent oxidative damage to sperms. To avail maximum fruit nutrition benefits eat organic produce many 'wild' varieties of berries and 'tropical tree' fruits that still have not been treated with any kind of fertilizers or chemicals and can be readily purchased from the local market. Organic fruits tend to be smaller, they feature special flavour and richness in vitamin and minerals and are stuffed with numerous antioxidants.

There is unnecessory pressure on our hospitals as many of today's illnesses are related to diet and could be avoided if people looked after themselves and ate more sensibly and healthily. Admittedly, a huge amount of money could be saved and thus spent where it is most needed. Fruit and vegetables are packed full of goodness and often contain a number of essential vitamins and minerals that cannot be found in other types of foods or they may contain higher levels of these nutrients than other foods. They are made up of water, melons up to 94 per cent, which is also essential for the body and their skin and seeds contain plenty of fibre, which our body needs to help cleanse and rid itself of waste and toxins. Fibre is needed to keep bowel movements regular, lower cholesterol, prevent constipation, bowel cancer and other illnesses of the bowel and intestine such as diverticulosis.

Fruits and vegetables should be included in your diet to maintain a healthy and balanced body, which then functions optionally thus leading a longer life span. Fruits and vegetables of different colours contain diverse mixtures of phyto-nutrients (protective plant compounds). These can act as powerful anti-oxidants, protecting the body from harmful free radicals and helping to protect against certain chronic diseases such as cancer. Some fruits and vegetables are labelled as 'super foods' because they contain high concentrations of some phyto-nutrients, particularly antioxidants which appear to be beneficial to health. About 80 per cent of our food intake should be alkaline vegetables and alkaline fruits and no more than 20 per cent of the entire food intake should be acidic. Alkaline vegetables help maintain the ph balance of the body at 7.3. They can be included to form part of your diet and should be substituted for most of the acid food. The greener the vegetable, the more alkaline it is said to be. So when looking for alkaline vegetables, choose greens more. Vegetable juices make you feel active, lighter, your skin and hair feel better. Vegetable juices can be made from every vegetable, unlike fruit juices. They help you absorb all the nutrients from the vegetables and allow you to consume optional amount of vegetables in an efficient manner. A study conducted at the Baylor College of Medicine (USA) has revealed that drinking at least one glass of low sodium vegetable juice daily may help over weight people with metabolic syndrome lose weight (Anonymous, 2009). Metabolic syndrome is defined by a cluster of risk factors, including excess body fat in the midsection, high blood pressure, high blood sugar and abnormal blood lipids.

There are a variety of fruits that you can choose from if you do not drink water that much. The best pick is a watermelon which has the maximum water content in it.

Fresh fruits will give you the needed hydration to feel better and be healthier. Some more examples are: apples, kiwi, grapefruit, coconut, strawberries, and any type of berries. Eat plenty of foods rich in vitamin C such as blackcurrants, blueberries, broccoli, guava, kiwi, oranges, papaya, strawberries and sweet potatoes- all help to produce collagen that strengthens the capillaries that feed the skin. Similarly eat foods rich in vitamin E, such as almonds, avocado, hazelnuts, pine nuts, sunflower and corn oils. Zinc rich foods like pumpkin seeds help to repair skin damage. Foods rich in vitamin A help new skin to grow. Carrots are also high in beta-carotene which the body converts to vitamin A. Vitamin A is one of the most important nutrient for healthy skin. Skin with vitamin A deficiency appears dry and dull. Carrots are also rich in antioxidants to help prevent free radical skin damage, vitamin B6, C, K, biotin, thiamine, potassium and fibre. Sulphur rich foods such as onions and garlic help to keep the skin sooth. Eating a few dried apricots, rich in iron daily help improve skin tone. Blueberries rich source of vitamin E, C, manganese and soluble fibre and in antioxidants that inhibit free radicals from damaging skin.

The green fresh vegetables provide nutrients that help the body to stay hydrated and not get too heavy with other foods. It is better to eat the vegetables raw. Cooking them will take out the natural moisture and hydration. Greens like spinach, amaranth, fenufreek leaves, argula not only have high iron content, but are horders of Vitamin K that helps maintain strong bones and healthy joints. A study shows that older adults with ample blood levels of K were less likely to develop osteoarthritis, compared to a low-in-K control group. However, Vitamin K also helps with blood clotting, so if you're taking blood thinners, check with your doc before boosting your K intake. A plant based diet with a variety of fruits, vegetables, nuts, grains and beans is the best organic way to fight cancer. To prevent cancer, crucuferous vegetables *viz.* cabbages. Broccoli, Brussels sprout and cauliflower are important as they contain powerful anti-cancer molecules but are destroyed while boiling them, so steam briefly or stir-fry rapidly in a little olive oil.

Green Leafy Vegetables

Packed with micronutrients, it is essential for building stamina and also to improve your RBC count. Green leafy veggies are rich in fibre and digest slowly while maintaining your blood-glucose level. We are blessed with foods that can help us maintain healthy hemoglobin levels without having to bust open the bank account. Beetroot, palak or spinach, green peas, rajma or kidney beans, cabbage, turnip, sweet potato, broccoli, lima beans, collards, black beans and cauliflower are some vegetables that are easily available in the market. Beetroot is the best natural remedy to boost blood count, it has the ability to regenerate iron content and activates red blood cells, supplying fresh oxygen to the blood. Raisins, prunes, dried figs, apricots, apples, grapes and watermelons not only get the red blood cells flowing but also improve the blood count. Citrus fruits like oranges, amla or Indian gooseberry, lime and grapefruit help to attract iron. They play a very important role in increasing blood count. Cherries have been found effective in promoting health by scavenging action against free radicals. Herbs like rosemary, thyme, oregano, basil and mint are rich in essential oils of the terpene. Watermelons, papayas and bananas (rich in potassium) increase blood flow by dilating arterioles (small blood vessels) and thus help in improving an erection.

Raw Vegetables

Raw vegetable is a diet,that is uncooked, unprocessed and often organic foods. If between 80 per cent to100 per cent of an individual's food consumption is raw, she or he is considered a raw foodist. The enzymes in raw foods (amylases, proteases and lipases) aid digestion. Uncooked foods populate the digestive tract with beneficial gut flora or microorganisms that are key in promoting the early development of the gut's mucosal immune system

Cooking is believed to kill the vitamins and minerals, but studies have found the opposite. While cooking may destroy some (but not all) vitamin C, the process boosts the uptake of disease-fighting nutrients — antioxidants. A 2008 study found that vegetables such as carrots, spinach, mushrooms, asparagus, cabbage and peppers supply more antioxidants when cooked than when eaten raw. This is because cooking breaks down vegetables' thick cell walls, making it easier for the body to absorb the nutrients they contain. Steaming is best, then gentle boiling. Frying preserves the least vitamins and minerals (Anonymous, 2013).

Our body possesses countless cells and tissues that have their own enzymes requirements and production to perform a whole range of tasks. The ones that are made availed from eating raw vegetables or juices work in conjunction with the already present ones. So one need not cut off raw vegetables completely. Striking the right balance is essential.

Benefits of Green Juices

Green juices are power-packed with nutrition. Green vegetables have the ability to transform sunshine into the food that all creatures consume. Greens produce chlorophyll, which oxygenate our body. This enables us to release stored toxins in the body. Hemoglobin (that has a similar molecular structure as that of your red blood cells) will get elevated, improving the blood circulation as a result and giving our body a boost of energy. A green juice will cleanse our digestive system, lungs, liver (which will rev up, if sluggish - causing us to retain weight) and uterus. So if you are a smoker, this is an excellent supplement to your diet.

Magnesium is essential for efficient calcium utilisation. The chlorophyll molecule has magnesium at its centre and its impact on the body's magnesium levels is highly significant. When green juices is include in daily diet, it helps absorb calcium better. All green plants have Vitamins A and C, which are important co-factors for calcium absorption. Chlorophyll foods also act as a form of 'stored sunshine', to regulate calcium and Vitamin D deficiency. So if you get no sunshine, then increase the greens in your diet. However, greater consumption of fruit juices is associated with a higher risk of type 2 diabetes, researchers led by Harvard School of Public Health (HSPH) found (Anonymous,2013).

What is a Vegan Diet?

A vegan diet is one that consists of only plant-derived foods, as vegans do not use or consume any animals or animal products, including flesh, eggs, and milk. Like non-vegans, vegans eat soups, stews, stir-fries, salads, and casseroles. They may consume a wide variety of ethnic foods, as well as vegan versions of traditional favourites such as pizza, tacos, burritos, lasagna, burgers, barbeques, loaves, chili, pancakes, waffles, sandwiches, and desserts. Green leafy vegetable like spinach, fenugreek, pak choy, radish leaves, lettuce, etc. are healthy and are known to reduce the risk of heart diseases and cancer as well. That's because they are extremely low in fat, calories and high in dietary fibre. They also contain folic acid, magnesium, calcium, potassium, etc. These minerals are beneficial for the optimum functioning of the heart. Studies have shown that one daily serving of green leafy vegetables can lower the risk of heart diseases by 11 per cent. However, certain vegetables like carrots, spinach, cabbage, cauliflower, onion, garlic, sweet potatoes and pumpkin are also said to cause allergic reactions like skin rashe, abdominal pain and gastric problems. Similarly among fruits, bananas, strawberries, oranges and grapes are principal offenders. Citrus fruits like lemon is said to be cause atopic allergy (Meghna Maukherjee, 2012).

What is a Healthful Vegan Diet?

A balanced vegan diet is made up of these four food groups: legumes, nuts, and seeds; grains; vegetables; and fruits. Because individual nutrient needs and energy requirements vary due to age, activity level, and one's state of health, this guide should only be considered a broad blueprint for a balanced vegan diet.

Legumes, Nuts, and Seeds (4+ servings per day)

The legume-nut-seed group includes beans, split peas, lentils, nuts, seeds, and soy products. These nutrient-dense foods are packed with protein, fibre, minerals, B vitamins, protective antioxidants, and essential fatty acids. Sample serving sizes from this group include: ½ cup cooked beans, 4 ounces of tofu or tempeh, 1 cup soy milk, 1 ounce of nuts or seeds, or 2 tablespoons of nut or seed butter.

Grains (4-6+ servings per day)

Whole grains provide B vitamins, fibre, minerals, protein, and antioxidants. They are preferable to refined grains because the refining process removes the healthiest nutrients. Also, intact whole grains—such as brown rice, oats, wheat berries, millet, and quinoa–are nutritionally superior to whole grain flours and puffed or flaked whole grains. A serving is one slice of bread, ½ cup cooked grain, or 1 ounce of ready-to-eat cereal. This group is fairly flexible with regard to servings per day. Vary your intake based on your individual energy needs.

Vegetables (4+ servings per day)

Eating a wide variety of colourful vegetables every day will ensure that you're getting an assortment of protective nutrients in your diet. A vegetable serving is ½ cup cooked, 1 cup raw, or ½ cup vegetable juice. For most vegetables, particularly calcium-rich leafy greens, it's nearly impossible to eat "too much."

Fruits (2+ servings per day)

Most fruits, especially citrus fruits and berries, are a great source of vitamin C; all fruits provide antioxidants and fibre. Choose whole fruits over fruit juices to get the most benefit, particularly from dietary fibre. A serving size is one medium piece, 1 cup sliced fruit, ¼ cup dried, or ½ cup of juice.

Eat Well and Feel Good

A healthy diet has shown experts that food could be the answer to combat diseases. Foods can be therapeutic and also make you feel good. Eat well not just for basic nutrition but to heal. After a lot of suffering, people realise it's better to eat well than pop in medicines. Food science introduces us to foods that heal rather than just provide basic vitamins. Increase your intake of yellow and orange vegetables to prevent infertility. Leafy vegetables are good for eyesight and detox the liver. If you want to avoid depression, just eat lots of seeds, nuts and watermelons for a better nervous system. Salads make the digestion stronger as they have enzymes. Eating fresh and dried fruit before a meal helps people feel fuller and eat slightly less during the main course, but being on fruit and vegetable heavy diet for months does not make difference on hunger and fullness.

Herbs Help Blood Pressure

Include herbs, green and herbal teas, and vegetables like celery, nettle (bichoo booti), lauki juice, cucumber, garlic, green coriander and parsley as they have diuretic effect, help in urine formation and control blood pressure. Include soya, nuts, fish, mustard seeds, flaxseeds, fenugreek seeds (methi) in the diet.

How Healthy is a Vegan Diet?

According to the American Dietetic Association's 2009 Position Paper on Vegetarian Diets, vegan diets "are healthful, nutritionally adequate, and may provide health benefits in the prevention and treatment of certain diseases." A healthy vegan diet helps reduce your risk of heart disease, cancer, obesity, and diabetes. Eat your vegetables as they are, by trying to retain as much of their original colour and nutrition as possible. Cook them lightly, stir-fry, pan cook, flavour with as little masala that is palatable and learn to love the natural taste of healthy vegetables. Once you love these subtle flavours, you can enjoy then without trying to mask their taste in anyway.

Helps in Weight Management

Being low in calories and fat content and high in fibre, micronutrients and antioxidants with a high satiety value, they help in weight management. They work as fillers and can be treated as free foods with the exception of potatoes, sweet potatoes, jimikand and arbi, which must be used in moderation.

Keeps BP in Control

Owing to their high potassium and low sodium content, intake of fruits and vegetables is recommended for BP patients. These natural foods protect you from heart disease by preventing accumulation of cholesterol in arteries. Green leafy vegetables are a rich source of Omega-3 fats, the kind present in fish, which prevents heart disease.

Maintain Blood Glucose

The fibre in fruits and vegetables plays an important role in maintaining blood glucose levels. Since diabetics are more prone to oxidative cell damage leading to complications of kidney, nerves and eyes, the antioxidants in vegetables and fruits prevent these. Diabetics need to restrict their intake of mangoes, bananas and potatoes.

Osteoporosis Cause

Calcium deficiency is not the cause of this condition. It's a diet that's either very rich in acidic foods or high in protein. Proteins are broken down into amino acids which are acid yeilding. And so, consuming large amounts of animal protein results in the body extracting calcium from the bones to neutralise the effect of acids. When the lost calcium is not restored, it causes osteoporosis over a period of time.

Sexdrive

There are fruits and vegetables that increase sex drive like asparagus-high in B-vitamin folate that increases the production of histamine, which in turn increases sex drive in men and women. Citrulline (watermelon) an amino-acid that relaxes blood vessels there by increaseing sex drive. Chillies contain capsaicin- a chemical which also increases blood flow and triggers the release of mood-enhancing endorphins that naturally pump up your libido. Similarly, avocado-loaded with potassium, boosts libido for both sexes and are rich in folic acid that provide energy and stamina- both of which are important once your libido is restored and pomogranate juice could also liven up libido. Almonds have a high quantity of a type of amino acid called arginine, which helps to relax blood vessels and improve blood circulation, therefore, for men who want a healthy sex life- start taking more almonds. Orange, green tea and pine apples are known to lower sperm acidity. Zinc which is present in bananas and almonds aids testostrone synthesis and increase sperm numbers, motility and volume. Nutritionists says foods with high content of vitamin A (carrot), vitamin C (orangs, strawberries, broccoli), vitamin E (peanuts, spinach, almonds, hazel nuts) optimum sperm health. Selenium rich foods such as garlic and lycopene, foods such as tomato and watermelon help prevent damage to sperms (Anand Holla, 2012).

The Scoop on some Important Nutrients

Like non-vegans, vegans need to be mindful of consuming all the nutrients they need in order to be healthy. Three nutrients that *everyone* needs to pay attention to are vitamin B_{12}, vitamin D, and Omega-3 fatty acids. Many common health problems can be prevented or alleviated with a healthy diet. Nutrients are important to lead a healthy life. Vitamins and minerals you get from your diet are just as important as carbohydrates, protein and fats; however, you only need them in small amounts. Vitamins and minerals usually function as co-enzymes, which mean they help chemical reactions in the body happen a lot faster. One of the most important vitamins for the body is Vitamin E. In order for the heart to stay healthy the body needs plenty of Vitamin E. Vitamin E is also an excellent option for helping limit bad heart conditions. The body is more prone to certain diseases if it doesn't have a proper amount of Vitamin E.

Vitamin B12

Vitamin B12 is necessary for proper red blood cell formation, neurological function, and DNA synthesis. It is manufactured by certain types of bacteria found in nature. Because plants vary widely in their levels of this bacteria (and most of us favour our food scrubbed squeaky clean), we cannot rely on plant foods to meet our B_{12} needs. We can ensure our dietary needs are met by consuming supplements and/or fortified foods. Our suggestion is to supplement with a vegan source of 2000 micrograms once a week or 10-100 micrograms a day (be advised that some B_{12} vitamins labelled as vegetarian are in a stomach base). Or, if you prefer not to use supplements, consume at least three servings of vitamin B_{12}-fortified food per day (each supplying at least 20 per cent of the Daily Value on the label), such as nondairy milks, breakfast cereals, meal replacement bars, beverage mixes, and Red Star Vegetarian Support Formula Nutritional Yeast (read labels to ensure B_{12} content).

Vitamin B

Vitamin B are vital, they help convert calories into energy and are components of enzymes that maintain normal skin function (including functioning of the oil-producing glands which keep skin moist and smooth). Poor intake of almost any B vitamin can cause dry or scaly skin. The best sources are bananas, whole grains and peanut butter etc.

Vitamin D

Vitamin D, the "sunshine vitamin", is also a hormone; our skin manufactures it from the ultraviolet rays of the sun. It plays an important role in bone health and supports normal neuromuscular and immune function. Good vitamin D status is linked to a lowered risk of osteoporosis, certain cancers, and other chronic diseases.

Vitamin D blood levels are an international public health concern. Getting enough of it is not as easy as we may think. The body's ability to produce vitamin D from sun exposure varies based on skin pigmentation, sunscreen, clothing, time of year, latitude, air pollution, and other factors, and the vitamin is found naturally in only a handful of foods. This is why all people—not just vegans—need to be mindful about vitamin D. The latest research suggest that getting even 100 per cent of the current Recommended Dietary Allowance (RDA) for vitamin D may be insufficient for many people. To ensure adequate vitamin D intake, take 1000-4000 International Units (IU) per day, depending upon your age and other individual needs.

Supplemental vitamin D comes in two forms: vegan D2 (ergocalciferol), usually synthetic or manufactured from yeast, and non-vegan D3 (cholecalciferol), manufactured from lanolin (from sheep's wool). A dose of sunlight along with almonds will benefit those who want a baby. Eat to beat infertility which is on the rise and have foods rich in iron like apples and radish. It's essential for menopausal women to take vitamins D and E, which are found in avocadoes, tomatoes, hazel nuts. Women can get vitamin D from butter and eggs too.

Omega-3 Fatty Acids

Make sure to include enough of omega-3 and omega-6, both essential fatty acids in the daily diet plan. A proper balance of essential fats is important for optimal brain function, heart health, and infant/child development. Alpha-linolenic acid (ALA) is an omega-3 fatty acid that partly converts to DHA and EPA in the body. It is present in several plant foods, including flax products, hemp products, canola oil, walnuts, and leafy green vegetables. Aim to consume 2 to 4 grams of ALA per day. Beat the blues by eating right, as certain foods can help cut the chances of depression. Make sure you're having enough Omega 3 fats from mustard oil, wheat, lobia, flax seeds, methi seeds, soybean and green leafy vegetables. Even oily fish can be good to beat the blues. "Wholegrains are important, everything from oats to wholewheat bread to barley are great sources of slow-release energy that will prevent your blood sugar from nosediving. Seeds like flaxseeds, pumpkin seeds, soybeans, chickpeas and sunflower seeds etc. are again friends of sexual hormones. "Consuming one tablespoon of flaxseed every day helps to increase the testosterone level in the body. These seeds are rich in Omega 3 fatty acids, a drop in which affects hormone levels leading to a plummeting sex drive. Pumpkin seeds are also rich in zinc which is a mineral needed to produce testosterone." An imbalance of Omega-3 fats in the diet can also aggravate acne. Omega-3 fats are known to counter the effect of inflammatory chemicals that lead to breakouts. It wards off cancer and cardio-vascular diseases. It also improves your immune system and reduces joint pain.

Be a Vegetarian, Live a Healthy Life

Today, an increasing number of people are turning towards a **vegetarian** diet. The reasons vary from religious, environmental, cruelty towards animals, to just the desire for a healthier lifestyle. A predominantly non-vegetarian diet poses many health risks. Here's why opting for a vegetarian diet achha hai.

The human body is anatomically and physiologically geared for a vegetarian diet. Some of the distinctions between vegetarians and carnivores are:

Natural Detox

Being vegetarian is a natural way to detox the body, because a vegetarian diet is more rich in fibre, vitamins, minerals and anti-oxidants, which help to cleanse the body's system. Meat and fish contain a lot of residue from toxic chemicals. It has been observed that the bodies of carnivorous animals contain 10 times more hydrochloric acid than that of herbivorous ones, but the human body should not have the same amount of hydrochloric acid. This establishes the fact that the human body is basically meant for a vegetarian diet. So digestion of vegetarian food is easier for our bodies. A vegetable diet contains dietary fibre (bottle gourd, pumpkin, spinach, cabbage) which flushes toxins out of the body. A diet containing only eggs, fish and mutton is a poor source of fibre.

Colourfully Appetising

Good food is not restricted to the taste, its visual appeal is equally important. In fact, if the food served to us is a visual treat, we are bound to eat and enjoy it even more. Vegetables come in an array in colours, which makes them more appetising and appealing to our eyes. Meat and fish usually come in boring shades of brown and beige. Therefore, using many coloured food items in your cooking is a great way to eat a variety of natural foods that will boost your health.

The best way to ensure that your diet is balanced is to colour code it. Pigmentation is what lends colour to fruits and vegetables, and also contains healthy nutrients. The deeper the colour, the higher the percentage of a specific nutrient the food item has. Purple coloured foods like plums and aubergines have phytochemicals called **anthocyanins** and **resveratrol** that fight ageing, reduce blood pressure and protect the body against liver diseases.

Orange-coloured foods such as oranges and carrots contain beta-carotene and beta-cryptoxanthin that keep the skin and eyes healthy, and also protect against arthritis. Yellow foods such as bananas and sweet corn contain carotenoids that boost the immune system and keep the heart healthy. White foods, including garlic, onions, cauliflower and mushrooms, are rich in vitamin C, and antibiotic and antifungal compounds that fight cell damage and keep the fighter white blood cells healthy. Red foods like tomatoes and watermelons contain the powerful antioxidant lycopene which acts as an internal sunscreen and protects you from sunburn, prevents osteoporosis and heart diseases.

Greens such as peas, broccoli and spinach are packed with sulforaphane, isothiocyanate, and indoles which stimulate the liver to keep cancer producing chemicals in check.

Good for Digestion

The most important difference between a vegetarian and a non vegetarian diet is that the former contains dietary fibres, whereas a non-vegetarian diet is lacking in fibre. This dietary fibres is very useful for the human body because people who have a diet rich in dietary fibre have low incidence of diseases like coronary heart diseases, cancer of intestinal tract, piles, obesity, diabetes, constipation, hiatus hernia, diverticulitis, irritable bowel syndrome, dental caries and gallstones. The food rich in this dietary fibre includes cereals and grains, legumes, fruits with seeds and citrus fruits. Complex carbohydrates in vegetarian foods are digested gradually providing steady source of glucose. Conversely, meats rich in fat and proteins are difficult to digest.

Boosts Cardio-vascular Health

Yes, it's true. A vegetarian diet boosts cardio-vascular health. Consumption of a vegetarian diet consisting of whole grains, legumes, vegetables, nuts, and fruits, and abstaining from meat and high-fat animal products, along with a regular exercise program, is consistently associated with lower blood cholesterol levels, lower blood pressure, less obesity, lower incidence of diabetes and consequently less heart disease. **Fruits and vegetables**, especially leafy greens and some legumes, are abundant in folic acid, which has been shown to help lower homocysteine levels (associated with heart disease). Many whole and unrefined plant foods are also important sources of minerals such as copper and magnesium, both of which can protect against cardiovascular disease. The many phytochemicals/flavonoids in fruits, vegetables, nuts and whole grains have properties that reduce the risk of heart disease. Various nuts are the source of heart **healthy** fatty acids (omega3, MUFA, PUFA). On the other hand, a non-vegetarian diet (mainly red meat) is associated with an adverse impact on cholesterol levels, and increases the incidence of heart disease. Freshly made vegetable juices contain natural plant pigments and enzymes that act as cancer protection agents. These juices should be made fresh and consumed immediately, unstrained to retain the fibre. The studies suggest that greater consumption of fruits and vegetables may reduce the risk of developing invasive bladder cancer in women. The study also suggested that women with the highest intake of vitamins A, C and E had the lowest risk of bladder cancer (Anonymous, 2013).

Reduces Risk of Cancer

A diet with adequate portions of green leafy vegetables and fruits ensures an intake of roughage or fibre. This has been shown to have beneficial effects on a number of cancers, notably cancers of the colon and rectum. A non vegetarian diet, especially one rich in red meats and animal fats, has been shown to have a carcinogenic effect. Carcinogens are substances or agents that are directly involved in causing cancer.

Longevity

A vegetarian diet is associated with lower blood cholesterol levels, lower blood pressure, less obesity, less heart disease, less stroke, less diabetes, less cancer, and hence vegetarians have a chance of living a longer life. But based on statistics, it certainly appears that a vegetarian diet is far healthier than a non-vegetarian one. Studies have shown that people who eat mostly plant-based **foods** look younger than their real physical age. And if that is true, it sure makes sense to include plant-based, **anti-ageing** foods in our daily diet to get that ageless, young skin and delay the inevitable effects of natural ageing by improving skin elasticity and thereby staving off wrinkles.

Red meat requires maximum digestion effort for your body to process. Over-eating will lead to issues liker perspiration and unwanted sweat. While vegetarians might not have such issues with their limited food intake, fenugreek and other strong flavoured plants cause similar results.

Stronger Bones

Gorging on meat can lead to protein over load. This can tax our kidneys, interfere with the absorption of calcium and prompt the body to extract existing calcium from the bones. Such calcium extraction is rare amongst the vegetarians. Our body is mostly alkaline. The human body's natural pH level stands at 7.45, making it alkaline. All fruits and vegetables, natural juices like coconut water are alkaline, which make them a better nutrition fit than say animal proteins, tea, coffee, sugar, alcohol, and packaged foods that have high acidic levels. An imbalance in pH levels, can lead to a disruption in cellular activity, causing health problems that range from heartburn to heart disease.

Carbohydrate Deficiencies

A non-vegetarian diet is a poor source of carbohydrat. Carbohydrates deficiency can lead to ketosis- a condition where the body starts breaking fat (instead of carbs) as a source of energy.

Healthy Skin

Eating beet root, tomato, pumpkin and bitter gourd can clear off blemishes. And guava, apples and peaches, eaten along with their peel, promise a glowing complexion.

Weight Management

Avoiding meat is the simplest way to reduce fat intake. Instead, eating whole grains, legumes, vegetables, nuts and fruits, lowers cholesterol levels, blood pressure and obesity.

Easy on the Teath

Our molars are more suitable for grinding grains and vegetables than tearing flesh. Digestion begins with the saliva, which can only digest complex carbohydrates present in plant foods.

Phyto Nutrients

Diabetes, cancer, kidney disease, stroke and bone loss are partially preventable with a good intake of phytonutrients. As these are present only in vegetarian diet, the non- vegetarians are at a loss.

Fruit compared with animal food.

Energy

When you look at this table you see that one banana and a few strawberries give you as much as energy as an entrecote! You will not feel as fulfilled, but your body really can extract as much of energy out of fruit.

Proteins–Fat–Sugar

As the energy fruit contains consists of sugars the fruit is digested in 30 minutes. This chemical process has no toxic waste products that are difficult to remove from the body and it even stimulates the removing of toxic elements. The energy animal products contain consists of fat and proteins. Because of the high amount of proteins it is digested in 6-8 hours. This chemical process does produce toxic waiste-products that have to be removed from your body. Protein is an important nutrient for the growth, development and repair of muscle and body tissues. Protein has a higher metabolic rate than fat, so an individual can burn more calories. It is also said to increase satiety, which prevents a person from overeating. Healthy sources of protein include lean chicken, fish, eggs and nuts.

Fibres

Another remarkable difference is that fruit contains a substantial amount of fibres while animal based products don't contain any. Fibres are very important. Your body does need dietary fibre, which might not carry nutritional value but acts as a cleaner for the digestive tract and intestines.

Water

The water percentage of fruit (80 per cent) is higher than that of meat (15 per cent) and comes more near the water percentage of the human body (80 per cent).

Vegetarians sweat through their skin to get rid of excess heat and impurities, where as carnivores pant or breathe rapidly and extend their tongues.

Vegetarians do not have long canines and claws, whereas meat-eating animals do for killing and tearing the flesh of prey.

The saliva of vegetarians contains the enzyme ptyalin to predigest starch whereas meat-eating animals do not have this enzyme.

Vegetarians secrete less hydrochloric acid during digestion in the stomach compared to flesh –eating animals, who secrete large quantities of it to digest meat and bones.

The jaws of vegetarians move up and down and also sideways for additional chewing and grinding, whereas those of carnivores open in an up and down motion only.

Vegetarian drink by suction through their teeth, whereas carnivores lap up liquid liquid with their tongue.

Vegetarians naturally salivate on seeing fruits and vegetables, but do not do so on seeing live chickens, pigs or other animals. Carnivares, on the contrary, naturally salivate on seeing its animal prey, but do not do so on seeing fruits and vegetables.

Therefore meat- eating is incompatible with the human body, causing internal damage and disease and also result in much damage to the eco-system. Experts have warned that eating three eggs a week could significantly increase a man's chance of dying from prostate cancer. The researchers found that men who consume more than two and a half eggs on a weekly basis were up to 81 per cent more likely to be killed by the disease. They suggested the damage might be done by the large amounts of cholesterol or choline - a nutrient that help cells to function properly - that are found in eggs.

The team at the Harvard School of Public Health in Boston examined the eating habits of 27,000 men over a 14-year period. They found that a higher number of cancer deaths among those who admitted consuming lots of eggs.

In one of the largest studies conducted till date, it has been found that vegetarians live longer than meat eaters. After following more than 70,000 men and women over six years, researchers have concluded that those who followed vegetarian lifestyles enjoyed 12 per cent lesser risk of death than non-vegetarians. The participants were divided into five dietary groups — non-vegetarians, semi-vegetarians, pescatarians, lacto-ovo-vegetarians and vegans. It was found that people who are vegetarian tended to be older, more highly educated and more likely to be married. They were also more likely to drink less alcohol, smoke less, exercise more and be thinner. According to researchers, all of these factors could play a role in their lower risk of death.

Interestingly, the study has found that the association between vegetarian diets and lower mortality was greater in men than in women. Men had a lower rate of cardiovascular disease and death from heart-related conditions.

Previously too, vegetarian diets have been linked to a lower likelihood of developing chronic diseases like heart disease or diabetes while red-meat consumption, because of its high levels of cholesterol and saturated fat, is likely to clog up arteries (Anonymous, 2013).

Freshly squeezed juices from fruits and vegetables are excellent sources of minerals and vitamins which catalyze chemical reactions occurring in the body. These enzymes also produce the energy needed for digestion, absorption, and conversion of food into body tissues. An increased intake of fruit and vegetable juices ensures that the body will efficiently absorb more minerals and vitamins.

Pure orange juice, for example, is very rich in Vitamin C, potassium, and folic acid. Other fruit juices that are rich in antioxidants, vitamins, and anthocynanins (water soluble pigments found in plants which act as antioxidants) include grape juice (purple), pomegranate juice, and cranberry juice. It is believed that juices that are made from fruits with rich colours such as grapes, pomegranates, cranberries, and blueberries are rich in vitamins and antioxidants.

A: Fruits and Vegetables Recommended in Rheumatism Diseases

- ☆ *Fruits* - gooseberry, cherry, red currant, lemon, apple, plum, grape.
- ☆ *Vegetables* - paprika, onion, pumpkin, bean, parsnip, salad, celery, garlic, cabbage.

B: Fruits and Vegetables Recommended in Nervous System

Tranquilizers

- ☆ Fruits - Cherry, mandarin, apple, plum.
- ☆ Vegetables - salad, cabbage.

Anaphrodisiac (The Opposite of Aphrodisiac)

- ☆ Vegetables - salad.

Hypnotics

- ☆ Fruits - Apple, apricot.
- ☆ Vegetables - onion, salad.

Headaches

- ☆ Fruits - cherry, lemon.
- ☆ Vegetables - cabbage, parsley.

C: Fruits and Vegetables Recommended in Beauty

Face Beauty

- ☆ Fruits - plum, lemon, strawberry.
- ☆ Vegetables - parsley, cucumber, tomato.

For Dry Skin

- ☆ Fruits - strawberry, quince.
- ☆ Vegetables - cucumber, salad

Wrinkled Skin

- ☆ Fruits - cherry, orange, almond.
- ☆ Vegetables - cucumber.

Hands Beauty

- ☆ Fruits- lemon.
- ☆ Vegetables - tomato.

D: Fruits and Vegetables Recommended in Other Diseases

Alcoholism

- ☆ Vegetables - paprika, cabbage.

E: Fruits for Diabetics

Diabetics can have fruits, provided the sugar level of the patient is in control, but these fruits must be consumed in a limited quantity. To help diabetic patients, some fruits which will not affect blood sugar level, provided they're consumed in moderation are as follows (**Renita Tisha Pinto, 2013**).

- ☆ **Kiwi:** Many researchers have shown a positive correlation between kiwi consumption and lowering of blood sugar level
- ☆ **Black jamun:** This is one of the best fruits for diabetics. It is known to improve blood sugar control. Seeds of these fruit can be powdered and consumed by patients to control diabetes.

- **White jamun:** Diabetic patients can consume jamun fruit daily, to control their sugar levels. White jamuns are also high in fibre.
- **Starfruit:** Similar to jamuns, starfruits are good for diabetics too as they help improve blood sugar control. But caution needs to be exercised in case the person has diabetes nephropathy.
- **Guava:** Guava controls diabetes and it is good for constipation. Guavas are high in vitamin A and vitamin C and contain high amounts of dietary fibre. This fruit has a reasonably low GI.
- **Cherries**: Their GI value is 20 (or even less in some varieties) which makes it a good healthy snack for diabetes patients at any time of the day.
- **Peaches:** These tasty fruits are a great healthy treat, with a low GI and good for diabetics.
- **Berries:** Numerous variety of berries are available throughout the world and almost all seem to be a rich source of antioxidants. Diabetics can include a serving of different berries to keep their sugars in check. To name a few: Strawberries, blueberries and blackcurrants, raspberries, cranberries, chokeberries and blackberries are good for diabetes patients.
- **Apples:** Apples contain antioxidants, which help to reduce cholesterol levels, cleanse the digestive system, and boost the immune system. Apples also contain nutrients that help in the digestion of fats.
- **Pineapples**: Good for diabetics, pineapples also benefit the body as they are rich in anti-viral, anti-inflammatory and anti-bacterial properties.
- **Pear**: These delicious fruits are a good snacking option for diabetics as they are rich vitamins and fibre.
- **Papaya**: They are good for diabetics because they are rich in vitamin and other minerals.
- **Figs**: Their richness in fibre helps with insulin function in diabetes patients.
- **Oranges**: These citrus fruits can be consumed on a daily basis by diabetics, as they are rich in vitamin C.
- **Watermelon**: Although watermelons have a high GI value, their glycemic load is low, making them good fruits for diabetes patients. However, consume in moderation.
- **Grapefruit**: It is a good option for diabetics as it slows down the blood sugar peak.
- **Pomegranate**: These tiny red rubies help diabetic people improve their blood sugar statistics.
- **Cantaloupe:** This fruit is high in the Glycemic Index, but has a good amount of fibre. Therefore, within moderation, it can easily be included in the diet of a diabetic.
- **Jackfruit**: Jackfruit contains vitamin A, vitamin C, thiamin, riboflavin, niacin, calcium, potassium, iron, manganese and magnesium among many other nutrients. Good for diabetes as they improve insulin resistance.

- ☆ **Amla**: A fruit loaded with vitamin C and fibre, Amla is a healthy addition to the diabetic diet.

F: Foods to Battle Cancer

One can keep cancer at an arm's length if they incorporate certain foods in their diet every day and exercise regularly. Here is a list of 12 foods that prevents cancer (Sobiya N. Moghul, 2013)

- ☆ **Peanuts**: Rich in Vitamin E, peanuts reduces the risk of colon, lung, liver, and other cancers. Have a spoon of peanut butter on a slice of whole grain toast for a filling cancer-fighting snack.
- ☆ **Grapefruit:** Oranges, grapefruits, broccoli all of them are high in Vitamin C. This element prevents the formation of cancer-causing nitrogen compounds. Grapefruits, Oranges, grapefruits, broccoli and other fruits reduce the risk of esophagus, bladder, breast cancer, cervical cancer, and stomach and colon cancer. So remember to fill your plates with veggies and fruits every day.
- ☆ **Berries:** Berries top the charts when it comes to cancer fighting foods. Raspberries, blueberries, and cranberries are very promising potential to help prevent cancer. Pterostilbene, an antioxidant found in berries is a storehouse of cancer fighting properties.
- ☆ **Sweet Potatoes**: Sweet potato is full of beta-carotene. Studies have revealed that consumption of high beta-carotene foods reduce the risk of cancers especially colon, breast, stomach and lung cancers. In a study including premenopausal women who ate a lot of vegetables that included beta-carotene like sweet potatoes had lowered their breast cancer risk by half.
- ☆ **Flaxseed:** Omega-3 fatty acids prevents in the growth of cancer by preventing the growth of cancer cells and interrupting the growth of cancerous tumor. Omega-3 fatty acids also help reduce inflammation. Along with fatty fish, one can also include a tbsp of flaxseed in your daily diet.
- ☆ **Curcumin:** Curcumin, the active ingredient in turmeric, acts as both an anti-inflammatory and an antioxidant, and prevents cancer by interfering with aspects of cellular signaling. Use it liberally in all your chicken and fish curries.
- ☆ **Cruciferous Vegetables:** Fruits and vegetables consists of small quantities of photonutrients and these naturally occurring nutrients are as important as minerals and vitamins. Cruciferous vegetables like cauliflower, cabbage contain phytonutrients which may help inhibit the metabolism of carcinogens and stimulate the body's production of detoxification enzymes.
- ☆ **Tomato:** According to sfgate. com, "A study published in 2009 in the Journal of Clinical Oncology shows that tomatoes, which contain lycopene, help prevent prostate cancer. Lycopene is a strong antioxidant that may also prevent other types of cancers, so Farquhar recommends three servings of tomatoes a week for men and women."

- **Pomegranates:** Pomegranates are rich in ellagic acid. According to a few studies, ellagic acid slows down cancer cell growth and deactivate cancer-causing compounds. So hog on this fruit in any form, a juice, as salad or milkshakes.
- **Onions:** Again, a food rich in anti-cancer substances including allicin, onions are a staple in Indian dishes. Ypui can also incorporate it through salads or just eating it raw.

G: Foods that Seduce

- **Bananas:** A banana is packed with vitamin A, B and C and also potassium. Vitamin B and potassium are known to increase sex hormone production in the body. Banana is also packed with Bromelain which increases the testosterone levels. High levels of sugar present in banana gives energy, which helps to last long.
- **Garlic:** Garlic has allicin which helps in blood circulation. So if there is enough blood going to your groin there surely wouldn't be any problem related to erection. Studies also show that garlic can stimulate the production of **nitric oxide** synthase which is responsible for erection.
- **Avocado:** This is one fruit which has been associated to both female as well as male sexuality. The fruit is voluptuous and feminine in shape but as the fruit hangs in pairs from the tree, they are often said to resemble the male testicles. Avocado is a rich source of beta carotene, magnesium, vitamin E, potassium and proteins — all of these are good for your sexual appetite.
- **Fig:** A fig when cut vertically seems to resemble the female sex organ. Since ancient times it has been associated with fertility. Figs contain vitamin A, vitamin B1, vitamin B2, calcium, iron, phosphorus, manganese, and potassium and these are known to decrease sexual weakness.
- **Asparagus:** Asparagus is a rich source of potassium, vitamin B6, vitamin A, C, thiamin and folic acid. It is said that folic acid boosts **histamine** production which help both male and females to reach orgasm. Folic acid also helps in reducing birth defects, so, asparagus is good for a pregnant woman. Asparagus is also known to increase blood circulation in the genito urinary system.
- **Basil:** This sweet smelling herb is called 'kiss me Nicholas' in Italy. It is believed that basil increases sex drive and fertility. Basil has magnesium, iron, vitamin A, C and K. It relaxes blood vessels and prevents clotting in arteries, hence increases blood circulation. Basil is also known to cure all kinds of 'headaches' so, next time if you get this excuse it is time to use basil.
- **Chilies:** From bell peppers to the red chilies – all are considered **aphrodisiacs**. Chilies have capsaicin which increases blood circulation and heartbeats, raises body temperature and also produces sweat. All these symptoms

occur while having sex so this might be another reason for calling it an aphrodisiac. Capsaicin also helps in releasing endorphins and stimulates the nerve endings which increases the pulse and makes the body sensitive.

H: Foods that Contain Oxalates

The following are some examples of the most common sources of oxalates in fruits and vegetable. It is important to note that the leaves of a plant almost always contain higher oxalate levels than the roots, stems, and stalks.

Fruits

Blackberries, blueberries, raspberries, strawberries, currant, kiwifruit, concord (purple) grapes, figs, tangerines, and plums.

Vegetables

Spinach, Swiss chard, beet greens, collards, okra, parsley, leeks and quinoa are among the most oxalate-dense vegetables.

Celery, green beans, rutabagas, and summer squash would be considered **moderately dense in oxalates.**

Nuts and Seeds

Almonds, cashews, and peanuts.

Legumes

Soybeans, tofu and other soy products.

Fruits

A fruit is generally a fleshy seed associated part of a particular plant; it is naturally and mostly edible and sweet in the raw state. Fruits are the natural and staple food of man. They are excellent source of minerals, vitamins and enzmes. They are easily disgested and exercise a cleaning affect on the blood and the digestive tract. Fruits and vegetables are lower in calories per gram compared to denser foods. They are alkalizing, energy boosters. The natural sugars in the fruits help to balance the pitta element of the body. These are natural cleansers and can act as an in-between snack for people who crave desserts or fried snacks. Fruits and vegetables are excellent sources of fibre. These are free of added colours, preservatives and additives when eaten raw and fresh.

To increase satiation, consume a cup of chopped water- rich fruits and vegetables, such as berries, citrus fruits, tomatoes, kiwi, celery, cucumber, bell peppers, leafy green vegetables and waterchest nuts. The research has shown that men and women whose skin has a yellow glow are thought to be particularly attractive and healthy - and yellow pigments called carotenoids, found in certain fruit and vegetables play a key role in giving the skin that hue. Scientists have found how micronutrients found in certain fruits and vegetables can have a rapid impact on your brain. They found that some foods like blueberries, cocoa and green leafy vegetables provide an instant boost to our mental sharpness and mood and help to ward off depression. "Certain micronutrients can have a big influence on moods", the Daily Star quated Dr Emma Williams of the British Nutrition Foundation as saying. Fruits are good source of vitamins and minerals and are a readily available source. A fruit will boost you instantly. It is something which can quench your thirst and satisfy your hunger at the same time. Fruits are guardian angels for all the busy bees out there, who have no time to even have their breakfast. Since its handy they can be eaten even on the move. Fruits are meant to be nutritious and are our sole companion which makes them our best friends! Make more fruit friends here. When it comes to the uses of fruits, it is never-ending! Fruits are consumed as food by millions of people in the form of salads, soups, juices, jams and pickles.

In this study, which involved over 100,000 women and men, researchers evaluated the effect of study participants' consumption of fruits; vegetables; the antioxidant vitamins A, C, and E; and carotenoids on the development of early ARMD or neovascular ARMD, a more severe form of the illness associated with vision loss. Food intake information was collected periodically for up to 18 years for women and 12 years for men.

While, surprisingly, intakes of vegetables, antioxidant vitamins and carotenoids were not strongly related to incidence of either form of ARMD, fruit intake was definitely protective against the severe form of this vision-destroying disease.

Three servings of fruit may sound like a lot to eat each day, but by simply tossing a banana into your morning smoothie or slicing it over your cereal, topping off a cup of yogurt or green salad with a half cup of berries, and snacking on an apricot, you've reached this goal.

In General

Keep a bowl of whole fruit on the table, counter, or in the refrigerator. Refrigerate cut-up fruit to store for later use. Buy fresh fruits in season when they may be less expensive and at their peak flavour.

For the Best Nutritional Value

Fruits are alkalizing, energy boosters. The natural sugars in the fruits help to balance the pitta element of the body. These are natural cleaners and can act as an in-between snack for people who crve desserts or fried snacks. They are also excellent sources of fibre. These are free of added colours, preservatives and additives.

Make most of your choices - whole or cut-up fruit rather than juice, for the benefits dietary fibre provides. Select fruits with more potassium often, such as bananas, prunes and prune juice, dried peaches and apricots, cantaloupe, honeydew melon, and orange juice.

When it comes to the uses of fruits, it is never-ending! Fruits are consumed as food by millions of people in the form of salads, soups, juices, jams and pickles.

Fruits = Taste + Flavour + Nutrition + Satiety!!

The natural antioxidants in fruits and vegetables will help keep your body working at its best, so consuming a diet that meets your daily recommended amount of fruits and vegetables is one of the best ways to give your body a strong defense against disease. Fruits and vegetables are protective to health as they're helpful at reducing the risk of coronary heart disease, stroke and some cancers. They're also low in calories, which helps prevent obesity, a significant risk factor for type 2 diabetes, cancer and cardiovascular disease.

Fruits are good source of vitamins and minerals and are a readily available source. A fruit will boost you instantly. It is something which can quench your thirst and satisfy your hunger at the same time.

Fruits are good source of vitamins and minerals and are a readily available source. A fruit will boost you instantly. It is something which can quench your thirst and satisfy your hunger at the same time. Fruits give you all the essential nutrients and are rich in vitamins like C and A. Eating more whole fruits, particularly blueberries, grapes and apples is significantly associated with a lower risk of type 2 diabetes, according to a new study.

Fruity Pleasure

All fruits, especially those rich in vitamin C can be vital in sexual gratifications. The body needs vitamin C to keep the sexual organs fine tuned. Tests reveal that consuming 500-1,000 milligrams of vitamin C in a day increases the number and quality of sperm produced. Some fruits which fall in this category include citrus fruits, specially kiwi, blackcurrants and strawberries.

A dose of sunlight along with almonds will benefit those who want a baby. Eat to beat infertility which is on the rise and have foods rich in iron like apples and radish. A study done at Harvard revealed women with high iron intake in their diet were more fertile. It's essential for menopausal women to take vitamins D and E, which are found in avocadoes, tomatoes, hazel nuts. Women can get vitamin D from butter and eggs too.

Foods that Help you Destress

Certain food items in our diet can actually help reduce stress in our bodies. While apricots and greens like spinach replenish the body with magnesium, which is known to relax the muscles and help with headaches, nuts and dark vegetables like almonds, pistachios, broccoli, and okra are rich in vitamins B and E that help lower blood pressure. Citrus fruits, which are a storehouse for Vitamin C, also help in controlling blood pressure. Other food items like sweet potatoes provide insulin to the body, provide energy, and help deal with fatigue in a stressful situation. Below is a list of foods recommend for stressed-out people:

- Cruciferous vegetables like cauliflower, cabbage, broccoli, etc
- Nuts like almonds, pistachios, etc
- Dark coloured vegetables like beetroot, okra, and peas
- Beans or lentils – Asparagus- Artichokes- Sweet potatoes- Leeks- Watercress

Fruit Nutrition Facts

Why Fruits?

Fruits are nature's wonderful medicines packed with vitamins, minerals, enzymes, anti-oxidants and many phyto-nutrients (Plant derived micronutrients). They are absolute feast to our sight, not just because of their colour and flavour but help body keep fit and healthy!They are the natural and staple food of man. They are easily digested and exercise a cleansing effect on the blood and the digestive tract.

1. Fruits are low in calories and fat and are a source of simple sugars, fibre, and vitamins, which are essential for optimizing our health.
2. Fruits provide plenty of soluble dietary fibre, which helps to ward of cholesterol and fats from the body and to get relief from constipation as well.
3. Fruits contain many anti-oxidants like *poly-phenolic flavonoids, vitamin-C, anthocyanins*. These compounds, firstly, help body protect from oxidant stress, diseases, and cancers, and secondly, help body develop capacity to fight against these ailments by boosting our immunity level. Many fruits, when *compared to vegetables and cereals, have very high anti-oxidant values which is* something measured by their "Oxygen Radical Absorbent Capacity" or ORAC.
4. Anthocyanins are flavonoid category of poly-phenolic compounds found in some "blue fruits" like blue-black grapes, mulberries, acai berry, chokeberries, blueberries, blackberries, and in many vegetables featuring blue or deep purple colour. Eating fruits rich in blue pigments offers many health benefits. These compounds have potent anti-oxidant properties, remove free radicals from the body, and thus offer protection against cancers, aging, infections etc. These pigments tend to concentrate just underneath the skin.
5. Fruit's health benefiting properties are because of their richness in vitamins, minerals, micro-nutrients, anti-oxidants which helps body prevent or at least prolong the natural changes of aging by protecting and rejuvenating cells, tissues and organs in the human body. The overall benefits are manifold! Fruit nutrition benefits are infinite!

   You are protecting yourself from minor problems like wrinkling of skin, hair fall, memory loss to major ailments like age related macular degeneration of retina in the eyes, Alzheimer's disease, colon cancers, weak bones (osteoporosis).. and Breaking the list of fruit nutrition benefits never ends.
6. No matter which beauty product rules the market, a fact that nobody can deny, is that fruits are nature's most valuable beauty enhancers.

Why Eat MORE Fruits and Veggies?

☆ **Colour and Texture.** Fruits and veggies add colour, texture ..and *appeal* ...to your plate.

- ✰ **Convenience.** Fruits and veggies are nutritious in any form – fresh, frozen, canned, dried and 100 per cent juice, so they're ready when you are!
- ✰ **Fibre.** Fruits and veggies provide fibre that helps fill you up and keeps your digestive system happy.
- ✰ **Low in Calories.** Fruits and veggies are naturally low in calories.
- ✰ **May Reduce Disease Risk.** Eating plenty of fruits and veggies may help reduce the risk of many diseases, including heart disease, high blood pressure, and some cancers.
- ✰ **Vitamins and Minerals.** Fruits and veggies are rich in vitamins and minerals that help you feel healthy and energized.
- ✰ **Variety.** Fruits and veggies are available in an almost infinite variety..there's always something new to try!
- ✰ **Quick, Natural Snack.** Fruits and veggies are nature's treat and easy to grab for a snack.
- ✰ **Fun to Eat!** Some crunch, some squirt, some you peel ..some you don't, and some grow right in your own backyard!
- ✰ **Fruits and Veggies are Nutritious and Delicious!**

When to Eat fruits?

The best time to eat fruits is first thing in the morning after a glass of water. Eating fruits right after a meal is not a great idea, as it may not be digested properly. The nutrients may not be absorbed properly either.

You need to leave a gap of at least 30 minutes between a meal and a fruit snack. Ideally, one should eat fruits an hour before the meal or two hours after, if they have diabetes or any other digestive problem like acidity. This is because, sometimes, diabetes is accompanied with digestive problems.

Can you mix fruits with other foods? You can mix fruit with yogurt or salt as long as you do not have any digestive problems like indigestion or acidity. You can also mix fruits like pineapple, oranges, melons or pomegranate with salads if you like. There is no harm in mixing berries and dry fruits with cereals either(Sumitra Nair, 2013).

Dried Fruits

In between your meals you snack on a packet of dried fruits instead of a processed fruits – you are unaware that even dried fruits can be harmful for our teeth because of their concentrated sugar content and stickiness. Dried raisins, prunes, apricots, anjeer are gummy. So, parts of them easily adhere to teeth and the sugar in them encourage bacteria in the mouth that erodes tooth enamel. Dried fruits are packed with non-soluble cellulose fibre, which can bind and trap sugars on and around the tooth, making it worse than sweets. Chewing regular fruits, however, is less harmful, because even though they have sugar too, they aren't as sticky and the saliva helps in cleaning up.

Chemical Constituents of Fruits and Nuts

Fruit	*Antioxidants*	*Chemical Compounds*
Apple		Quercetin, Epicatechin, Procyanidin B2, Elastin and Collagen
Apricot	Antioxidants	Lutein, Zeaxanthin, β-cryptoxanthin, Carotenes, Laetrile (B_{17}).
Avacoda	Monosaturated fatty acid Poly saturated fatty acid Flesh	Oleic acid and Palmitoleic acid, ω-3 –fatty acids Omega -6 –poly- unsaturated fatty acid, Tannin (poly phenolic compound). Cryptoxanthin, Lutein, Xeaxanthin, β and α carotenes
Almond	Unsaturated fatty acids Amino acid	Oleic, Palmitoleic acids, Alpha-tocophenol, Arginine. (helps relax blood vessels and improve blood circulation) and Folic acid.
Banana		Bromelain (increases libido), TNF, Vitamin B6(Pyridoxine).
Blueberry	Aroma Poly phenolic anthocyanidin Flanovoids	Amyl acetate, Pterostibene (remedy for colon and liver cancers) Chlorogenic acid!, Tannins, Myricetin, Quercetin, Kaempferol, Carotene β, Lutein, Zeaxanthin.
Black berries	Phenolic flavonoid phytochemical	Xylitol <-a calorie sugar substitute Anthocyanins, Ellagic acid tannins, Quercetin, Gallic acid, Cyanidins, Pelargonidins, Catechins, Kaempferol and Salicylic acid.
Brazil nuts	Mono un saturated fatty acid	Oleic acid, Palmitoleic acid and Selanium, Copper, Mg. Mn.
Cranberries	Anti-oxidant compounds	Oligomeric Proanthocyanidins (OPC'S), Flavonoids, Cyanidin, Peonidin and Quercetin.
Cashewnut		Oleic acid and Palmitoleic acid
Citrus Fruits	Bitterness Aromatic principae	Narangi and Limonoids (Flavonoid compounds) Limonene and other essential Oils
Cranberry		Proanthocyanidins (prevents plaque formation on teeth)
Chestnut		Oleic acid and Palmitoleic acid, Vit. C, Folate
Coconut	Saturated fatty acids	Lauric acid*, Cytokinins**
Cherry	Poly phenolic flavonoid compounds	Anthiocyanin glycosides, Melatonin***, Phenols, Quercetin
Cherimoya	Annonaceous acetogenins	Asimicin, Bullatacinare#
Eraneberries	Antioxidant	Proanthocyanidin
Dates	Anthocyanidin flavonoids	Cyanidin, Peonidin and Quercetin, β carotene, Tannins.
Drum stick		Terigospermin. Moringin
Fig	Poly phenolic flavonoid antioxidant	Chlorogenic acid, Carotenes, Luteins, Tannins, ω-3-fatty acids, ω-6-fatty acids. Lycopene, Cryptoxanthin.
Grapes	Polyphenolic Phyto-chemical Compound	Resveratrol, Anthocyanins, Catechins
Grape fruit	Antioxidant	Naringenin, Lycopene, β-carotene, Xanthin, Lutein, Pectin
Guava		Lycopene, Vit. A, Vitamin C, Potassium.
Jackfruit		Carotene β, Xanthin, Lutein, Cryproxanthin-β, Vitamin A, Vit. B6, FolateJambosine and Glycoside

Contd...

Table–*Contd...*

Fruit	*Antioxidants*	*Chemical Compounds*
Jamun		Jambolin or antimellin
Kiwi fruit		Omega-3-fatty acid, Serotonin, Manganese, Potassium, Vit. C and K
Kumquat	Essential Oils	Limonene, Pinene, α-bergamotene, Cryophyllene, α-humulene, α-muurolene, Vitamin C.
LemonLime	Phytochemical	Hesperetin and Naringenin, Ascorbic acid. Hesperidine
Liquat	Antooxidants	Chlorogenic acid, Neo-chlorogenic acid, Hydroxybenzoic acid, Feruloylquinic Acid, Protocatechuic acid, Epicatechin, Coumari acid, Ferulic acid
Litchi	Polyphenol	Oligonol+ Vit. C, Copper.
Mango	Flavonoids Unripe fruits Antioxidants	β-carotene, α-carotene, β- cryptoxanthin, Vit. A and C, Vit. B6 Anacardic acid Glutamine acid (An important. protein for concentration and memory)
Mulbery	Berries (Phyto-chemicals)	Anthocyanins, Resveratrol, Zeaxanthin
Orange	Phytochemicals	Hesperetin and Narigenin- Flavonoids.
Olive	Pungent/bitter taste	Oleuropein aglycone- a poly phenolOleuropein- anti-oxidant, anti- inflammatory, anti ageing, anti-cancer, anti-viral and anti-microbial and Oleocanthal-α phytonutrient.
Papaya	Proteolytic enzyme Alkaloid	Papain Carpaine++, β-carotene, Lutein, Zeaxanthin, Cryptoxanthins Potassium.
Peaches	Phenolic anti oxidants	β-carotene, lutein, Zeaxanthin, β-cryptoxanthin, K, Fluride, Fe.
Pears		Non Soluble Polysaccharide (NSP), cOPPER
Persimmon	Polyphenolic antioxidants	Catechins## and Gallocatechins, Betulinic acid, Zeaxanthin, Lycopene, Vitamin C.
Pineapple	Proteolytic enzyme	Bromelain, Folates, Copper, Manganese, Vit. C
Plums		Sorbitol, Isatin, β-carotene, Vit. A and C, Carotene, Oxalic acid
Pome-granate	Polyphenolic oxidant compound	Punicalagin and Ellagic acid
Peanuts	Polyphenolic antioxidants	P-coumaric acid^, Resveratrol^^, Vit. C, Folates, Niacin.
Peacannuts	Mono unsaturated fatty acid	Oleic acid
	Ployphenolic antioxidate	Ellagic acid
Pine nuts	Monounsaturated fatty acids	Oleic acid, (ω-6) Pinolenic acid^^^, Vitamin E.
Pistachio		Oleic acid, carotenes, Vitamin E, Copper, Iron.
Raspberry	Phenolic flavonoids phytochemicals	**Xylitol<,** Anthocyanins, Ellagic acid (a tannin), Quercetin, Gallic acid, Cyanidins, Pelargonidins, Catechins, Kaemferol, Salicylic acid, Vitamin acid.

Contd...

Table–*Contd...*

Fruit	*Antioxidants*	*Chemical Compounds*
Strawberry	Phenolic flavonoids	Anthocyanins, Ellagic acid, Vitamin C, Manganese
Sita phal	Seeds/Fruits	Anonine, michelabbine, oxoushinsuie, corydine, isocorydine, alanine, β-alanine,
Walnut+++	Monounsaturated fatty acid. Phytochemical	**Oleic acid** and **Omega -3-fatty acid** like linoleic acid, α-linolenic acid (ALA) and Arachidonic acid. Melatonin, Ellagic acid, Vit. E, Carotenes. Jugladic acid

<: Alow calorie sugar substitute present in the fruit fibres, absorbs more slowly than sugar and does not contribute to high blood sugar levels.

!: Helps lower blood sugar levels and controls blood glucose levels in type 11 diabetes mellitus condition.

*: Helps rise HDL cholesterol level in blood.

**: Anti ageing anti cercirogenic and anti thrombotic effects.

***: Can cross the blood brain barrier easily and produce soothing effects on the brain neurons, calming down nervous system irritibility which helps relieve neurosis, insomnia and headache condition.

#: Are powerful cytotoxins and have been found to have anti cancer, anti malerial and anti helminthes properties.

+: Antioxidant, anti –influenza. It improves blood flow in organs, reduces weight and protect skin from harmful UV Rays.

++: Which could be dangerous when eaten in high doses (unripe fruit, seeds, latex, leaves).

##: Have anti infective, anti inflammatory and anti hemorrhagic properties.

^: Is believed to reduce the risk of stomatch cancer by reducing the formation of carcinogenic nitrosamines.

^^: It reduces stroke risk by alteration of molecular mechanism in blood vessels and by increasing production of vasodilator hormone, nitric oxide.

^^^: Causes the triggering of hunger suppressant enzymes cholecystokinin and glucagon like peptide. In addition Pinolenic acid may have LDL lowering properties by enhancing hepatic LDL uptake.

+++: Walnut contains 13541 μmol TE (Trolex equivalents) of oxidant radical absorbantcapacity (ORAC).

Based on a study in 2010, reported by the Dietary Guidelines of America, you need 1.5 - 2 cups of fruits daily, but the amount depends on the number of calories per day recommended for your healthy weight. Also, it is recommended that you eat five servings of fruits and veggies a day, but it depends on your age and gender: Women between the ages of 19-50 should eat 7-8 servings, while men should eat 8-10 servings. And for those above 50, men and women should consume seven servings of fruits and vegetables a day (Trina Remedios, 2013).

Fruits help fight cancer, obesity and heart disease. Fruits provide nutritional vitamins. Certain fruits contain a good amount of natural sugar and fluids, which also keep us hydrated. Fruits are rich in fibre, and are great for bowel movement and weight loss. It is also advised that you consume fruits whole, instead of skinned or juiced. Always opt for fresh fruits as their nutritional value is higher and try different seasonal fruits from time to time.

Eating many of the numerous coloured fruits and vegetables, can supply your body with a wide range of vitamins and minerals like fibre, folate, potassium and vitamins A and C. Some examples include green spinach, orange, sweet potato, black beans, yellow corn, purple plums, red watermelon and white onion.

Orange, red and dark green vegetables like pumpkin, carrots, cantaloupe, kale and yellow corn are rich source of natural alpha- carotene, where as potatoes, carrots and pumpkins, spinach and leafy vegetables are rich in beta-carotene. The red- orange colour in carrots shows the presence of beta-carotene which helps boost your immune system and provides your body with antioxdants.

Green vegetables, such as spinach and kale contain lutein and zeaxanthin, which may help protect against age related diseases. Green foods are particularly good for the circulation of system. Green foods include minerals and vitamins B.

Red foods are specially for reducing internal damage. Red foods include phytochemicals so take Watermelon, Tomatoes, Strawberry.

Yellow foods are specially for reducing the risk of developing cataracts. So take Yellow Corn, Oranges, Pineapple.

The fibre and water content of fruits and vegetables help to maintain a healthy digeative tract and weight control (as they are lower in calories than most packed foods).

Diets that are high in insoluble fibre may offer the best protection against Diverticulosis.

Fruits and vegetables are high in cellulose–a type of insoluble fibre.

Studies have reported that high fruit and vegetable intake can reduce the risk of a stroke by up to 25 per cent.

Unadorned fruits and vegetables are naturally low-in calories, and may be an important way to prevent and treat obesity. One study concluded that diets that are high in fibre are associated with lower body weight.

The higher intake of fruits and vegetables seemed to increase the ventilation function of the lungs. All fruits are rich in vitamins and plant derived nutrients known as phytonutrients like beta-carotenes, flavonoids and anthocyanins. Vitamin A and C commonly found in most of the fruits have powerful antioxidant properties.

Another helpful benefit of fruits and fruit juices is their ability to promote detoxification in the human body. Fruits help to cleanse the body, especially those with high acid levels. Tomatoes, pineapples, and citruses such as oranges, red grapefruits, and lemons are known for their detoxifying properties. While these fruits promote cleansing, they still provide the body with a high boost of Vitamin C!

Vitamin A

Vitamin-A is also required for maintaining healthy mucus membranes and skin. Consumption of natural fruits rich in carotenes helps protect body from lung and oral cavity cancers. Vitamin A is a powerful antioxidant and is essential for vision. It is available in carrot, melon, mangoes, apricot etc. Carotenoids give fruits and vegetables their bright orange, red, or yellow colours, and are a source of dietary vitamin A. Findings suggest that consuming carotenoid-rich foods may help prevent or delay the onset of ALS (amyotrophic lateral sclerosis).

Vitamin C

Vitamin C is a powerful natural antioxidant. Regular consumption of foods rich in vitamin C helps body develop resistance against infectious agents and scavenge harmful, pro-inflammatory free radicals from the body. Further, the vitamin is required for collagen synthesis and wound healing, anti-viral, anti-cancer activity and helps prevent from neuro-degenerative diseases, arthritis, and cold/fever. etc by removing oxidant free-radicals from the body. Collagen is the main structural protein in the body required for maintaining the integrity of blood vessels, skin, organs, and bones. Vitamin-C is one of powerful natural anti-oxidant which has many essential roles like, anti-viral, anti-cancer activity. Vitamin C helps strengthen your immune system and protects you from infections like cold and cough, which sap your energy. Also, when you are outdoors, you are more susceptible to bacterial infections which hamper your health. Thus ensure that you consume foods and fruits rich in vitamin C.

Vitamin C

Banans, sweet potato and beans. Cure your crankiness and satiate your food carvings by increasing vitaminB in your diet.

Vitamin B-6

Vitamin B-6 or **pyridoxine** is required for GABA hormone production in the brain. It also controls homocystiene levels in the blood, which may otherwise be harmful to blood vessels resulting in CAD and stroke.

Vitamin E

Vitamin E is a powerful lipid soluble antioxidant, required for maintaining the integrity of cell membrane of mucus membranes and skin by protecting it from harmful oxygen free radicals. Present in avocados, soaked almounds

Vitamin K

Vitamin K is essential for many coagulant factors in the blood as well as in bone metabolism. Vitamin K has potential role in the increase of bone mass by promoting osteotrophic activity in the bone. It also has established role in Alzheimer's disease patients by limiting neuronal damage in the brain.

Carotene

Consumption of natural fruits rich in carotene is known to protect from lung and oral cavity cancers. Alpha carotene reduces the risk of lung cancer and boosts the immune system. Beta- carotene functions as a powerful antioxidant protecting the protein, fat and DNA in cells from free radicals damage.

Folic Acid

Folic acid is essential for DNA synthesis and cell division.

Resveratrol

This antioxidant has anti-ageing and disease preventing properties. It is very good for heart health, because it helps to reduce inflammation and bad cholesterol. Besides, it also helps to prevent cancer and Alzheimer's.

Zeaxanthin

Zeaxanthin, an important dietary carotenoid, selectively absorbed into the retinal macula lutea in the eye where it is thought to provide antioxidant and protective light-filtering functions; thus it protects eyes from "Age related macular degeneration" (ARMD) disease in the elderly. Zeaxanthin has photo-filtering effects on UV rays and thus protects eyes from age related macular disease in the elderly populations. Examples of fruits and vegetables that contain lutein and zeaxanthin are spinach, collards, kale and sweet corn.

Anthocyanins

This antioxidant possesses great anti-inflammatory and anti-carcinogenic property. It helps in lowering the risk of cardiovascular diseases, obesity and diabetes. Anthocyanins are red, purple or blue pigments found in many fruits and vegetables, especially concentrated in their skin, known to have powerful anti-oxidant properties. They are another class of polyphenolic anti-oxidants present abundantly in the red grapes. These phyto-chemicals have been found to have anti-allergic, anti-inflammatory, anti-microbial, as well as anti-cancer activities. Anthocyanin is a naturally occuring bioactive compound in berries and other fruits and vegetables including brinjals, blackberries, blackcurrants, are most likely to reduce the risk of developing parkinson's disease(Anonymous, 2012). Anthocyanins which give colour to most red, purple and blue fruits and vegetables, can possibly slow the growth of colon cancer cells.

Lycopene

Lycopene is the most powerful flavonoid antioxidant, studies have shown that it protects skin damage from UV rays, and offers protection against prostate cancer. Carrots, yams, sweet potatoes, squash, tomatoes, apricots, beets etc. contain Vitamin A and Lycopene, which slow down the growth of aggressive cancers.

Omega-3-fatty acids

Consumption of foods rich in omega-3 fatty acids may reduce the risk of coronary heart disease, stroke, and help prevent development of ADHD, autism, and other developmental differences in children. For omega-6 use safflower, sunflower and corn oils. Omega-3 fatty acids help prevent toxins from entering the cell membranes, aids nutrients in entering and helps in excretion of waste products. Omega -3 also helps prevent inflammation which can damage skin. Lack of Omega 3 may be linked to pre mature birth or low weight babies. Fish is the best source of Omega 3 but there are vegetarian sources of Omega 3 like flax seeds and walnuts. Omega 3 is responsible for regulating hormones and a healthy uterus is essential for improving fertility. Omega 3 also improves the blood supply to reproductive organs (Trina Remedios, 2013).

Folate

Folate helps prevent birth defects such as spina bifida. Fruits and vegetables such as oranges, spinach, broccoli, and dried beans are good sources of folate.

Antioxidants

Antioxidants are essential to optimizing health by helping to combat the free radicals that can damage cellular structures as well as DNA.

Melatonin

Melatonin can cross the blood-brain barrier easily and produces soothing effects on the brain neurons, calming down nervous system irritability which helps relieve neurosis, insomnia and headache conditions.

Pectin

Pectin, which by acting as bulk laxative helps to protect the colon mucous membrane by decreasing exposure time to toxic substances in the colon as well as binding to cancer causing chemicals in the colon. Pectin has also been shown to reduce blood cholesterol levels by decreasing its re-absorption in the colon by binding bile acids resulting in its excretion from the body.

Tannins

Tannins have shown to lower blood pressure and stimulate the immune system, while anthocyanins help protect blood vessels and reduce inflammation.

Fluoride

Fluoride is a component of bones and teeth and is important for prevention of dental caries.

Calcium

Calcium is an important mineral that is an essential constituent of bone and teeth, and required by the body for muscle contraction, blood clotting, and nerve impulse conduction.

Potassium

Potassium is a heart-healthy mineral, it is an important component of cell and body fluids helps controlling heart rate and blood pressure; thus counters the bad influences of sodium, thus offers protection against stroke and coronary heart diseases. Yet another mineral that has a significant effect on your PMS mmod swings. Fresh fruit salads, with oranges and apples are good options for an early evening snacks.

Manganese

Manganese is used by the body as a co-factor for the antioxidant enzyme, *superoxide dismutase*. It is essential for bone strengthening and has cardiac-protective role as well. Adding magnesium (Spinach, beans, cashews) to your diet reduces PMS symptoms drastically. Foods that are rich in magnesium help regulate serotonin, which in simple terms is the mood balancer of the brain. Serotonin, a brain chemical, plays a vital role in mood fluctuatrions and pain sensitivity. Thus keeping it under control, significantly influences the onset of PMS. Increasing your magnesium intake will help cure symptoms like breast tenderness and abdomnal bloating. Magnesium is essential for bone growth.

Copper

It is required in the bone metabolism as well as in production of white and red blood cells. Copper is a co-factor for many vital enzymes, including cytochrome c-oxidase and superoxide dismutase (other minerals function as co-factors for this enzyme are manganese and zinc

Iron

Iron is essential for red blood cell production. Lack of iron in our diet can cause loss of stamina. You can get iron from meat, beans, nuts and some vegetables such as broccoli and spinach. Iron is required for red blood cell formation as well for cellular oxidation. Iron, being a component of hemoglobin inside the red blood cells, determines the oxygen carrying capacity of the blood.

Chromium

Chromium, the trace mineral that helps tissue cells respond appropriately to insulin levels in the blood; thus helps facilitate insulin action and control sugar levels in diabetes.

What are Some of the Good Things in Fruits and Vegetables?

- ☆ **Vitamins**
- ☆ **Minerals**
- ☆ **Flavonoids** - plant chemicals that act like antioxidants
- ☆ **Saponins** - plant chemicals that have a bitter taste
- ☆ **Phenols** - organic compounds in foods
- ☆ **Carotenoids** - vitamin A-like compounds
- ☆ **Isothiocyanates** - sulphur-containing compounds.
- ☆ Several types of **dietary fibre**

Functions of Antioxidants and Phytonutrients

- ☆ They are scavengers of free radicals
- ☆ Anti-cancer properties

- ☆ Anti-ageing properties
- ☆ Anti-allergic properties
- ☆ Natural immune boosters.

Vegetables and Fruits

1. The orange foods help to prevent cancer. Carrots (Gajar), Pumpkin (Kaddu), Sweet Potatoes (Shakar kand), Apricots (Khubani), Cantaloupe (Kharbuja) and other orange foods are particularly good for our eyes. The orange foods contains vitamin A.
2. White and green foods contain a variety of minerals. Include Garlic, Onions in your meals which are anti-bacterial and anti fungal.
3. Red and Blue foods are protecting against heart disease by improving circulation of blood. Add foods in your meals like Apple, Beets (Chukandar), black grapes. Overall, blue colour foods help in boosting overall health, by preventing diseases and promoting longevity.They are also good memory boosting foods. The best blue foods are blueberries, blackberries, figs, plums, raisins and purple cabbage. Food is blue in colour because of its richness in health-boosting phytochemicals known as anthocyanins and resveratrol.

Apple (*Malus domestica*)

An apple a day keeps the doctor away is cliched, but its health benefits are undisputable. Apple's antioxidant property prevents cell and tissue damage. Studies by nutritionists have shown that apples contain abundant amounts of **elastin** and **collagen** that help keep the skin young. Applying a mixture of mashed apple, honey, rose water and oatmeal can act as a great exfoliating mask on your skin. An apple a day, will surely keep the doctor away because they contain **guercetin**, a photochemical containing ant-inflammatory properties. It also helps in prevention of blood clots. Eat apples for breakfast with your cereal or eat them as a snack when you're hungry instead to pigging out on deep-fried chips. The apple is a highly nutritive food. It contains minerals and vitamins in abundance. The food value of the apple is chiefly constituted by its contents of sugar which ranges from 9 to 51 per cent. Of this, fruit sugar constitutes 60 per cent and glucose 25 pet cent and cane sugar only 15 per cent. Apples are a source of both soluble and insoluble fibre. Soluble fibre such as pectin actually helps to prevent cholesterol buildup in the lining of blood vessel walls, thus reducing the incident of atherosclerosis and heart disease. The insoluble fibre in apples provides bulk in the intestinal tract, holding water to cleanse and move food quickly through the digestive system. Although an apple has a low vitamin C content, it has antioxidants and flavonoid which enhances the activity of vitamin C thereby helping to lower the risks of colon cancer, heart attack and stroke.

It is a good idea to eat apples with their skin. Almost half of the vitamin C content is just underneath the skin. Eating the skin also increases insoluble fibre content. Most of an apple's fragrance cells are also concentrated in the skin and as they ripen, the skin cells develop more aroma and flavour. The skin of apple should not be discarded when taking it in raw form as the skin and the flesh just below it contain more vitamin C than the inner flesh. The vitamin content decreases gradually towards the center of the fruit. The skin also contains five times more vitamin A than the flesh.

They have been used as a beauty aid for decades. A cup of apple juice added to your bath will cleanse and smoothen your skin. Apple juice applied to your scalp helps prevent dandruff. A final rinse of your hair with apple juice after shampooing brings about an added shine. Apples are also a good conditioner and toner, and help fight acne.

Peeling apples before eating can cause loss of many vital nutrients. Laboratory analysis has shown that 100 gm of unpeeled apple contains about 16 mg of vitamin C. The same quantity of peeled fruits contain only 8 mg this vitamin. Nearly half of an apple's vitamin C content lies beneath its skin. It contains as calcium, potassium, phosphorus, folate and iron. The peel is a significant source of both insoluble and fat –soluble fibre called pectin. It accounts for around 50 per cent of total fibre of the apple. Peels are rich in flavonods and phenolic acids. It has triterpenoids. These compounds can inhibit or kill cancer cells.

Health Benefits of Apple

Apples combine certain nutrients in a way that sets them apart from all other fruits and makes them a food of choice for achieving several health goals. Here's what apples can do for you when it comes to your health. Delicious and crunchy apple is one of the popular fruit that contain an impressive list of essential nutrients, which are required for normal growth and development and overall nutritional well-being (Sobiya, 2013).

Unique Support for Heart Health

Apples are a rich source of phytonutrients known as **quercetin**, **catechin**, **phloridzin** and **chlorogenic acid**. These unique phytonutrients help protect the cardiovascular system from oxygen-related damage. In addition, the high fibre content (pectin) in apples helps keep your LDL ("bad") cholesterol levels low and under control. People who eat two apples per day may lower their cholesterol by as much as 16 per cent.

Alzheimer's Prevention

A study on mice at Cornell University found that the quercetin in apples may protect brain cells from the kind of free radical damage that may lead to Alzheimer's disease.

Lung Cancer Prevention

According to a study of 10,000 people, those who ate the most apples had a 50 per cent lower risk of developing lung cancer. Researchers believe this is due to the high levels of the flavonoids quercetin and naringin in apples.

Breast Cancer Prevention

A Cornell University study found that eating one apple per day can reduce the risk of **breast cancer** by 17 per cent. By eating three apples per day can reduce the risk by 39 per cent and those who eat six apples per day can reduce their risk by 44 per cent.

Colon Cancer Prevention

One study found that rats fed an extract from apple skins had a 43 perc ent lower risk of colon cancer. Other research shows that the pectin in apples reduces the risk of colon cancer and helps maintain a healthy digestive tract.

Liver Cancer Prevention

Research found that rats fed an extract from apple skins had a 57 per cent lower risk of liver cancer.

Diabetes Management

The pectin in apples supplies galacturonic acid to the body which lowers the body's need for insulin and may help in the management of diabetes.

Weight Loss

A Brazilian study found that women who ate three apples or pears per day lost more weight while dieting than women who did not eat fruit while dieting.

Bone Protection

French researchers found that a flavanoid called phloridzin that is found only in apples may protect post-menopausal women from osteoporosis and may also increase bone density. Boron, another ingredient in apples, also strengthens bones.

Asthma Help

One recent study shows that children with asthma who drank apple juice on a daily basis suffered from less wheezing than children who drank apple juice only once per month. Another study showed that children born to women who eat a lot of apples during pregnancy have lower rates of asthma than children whose mothers ate few apples.

- ☆ Apples are low in calories; 100 g of fresh fruit slices provide only 50 calories. The fruits contain no saturated fats or cholesterol; but rich in dietary fibre, which helps, prevent absorption of dietary LDL cholesterol in the gut. The dietary fibres also help protect the mucous membrane of the colon from exposure to toxic substances by binding to cancer causing chemicals in the colon.
- ☆ It contains good quantities of vitamin-C and beta-carotene. Vitamin C is a powerful natural antioxidant. Consumption of foods rich in vitamin C helps body develop resistance against infectious agents and scavenge harmful, pro-inflammatory free radicals from the body.
- ☆ They are rich in antioxidant phyto-nutrients flavonoids and polyphenols. The important flavonoids in apples are **quercetin, epicatechin,** and **procyanidin B2**.
- ☆ Apples are also good in **tartaric acid** that gives tart flavour to them. These compounds help body protect from deleterious effects of free radicals.

- Apples are good source of B-complex vitamins such as riboflavin, thiamin, and pyridoxine (vitamin B-6). Together these vitamins help as co-factors for enzymes in metabolism as well as in various synthetic functions inside the body.
- Apple also contains small amount of minerals like potassium, phosphorus, and calcium.
- Apple is a fair source of fibre.
- Apple contains Pectin and Malic acid.
- Iron contained in the apple helps in formation of blood.
- Raw apples are good for constipation.
- Cooked or baked apples are good for diarrhea.
- Apples have been found useful in acute and chronic dysentery among children. Ripe and sweet apples should be crushed into pulp and given to the child several times a day.
- Apples are of special value to heart patients. They are rich in potassium and phosphorus but low in sodium. It is also useful for patients of high blood pressure.
- Apple is also said to be beneficial to gout patients caused by increase of uric acid in blood.
- Apples, boiled to a jelly, make a very good liniment for rheumatic pains. They should be rubbed freely on the affected area.
- Sweet apples are valuable in dry hacking cough.
- Apples are useful in kidney stones.
- The apple peel water is an excellent medicine for the inflamed eyes as an eye wash. The over-ripe apples are useful as a poultice for sore eyes. The pulp is applied over the closed eyes.
- Tooth-decay can be prevented by regular consumption of apples as they possess a mouth cleansing property.
- The apple is the best fruit to tone up a weak and run-down patient. It removes deficiencies of vital organs and makes the body stout and strong. It tones up the body and the brain as it contains more phosphorus and iron than any other fruit or vegetable. Its regular consumption with milk promotes health and youthfulness and helps build healthy and bright skin.

Here are some serving tips

Eat apple fruit as they are with skin to get maximum health benefits.

- Sliced apple turns brown (enzymatic brownish discolouration) on exposure to air due to conversion in iron form from *ferrous oxide* to *ferric oxide*. If you have to serve them sliced, rinse slices in water added with few drops of fresh lemon.
- Cloudy apple juice is a good drink with dinner.
- Apple fruit is also used in the preparation of fruit jam, pie, and fruit sala

Here are five things you may not know you can do with an apple:

- ☆ Ease your allergies. Apples are abundant in the plant compound **quercetin**. The compound seems to slow down the secretion of histamine, a chemical your body releases during an allergic reaction. A 2000 study in the journal Thorax also found that men who ate five or more apples a week enjoyed better lung function.
- ☆ Mitigate a migraine. Some studies have found that the smell of green apples can reduce the brain-thumping symptoms of a migraine, and can shorten a migraine episode. Researchers are also trying to figure out if it could help alleviate joint pain. If you feel a headache coming on, you can cut a green apple in half and smell it. You be the judge if it works or not! Just don't sniff artificial apple fragrances–they could cause a headache!
- ☆ Ripen other apples, and other fruit. Apples, like peaches, pears, and bananas, release off ethylene gas, which speeds ripening. If you want to ripen certain fruits fast, put them in a paper bag with a ripened apple. On the flipside, if you don't want your fruit to ripen too quickly, make sure you don't store these fruits close to each other in the fridge.
- ☆ Clean your hair (with apple cider vinegar). To combat oily hair, put a teaspoon of apple cider vinegar in a pint of water, and rinse it through your hair while taking a bath or shower. The vinegar solution will take out soap residue that can cling to and weigh down oily hair. The vinegar scent will go away when your hair dries.

Apricot (*Prunus armeniaca*)

It can be easily categorized as one of the most beneficial and healthy fruits. The shape of the stone fruit, apricot, is quite similar to that of a peach and has a velvety skin, which is golden orange in colour. It is slightly smaller in size when compared to a peach and has a seed nut in it. Though it is not appropriate for making juice, the fruit can definitely be blended with other fruit juices. The taste is very fruity and feels like a cross between peach and plum. Apricot is widely used in jams, making pastry and also distilling into brandy and liquor. The fruit, when taken raw, is a bit acidic in nature. However, the acidity decreases as it ripens and the sugar content increases. Read on to find nutritional value and benefits of eating apricot. The mineral content in apricots is high, which makes them helpful in overcoming disorders like anemia, tuberculosis, asthma, bronchitis, etc. The high copper and iron content in apricots increases the hemoglobin in the blood. The vitamin A content in apricots is high and helps in reducing skin disorders like pimples and zits. Vitamin A also aids in improving eye-sight. A certain substance called lycopene is present in apricots, which helps in the prevention of cancer and also protects the body from high cholesterol, thereby preventing heart disease. Eating apricots helps in removing gallstones and also destroys intestinal worms. It also aids in digestion, since it reacts as an alkaline, breaking down food particles. The fruit is considered to be a laxative and consuming it regularly will help you get rid of constipation. It is very helpful in getting rid of the waste in the body, due to its high fibre content.

Nutrients in apricots can help protect the heart and eyes, as well as provide the disease-fighting effects of fibre. The high beta-carotene content of apricots makes them important heart health foods. An apricot in its raw state is somewhat acidic but the acidity decreases as it ripens and its sugar content increases. When it ripens, the vitamin A within also doubles.

Apricots contain nutrients such as vitamin A that promote good vision. Vitamin A, a powerful antioxidant, quenches free radical damage to cells and tissues. Free radical damage can injure the eyes' lenses.

The degenerative effect of free radicals, or oxidative stress, may lead to cataracts or damage the blood supply to the eyes and cause macular degeneration. Researchers who studied over 50,000 registered nurses found that women who had the highest vitamin A intake reduced their risk of developing cataracts by nearly 40 per cent. Apricots are a good source of fibre, which has a wealth of benefits. These include preventing constipation and digestive conditions such as diverticulosis. A healthy, whole-foods diet should include apricots as a delicious way to add to your fibre intake of the body.

The apricot has natural sugars that are easily digestible, vitamins A and C, Riboflavin and Niacin, and also minerals like calcium, phosphorus, iron and traces of sodium, sulphur, manganese, cobalt and bromine., Interestingly, when the fruit ripens, its vitamin A content doubles. When the fruit is dried not only does it double in calories, but its calcium, phosphorus and iron content also increases.

Nutritional Benefits

- ☆ The apricot has highly health-building virtues. The fresh fruit is rich in easily-digestible natural sugars, vitamins A and C, riboflavin (B2) and niacin (B3).
- ☆ It is also an excellent source of minerals like calcium, phosphorus, iron and traces of sodium, sulphur, manganese, cobalt and bromine. Apricots are often dried, cooked into pastry or eaten as jam. The calories in apricots multiply many times over when dried, but the amount of calcium, phosphorus and iron also increased significantly.
- ☆ The beta-carotene and lycopene in this golden fruit helps protect the LDL cholesterol from oxidation, which in turn helps prevent heart disease. The apricot seed is a nut that is rich in protein and fat like any other nuts. It also has an extremely high content of vitamin B17 which is known as Laetrile. Daily consumption of this seed is claimed to be highly effective in preventing cancer. Cancer patients on Laetrile Cancer Therapy have reported that their tumors have shrunk with high doses of vitamin B17.
- ☆ These bitter seeds may be chopped up or ground and swallowed with a teaspoon of honey.
- ☆ Studies show that the fruit, kernel, oil and flowers of the apricot have been used in medicine and medical treatment from ancient days.
- ☆ Its high iron content makes it beneficial for people suffering from anemia.

- ☆ Its cellulose and pectin content make apricot a gentle laxative and therefore, useful to treat constipation.
- ☆ It aids digestion.
- ☆ Its high Vitamin A content improves eyesight.
- ☆ The beta-carotene and lycopene in this golden fruit help protect LDL or good cholesterol from oxidation, which in turn helps prevent heart disease.

Medicinal and non-food uses

- ☆ Due to their high fibre to volume ratio, dried apricots are sometimes used to relieve constipation or induce diarrhea. Effects can be felt after eating as few as three.

Health Benefits

- ☆ The fruit, kernel (inner softer part of the seed), oil and flowers of the apricot have always been used in medicine and medical treatment from ancient days.
- ☆ The kernel yields an oil that is similar to that of the almond and is widely used for their sedative, anti-spasmodic relief to strained muscles. It is also useful for healing of wounds, expelling worms and as a general health tonic.
- ☆ **Anemia:** The high content of iron in apricot makes it an excellent food for anemia sufferers. The small but essential amount of copper in the fruit makes the iron available to the body. Liberal consumption of apricot can increase the production of hemoglobin in the body. This is ideal for women after their menstrual cycle, especially those with heavy flow.
- ☆ **Constipation:** The cellulose and pectin content in apricot is a gentle laxative and are effective in the treatment of constipation. The insoluble cellulose acts as a roughage which helps the bowel movement. The pectin absorbs and retains water, thereby increasing bulk to stools, aiding in smooth bowel movement.
- ☆ **Digestion:** Take an apricot before meal to aid digestion, as it has an alkaline reaction in the digestive system.
- ☆ **Eyes/Vision:** The high amount of vitamin A (especially when dried) is essential to maintain or improve eyesight. Insufficiency of this vitamin can cause night blindness and impair sight.
- ☆ **Fever:** Blend some honey and apricots with some mineral water and drink to cool down fevers. It quenches the thirst and effectively eliminates the waste products from the body.
- ☆ **Skin Problem**: Juice fresh apricot leaves and apply on scabies, eczema, sun-burn or skin itchiness, for that cool, soothing feeling.
- ☆ Nutrients in apricots can help protect the heart and eyes, as well as provide the disease-fighting effects of fibre. The high beta-carotene content of apricots makes them important heart health foods. Beta-carotene helps

protect LDL cholesterol from oxidation, which may help prevent heart disease.

- Apricots contain nutrients such as vitamin A that promote good vision. Vitamin A, a powerful antioxidant, quenches free radical damage to cells and tissues. Free radical damage can injure the eyes' lenses.
- The degenerative effect of free radicals, or oxidative stress, may lead to cataracts or damage the blood supply to the eyes and cause macular degeneration.
- Apricots are a good source of fibre, which has a wealth of benefits including preventing constipation and digestive conditions such as diverticulosis.
- Nutrients in **apricots** can help protect the heart and eyes, as well as provide the disease-fighting effects of fibre. The high beta-carotene content of apricots makes them important heart health foods.
- An apricot in its raw state is somewhat acidic but the acidity decreases as it ripens and its sugar content increases. When it ripens, the vitamin A within also doubles.

Consumption Tips

- Apricots are usually picked when they are still firm. An unripe apricot is often yellow and hard. When ripe and soft, its colour turn a consistent golden-orange hue. At this time, handle the fruit with care as it is easily bruised.
- Stored in the fridge, these fruits can last for three or four days. When overripe, the fruit turns soft and mushy.

Caution

- Fresh apricots contain a small amount of oxalates. Individuals with a history of calcium oxalate-containing kidney stones should not consume too much of this fruit.
- Whereas dried apricots contain sulphur-containing compounds such as sulphur dioxide. These compounds may cause adverse reactions in people who suffer from asthma.
- Apricot seeds are best known for the controversy surrounding their ability to purportedly cure cancer. The seeds, or kernels, are found inside pits of apricots. Aside from the controversial substance in them called amygydalin, the kernels have monounsaturated fats, protein and carbohydrates.

The Good

This food is very low in Saturated Fat, Cholesterol and Sodium. It is also a good source of Dietary Fibre and Potassium, and a very good source of Vitamin A and Vitamin C. But very high in sugar

Banana *(Musa acuminata colla)*

Banana is a tropically grown and is a very beneficial fruit. Though banana is a very nutritious food, it's significance is often underestimated as it is easily available, can grow in plenty, and are considerably cheap compared to other fruits, but this does not change the nutrition facts. Calcium can be considered as the main nutrition that a banana provides and as we know calcium helps in strengthening the bones and that is just a tip of the iceberg, banana has ample benefits both in terms of nutrition and energy. The fact that banana is very high in calories, but at the same time low in terms of fats is very less known and this makes it a very good option for those who are very much concerned about their weight and fitness level. It is also easy to digest making it a good food for small kids because they do not have trouble digesting it. Banana also is rich in anti-oxidants and it is a well known fact that anti-oxidants help in increasing immunity, increase energy, reduce high blood pressure, etc. Not only this, one of the several banana nutrition are minerals. Banana is loaded with trytophan that increases serotonin levels in the body. Apart from this, the fruit has many important acids, enzymes and physiologically important chemical compounds. Th banana develops its particular aroma on ripening because of the presence of a chemical compound called **Amylacetate.** Iron and potassium in banana are wholly available. Its intake help children to retain many mineral nutrients. The fruit is a fair source of B vitamins and calcium. Banana is a very energy rich fruit, *i. e.* two fruits can provide fuel for body for two hours. That is why banana is favoured fruit of sportspersons. Besides energy, bananas are one of the best sources of potassium, an essential mineral for maintaining normal blood pressure and heart function. An average banana contains a whopping 467mg of potassium and only 1 mg of sodium. So a banana a day helps in preventing high blood pressure and protect against hardening of arteries (Parmar, 20012). Bananas are a particularly good energy source and these are best eaten a few hours before you intend to exert yourself. Along with being great sources of carbs, they are also effective because they trigger the release of 'dopamine' – a chemical that builds your concentration and focus.

1. The fully ripe banana produces a substance a called TNF which has anti cancer properties. The degree of anti-cancer effect corresponds to the degree of ripeness of the fruit. Bananas increase the number of white blood cells, enhance the immunity of the body and produce TNF. Eating one to two bananas a day increases the body's immunity to diseases like cold, flu and others. The bananas with dark spots are eight times more effective in enhancing the property of white blood cells than the green one (Charnjit Parmar, 2012).
2. The fruit **banana** can be called the perfect food item for a human body. **Banana nutrition facts** tell us that it has all the components that are needed to make health, sound and playful. It helps our body to fight against **diseases** and provides useful **minerals**, like **potassium**. Banana makes your body to **generate probiotics**, reduces the side effect of salt and helps to **retain calcium** in the body. This tropical fruit can do wonders to our body nutrition system with its **minerals, vitamins** and **micro nutrition**.
3. This is one fruit that's abundantly available in India all through the year. We know it's a good source of iron, magnesium and potassium and helps reduce menstrual cramps. The effect of banana on skin too is not something that can be ignored. Banana is rich in vitamin A, B and E and hence works as an anti-aging agent. A fresh mashed banana facial can do wonders to your skin.
4. Boasting three naturally-occurring sugars - sucrose, fructose, and glucose - bananas are an obvious choice for a quick energy boost.
5. Bananas also contain fibre, which will help sustain the glucose levels in your blood, thus giving you a steady source of energy over a longer period of time.
6. Bananas are a great sex food. "Bananas are ideal to give you endurance in the bedroom as they are a rich source of vitamin B that converts carbohydrates into energy. It also helps your body to produce sex hormones such as testosterone. Having a banana a few hours before getting into action can keep you pepped up! Sexual pleasure and stimulus aside, here's a thought not many have had - why not nibble on quick energy foods so that you can last longer in bed? After all, you don't want your energy dipping just when your partner starts reciprocating. Nibble on this storehouse of B

vitamins and potassium slowly and it is sure to give you that extra kick of energy.

7. Bananas are known to contain bromelain enzyme which is known to give a boost to your libido. Bananas contain the **bromelain** enzyme, that increases libido and reverse impotence in men. Also, they are potent sources of potassium and B vitamins like riboflavin, which increase the body's overall energy levels. A great way to incorporate bananas into your sex fantasy is to place cut banana pieces on your lady love's body and slowly eat them.

- ☆ Researchers have found that eating three bananas cuts the risk of a stroke.
- ☆ In the latest research, scientists analysed data from eleven different studies - dating back to the mid-Sixties - and pooled the results to get an overall outcome. They found a daily potassium intake of around 1,600 mg, less than half the UK recommended daily amount for an adult of 3,500mg, were enough to lower stroke risk by more than a fifth.
- ☆ The average banana contains around 500 milligrams of potassium, which helps to lower blood pressure and controls the balance of fluids in the body.
- ☆ Too little potassium can lead to an irregular heartbeat, irritability, nausea and diarrhoea.
- ☆ If consumers ate more potassium-rich foods and also reduced their salt intake, the annual global death toll from strokes could be cut by more than a million a year.
- ☆ They said that having one banana for breakfast, one for lunch and one in the evening would provide enough potassium to reduce the chances of suffering a blood clot on the brain by around 21 per cent.
- ☆ The findings suggest that thousands of strokes could be prevented by the consumption of other potassium-rich foods such as spinach, nuts, milk, fish and lentils.
- ☆ Although some previous studies have suggested bananas could be important for controlling blood pressure and preventing strokes, results have not always been consistent.
- ☆ The nutrition that can be found in an average banana has many useful benefits to our body. As the fruit contains high vitamin C, this vitamin can protect a human body from infections. Not only this, the vitamin is useful for the combination of connective type of tissues. Vitamin C helps to retain iron in our body and enhances the blood formation system. Banana does not help us by Vitamin C only; it supplies another good mineral, which is Potassium.
- ☆ Potassium makes our body strong, and helps the growth of body muscles. Not only this, banana provides Potassium, which reduce high stoke diseases, *i. e.* hypertension. The mineral combination of Banana is really

helpful for body system. It contains only potassium and no sodium. This combination of nutrition can reduce high blood pressure as well. Banana frees us from the risks such as formation of red blood, as it contains Vitamin B6. Vitamin B6 also keeps the center of nerve functioning well.

Banana Nutrition Facts

- ☆ Vitamin C which is found in bananas, helps the body to defend and heal against infections. This vitamin also proves valuable in the synthesis of the connective tissue, absorption of iron and the formation of blood.
- ☆ Not only are bananas rich in vitamin C, they also contain potassium. Potassium is a mineral that helps in the building of muscles and protein synthesis. This is done as potassium stimulates nerve impulses for muscle contraction.
- ☆ A diet rich in potassium is said to reduce the risk of hypertension and stroke. As bananas are free from sodium and very rich in potassium, they can be included in the diet to reduce the risk of high blood pressure.
- ☆ Bananas contain three natural sugars, sucrose, fructose and glucose, along with fibre. A banana thus gives an instant and substantial boost of energy.
- ☆ A banana contains vitamin B6, which helps in the synthesis of antibodies in the immune system, apart from red blood cell formation, protein metabolism and functioning of the central nervous system.
- ☆ No other fruit contains more digestible carbohydrates than bananas. This is advantageous, because the body burns off calories from carbohydrate more easily and quickly than calories from fat or protein.
- ☆ Sweet banana, can provide you with the necessary energy to work all day long. This fruit gives our body immediate energy supply, because it has a sufficient amount of sugar and glucose. We know that banana contains carbohydrates, but it is a digestible carbohydrate. This type of micronutrient can help in reducing the amount of calories from our body. Eating a banana will help us in balancing the body fluid and getting dehydrated.
- ☆ Now we have a clear idea about all the nutrition's that a banana can supply to our body. All the minerals and vitamins that banana contains help us to build our body properly and helps our system to fight against the harmful diseases.
- ☆ Banana benefits our body by treating constipation. The fibre and minerals in a banana help the bowel move in a healthy way. Furthermore, when we eat food, the extra water needs to be absorbed to digest the foods properly. Banana provides pectin, which absorbs the water and helps in reducing constipation. Moreover, as we have discussed earlier, the potassium helps to prevent heart related diseases and bananas are a good source of potassium.
- ☆ Banana can be used in making remedial foods. It will reduce the acids in the stomach, which in turn results in avoidance of ulcers. Regular eating of

bananas does not only ensure the body growth, but also enhance the body immune. This provides you with a clear picture of banana nutrition facts.

- Rich in potassium and vitamins A, C and E, bananas are ideal for skin and hair (they help restore dull and damaged hair). While vitamin A restores natural oils of the skin, vitamin Erepairs damaged skin and lightens age spots. Vitamin C, on the other hand, prevents cell oxidation and wrinkle
- You don't need to eat bananas for the potassium! (Although it is present in bananas, potassium is the predominant nutrient among most all fruits and vegetables.)
- Bananas are high in sugar, so they should not be eaten if you have blood sugar problems.
- Don't eat bananas on an empty stomach; combining them with a bit of protein will help to normalize the insulin response caused by the sugar in the banana.
- Green-tipped bananas are better for your health than over-ripe bananas.

Health Benefits of Banana Fruit

- Banana fruit is rich in calories, but very low in fats. The fruit contains good amounts of health benefiting anti-oxidants, minerals, and vitamins.
- Banana pulp is composed of soft, easily digestible flesh with simple sugars like fructose and sucrose that when eaten replenishes energy and revitalizes the body instantly; thus, for these qualities, bananas are being used by athletes to get instant energy and as supplement food in the treatment plan for underweight children.
- The fruit contains good amount of soluble dietary fibre that helps normal bowel movements; thereby reducing constipation problems.
- It contains many health promoting flavonoid poly-phenolic antioxidants such as lutein, zeaxanthin, beta and alpha carotenes in small amounts. These compounds help act as protective scavengers against oxygen-derived free radicals and reactive oxygen species (ROS) that play a role in aging and various disease processes.
- Banana is a very good source of vitamin-B6 (pyridoxine), provides about 28 per cent of daily-recommended allowance. Pyridoxine is an important B-complex vitamin that has beneficial role in the treatment of neuritis, anemia, and decreasing homocystine (one of the causative factor for coronary artery disease (CHD) and stroke episodes) levels in the body.
- The fruit is also good source of vitamin-C. Consumption of foods rich in vitamin-C helps body develop resistance against infectious agents and scavenge harmful oxygen free radicals.
- Fresh bananas provide adequate levels of minerals like copper, magnesium, and manganese.

- ☆ Fresh banana is a very rich source of potassium. 100 g fruit provides 358 mg potassium. Potassium is an important component of cell and body fluids that helps control heart rate and blood pressure, countering bad effects of sodium.
- ☆ Once ripened, bananas are very fragile and start decaying in short time. In the field, bananas are usually harvested while they are green as it is easy to transport when the fruits are raw and firm.
- ☆ Ready to eat bananas should be quite firm, bright yellow in colour and emanate rich fragrance, and the skin should be peeled off easily. Ripened, fresh bananas are nutritionally enriched and sweeter in taste.
- ☆ Avoid mushy or damaged bananas, as they are un-appealing.

Now we have a clear idea about all the nutrition's that a banana can supply to our body. All the minerals and vitamins that banana contains help us to build our body properly and helps our system to fight against the harmful diseases.

- ☆ Bananas comes with nature gifted protective outer layer of skin, therefore are less likely contaminated by germs and dust.
- ☆ Just remove the peel and enjoy!
- ☆ Banana fruit sections are a great addition to the fruit salads.
- ☆ Fresh "banana-milkshake" with sugar syrup is a delicious drink.
- ☆ Bananas have also been used in the making of fruit jams.
- ☆ Banana fritters can be served with ice cream as well.
- ☆ Banana chips are a snack produced from dehydrated or fried banana or plantain slices.
- ☆ Mash ripe banana fruits and add to cakes, muffins, bread pudding etc.
- ☆ Plantain is raw unripe banana that is used as vegetable in recipes.

Caution

- ☆ Banana fruits are sometimes known to cause skin and systemic allergic reactions. In "oral allergy syndrome" which causes itching and swelling in the mouth or throat within hours after ingestion and is related to birch tree and other pollen allergies.
- ☆ The other type of reaction is related to latex allergies and causes urticaria and potentially serious gastrointestinal symptoms like nausea, vomiting, and diarrhea.

Cherries (*Prunus cerasus*)

Cherries (*Prunus cerasus*) are certainly one of today's most popular dessert fruits, but they have been recognized for their medicinal purposes since the 1400's. One cup of sweet cherries has just 90 calories and is a good source of fibre and vitamin C. Cherries are packed with antioxidant value and vitamins. They are slightly high on natural sugar content, but they're still very low-carb and are an excellent source of important fibre. Cherries are amazing with feta and greens. With a glycemic index of just 22, one should go for a daily serving of cherries, when they are in season locally. They are sweet, sour and astringent fruits. It cleansees the system and is good for dental problems. It is useful for lack of sleep. Cherries rank among the top 20 foods with the highest concentration of antioxidants. In fact, the standard one-cup serving of cherries has the capacity to carry 4,873 antioxidants! Antioxidants are substances found in foods that may protect cells from damage caused by unstable molecules, known as free radicals. Cherries are especially rich in a phytochemical called anthocyanin. They also contain melatonin, phenols and quercetin. Melatonin is a harmone that is naturally present in the human body and aids sleep. Like many fruits, raw sweet cherries provide carbohydrates, potassium and vitamin C. They also are a source of cholesterol-lowering phytosterols.

Researchers tracked the participant's sleep habits and after drinking the cherry juice, they found significant improvements in sleep behaviour, most notably longer sleep time, less daytime napping and increased overall sleep efficiency. They attribute the sleep benefits to the **melatonin** content of the red super fruit - a powerful antioxidant critical for sleep-wake cycle regulation. Each serving of the tart cherry juice concentrate was estimated to contain the equivalent of 90-100 tart cherries, providing a significant level of melatonin in the juice and ultimately in the bodies of the participants (Anonymous, 2012).

Carbohydrates and Sugar—One cup of raw sweet cherries contains 87 calories, most of which are from the 22.1 g carbohydrates per serving. One cup of cherries contains 17.7 g sugar, most in a combination of the 7.4 g fructose and 9.1 g glucose per cup. The remaining sugars are a combination of less than 1 g each of sucrose, maltose and galactose. One cup of cherries contains no starch and 2.9 g fibre, which is just over 10 per cent of the DRI.

Protein-Cherries contain very little protein—one cup contains only 1.46 g. This protein is made up of small amounts of 18 different amino acids, but like all plant foods, cherries do not contain all of the essential amino acids and are not a complete protein.

Fat-Cherries also contain very little fat, with one cup containing only.3 g. This fat is a combination of trace amounts each of saturated, polyunsaturated and monounsaturated fats. Cherries contain no cholesterol and, in fact, contain phytosterols, which can help create lower cholesterol levels by competing for absorption with cholesterol.

Minerals-Cherries contain significant amounts of some minerals, while supplying very little of others. One cup of cherries has 15.3 per cent of the DRI for potassium and 9.2 per cent of the DRI for copper, but less than 5 per cent of calcium, magnesium, manganese, phosphorus, selenium but have no sodium.

Water-Soluble Vitamins- Cherries also supply varied amounts of water-soluble vitamins. One cup provides 12.9 per cent of the DRI for vitamin C for women and 10.8 per cent for men. Vitamin C acts as an antioxidant and is important for iron absorption. One cup also supplies approximately 5 per cent of pantothenic acid and B6, but contains less than 4 per cent of all other water-soluble vitamins, all of which are B vitamins.

Fat-Soluble Vitamins-Cherries supply very little fat-soluble vitamins. One cup provides 3.2 per cent of the DRI for vitamin K and less than 1 per cent of vitamins D, E and A.

Lutein-Cherries also contain lutein, a pigment in many plants that works as an antioxidant in the body. Lutein is thought to promote eye and cardiovascular health.

Cherries and Melatonin

Melatonin is a naturally occurring hormone produced by the pineal gland in the brain. It plays a key role in regulating the body's internal clock and helps determine when we fall asleep and when we wake up. Eating a handful of cherries just before bed is a great way to naturally regulate your sleep cycle. According to a study Cherries grown in sun-rich Central Otago are natural source of melatonin at levels 30 times higher than their norther hemisphere counterparts.

Cherries, Arthritis and Pain Relief

Great news for arthritis sufferers! A bowl full of cherries may help alleviate pain and inflammation associated with arthritis and gout, the most severe form of arthritis. A gout attack occurs when excessive amounts of uric acid (waste product found in the blood) accumulate in the joints, and cause inflammation and pain.

Cherries and Heart Health

- ☆ **Cardiovascular disease**- One of the many health benefits of cherries is that they contain powerful antioxidants called anthocyanins, which may reduce a person's risk of developing cardiovascular disease.
- ☆ **Just like red wine**, anthocyanins give cherries their deep red colour and also protect cells from damage during an interaction with oxygen. This important process also serves to protect the heart and surrounding tissue, inhibit plaque formation and reduce inflammation.

Cherries and Brain Health

- ☆ Cherries are one of the few foods that contain melatonin. In addition to helping regulate sleep patterns, melatonin is an important antioxidant that helps maintain optimum brain functioning and may deter the onset of age-related chronic diseases like Alzheimer's. Research also suggests that the anthocyanins found in cherries further protect neural cells and promote brain health.

A Rich Source of Antioxidants

Cherries are a rich source of antioxidants and they prevent or repair the damage that is done to your body's cells by free radicals. They are also known to be potent anticancer agents. Apart from this, they relieve arthritis pain which is on the rise during the rainy season. Cherries also relieves stress and is known to prevent premature ageing.

Health Benefits

- ☆ Cherries are one of the very low calorie fruits; yet are rich source of nutrients, vitamins and minerals. Both sweet as well as tart Cherries are packed with numerous health benefiting compounds that are essential for wellbeing.

- Cherries are pigment rich fruits. These pigments are in fact polyphenolic flavonoid compounds known as anthocyanin glycosides. Anthocyanins are red, purple or blue pigments found in many fruits and vegetables, especially concentrated in their skin, known to have powerful anti-oxidant properties.
- Anthocyanins in the cherries are found to act like anti-inflammatory agents by blocking the actions of cycloxygenase-1 and 2 enzymes. Thus consumption of cherries has potential health effects against chronic painful episodes such as gout arthritis, fibromyalgia (painful muscle condition) and sports injuries.
- Studies also suggest that tart cherries are help body to fight against cancers, aging and neurological diseases and pre-diabetes.
- Cherry fruits are very rich in stable anti-oxidant melatonin. Melatonin can cross the blood-brain barrier easily and produces soothing effects on the brain neurons, calming down nervous system irritability which helps relieve neurosis, insomnia and headache conditions.
- Cherries are good source of minerals such as potassium, iron, zinc, copper and manganese.
- The fruits, especially tart cherries are exceptionally rich in many health promoting flavonoid poly phenolic anti-oxidants such as lutein, zeaxanthin and beta-carotene.
- Cherries' anti-inflammatory functions, effective in reducing heart disease risk factors by scavenging action against free radicals.

Coconut (*Cocuc nucifera*)

Coconut (*Cocuc nucifera*) is highly nutritious and rich in fibre, vitamins, and minerals. It is classified as a "functional food" because it provides many health benefits beyond its nutritional content. Coconut oil is of special interest because it possesses healing properties far beyond that of any other dietary oil and is extensively used in traditional medicine among Asian and Pacific populations. Pacific Islanders consider coconut oil to be the cure for all illness. The coconut palm is so highly valued by them as both a source of food and medicine that it is called "The Tree of Life." For thousands of years coconut products have held a respected and valuable place in local folk medicine. Coconut is used to treat a wide variety of health problems including the following: abscesses, asthma, baldness, bronchitis, bruises, burns, colds, constipation, cough, dropsy, dysentery, earache, fever, flu, gingivitis, gonorrhea, irregular or painful menstruation, jaundice, kidney stones, lice, malnutrition, nausea, rash, scabies, scurvy, skin infections, sore throat, swelling, syphilis, toothache, tuberculosis, tumors, typhoid, ulcers, upset stomach, weakness, and wounds. Young coconuts are good for most people, but are particularly good for carb and mixed types. Mature coconuts, on the other hand, are best for protein types.

Health Benefits

Coconuts provide vitamins, minerals and fibre that prevent diseases like gall blader stones, and liver diseases, inflammation and skin diseases, to name a few. If you want proteins, coconut milk is a good source. It also provides iron, selenium, sodium, magnesium, phosphorus and potassium; vitamin C, E, B1, B3, B5 and B6. Surprisingly coconut milk is far richer in calcium, than regular milk and supports the immune system, helps in weight loss and is great for digestion. less

- ☆ Coconuts can add flavour, variety and best of all—healthy nutrients to your diet.
- ☆ Coconuts are rich in lauric acid, which is known for being antiviral, anti-bacterial and anti-fungal, and boosts the immune system.
- ☆ Fresh coconut juice is one of the highest sources of electrolytes known to man, and can be used to prevent dehydration, for instance in cases of diarrhea or strenuous exercise, instead of a sports drink.
- ☆ Help you lose weight, or maintain your already good weight.
- ☆ Reduce the risk of heart disease. Lower your cholesterol, improve conditions in those with diabetes and chronic fatigue, improve Crohn's, IBS, and other digestive disorders
- ☆ Prevent other disease and routine illness with its powerful anti-bacterial, anti-viral and anti-fungal agents.
- ☆ Increase metabolism and promotes healthy thyroid function, boost your daily energy, rejuvenate your skin and prevent wrinkles. Modern medical science is now confirming the use of coconut in treating many of the above conditions and may provide a wide range of health benefits. Some of these are summarized below:
- ☆ Kills viruses that cause influenza, herpes, measles, hepatitis C, SARS, AIDS, and other illnesses.
- ☆ Kills bacteria that cause ulcers, throat infections, urinary tract infections, disease and cavities, pneumonia, and gonorrhea, and other diseases.
- ☆ Kills fungi and yeasts that cause candidiasis, ringworm, athlete's foot, thrush, diaper rash, and other infections.
- ☆ Expels or kills tapeworms, lice, giardia, and other parasites.
- ☆ Provides a nutritional source of quick energy.
- ☆ Boosts energy and endurance, enhancing physical and athletic performance.
- ☆ Improves digestion and absorption of other nutrients including vitamins, minerals, and amino-acids.
- ☆ Improves insulin secretion and utilization of blood glucose.
- ☆ Relieves stress on pancreas and enzyme systems of the body.
- ☆ Reduces symptoms associated with pancreatitis.
- ☆ Helps relieve symptoms and reduce health risks associated with diabetes.
- ☆ Reduces problems associated with malabsorption syndrome and cystic fibrosis.
- ☆ Improves calcium and magnesium absorption and supports the development of strong bones and teeth.
- ☆ Helps protect against osteoporosis.

- ☆ Helps relieve symptoms associated with gallbladder disease.
- ☆ Relieves symptoms associated with Crohn's disease, ulcerative colitis, and stomach ulcers.
- ☆ Improves digestion and bowel function.
- ☆ Relieves pain and irritation caused by hemorrhoids.
- ☆ Reduces inflammation.
- ☆ Supports tissue healing and repair.
- ☆ Supports and aids immune system function.
- ☆ Helps protect the body from breast, colon, and other cancers.
- ☆ Is heart healthy; improves cholesterol ratio reducing risk of heart disease.
- ☆ Protects arteries from injury that causes atherosclerosis and thus protects against heart disease.
- ☆ Helps prevent periodontal disease and tooth decay.
- ☆ Functions as a protective antioxidant.
- ☆ Helps to protect the body from harmful free radicals that promote premature aging and degenerative disease.
- ☆ Does not deplete the body's antioxidant reserves like other oils do.
- ☆ Improves utilization of essential fatty acids and protects them from oxidation.
- ☆ Helps relieve symptoms associated with chronic fatigue syndrome.
- ☆ Relieves symptoms associated with benign prostatic hyperplasia (prostate enlargement).
- ☆ Reduces epileptic seizures.
- ☆ Helps protect against kidney disease and bladder infections.
- ☆ Dissolves kidney stones.
- ☆ Helps prevent liver disease.
- ☆ Is lower in calories than all other fats.
- ☆ Supports thyroid function.
- ☆ Promotes loss of excess weight by increasing metabolic rate.
- ☆ Is utilized by the body to produce energy in preference to being stored as body fat like other dietary fats.
- ☆ Helps prevent obesity and overweight problems.
- ☆ Applied topically helps to form a chemical barrier on the skin to ward of infection.
- ☆ Reduces symptoms associated the psoriasis, eczema, and dermatitis.
- ☆ Supports the natural chemical balance of the skin.
- ☆ Softens skin and helps relieve dryness and flaking.
- ☆ Prevents wrinkles, sagging skin, and age spots.

- ☆ Promotes healthy looking hair and complexion.
- ☆ Provides protection from damaging effects of ultraviolet radiation from the sun.
- ☆ Helps control dandruff.
- ☆ Does not form harmful by-products when heated to normal cooking temperature like other vegetable oils do.
- ☆ Has no harmful or discomforting side effects.
- ☆ Is completely non-toxic to humans.

Coconut water has many health benefits and one of them is losing weight quickly. Thus 'coconutt' is packed with anti-aging properties and has the right number of calories to keep you going. The water inside the green coconuts is ones to be had. One cup of this water has around 46 calories compared to 550 calories that coconut milk has. The water contains no fat and consists mostly of natural sugars and minerals. Coconut water is very rich in magnesium and potassium, as well as being very high in fibre.

- ☆ It is so hydrating because of the electrolyte composition, which is very similar to human plasma. Staying hydrated will help you feel full for longer, which in turn will reduce hunger and sugar cravings. This will effectively lead to you eating less, which means your total calorie consumption will go down. Drinking this water will known to flush out toxins which will help you to lose weight faster and also reduce those cravings and hunger pangs.
- ☆ So, the next time you feel thirsty or feel a hunger/craving pang, go get this nature's pure drink and sip it. It is sweet, oily fruit. It is considered as an auspicious fruit and is used in almost every religious ritual
- ☆ Coconut water contains organic compounds that have been known to help.

Benefits of Coconut Water

- ☆ The coconut is known for being a light moisturizer that removes the excess oil from the skin. You can also add it in your bath too. And, can be used as a conditioner or shampoo for oily hair.
- ☆ Regular comsumption of the coconut water helps in blood circulation and prevents from excessive heat burn problems and also controls acidy.
- ☆ It is best for acne and blackheads and also a natural toning and leansing mask for oily skin.
- ☆ Coconut water is a rich source of natural fibre, natural calcium, magnesium, and natural potassium.
- ☆ Coconut water is anti-viral, anti-fungal.
- ☆ Coconut can be used as medicine. It helps in the problems of kidney, reduces the risk of cancer, and diabetes. It also helps in digestion.
- ☆ Coconut water is a rich source of natural fibre, natural calcium, magnesium, and natural potassium- making it a healthy electrolyte drink.

Coconut Oil

While coconut possesses many health benefits due to its fibre and nutritional content, it's the oil that makes it a truly remarkable food and medicine. Once mistakenly believed to be unhealthy because of its high saturated fat content, it is now known that the fat in coconut oil is a unique and different from most all other fats and possesses many health giving properties. Coconut oil has been described as "the healthiest oil on earth."What makes coconut oil so good? All fats and oils are composed of molecules called fatty acids. There are two methods of classifying fatty acids. The first, is based on saturation (saturated, mono- unsaturated, and polyunsaturated fats). Another system of classification is based on molecular size or length of the carbon chain within each fatty acid. Fatty acids consist of long chains of carbon atoms with hydrogen atoms attached. In this system you have short-chain fatty acids (SCFA), medium-chain fatty acids (MCFA), and long-chain fatty acids (LCFA). Coconut oil is composed predominately of medium chain fatty acids (MCFA), also known as medium-chain triglycerides (MCT). The saturated fatty acids in coconut oil are predominately medium-chain fatty acids. Both the saturated and unsaturated fat found in meat, milk, eggs, and plants (including most all vegetable oils) are composed of LCFA. MCFA are very different from LCFA. They do not have a negative effect on cholesterol and help to protect against heart disease. MCFA help to lower the risk of both atherosclerosis and heart disease. It is primarily due to the MCFA in coconut oil that makes it so special and so beneficial. There are only a very few good dietary sources of MCFA. By far the best sources are from coconut and palm kernel oils.

Benefits of Coconut Oil

- ☆ Apart from being good for the skin and hair of a person, coconut oil has been found to be beneficial in case of the following ailments. Stress, Heart Diseases, High Cholesterol Levels, Too Much Weight, Kidney Problems, Poor Digestion, Low Metabolism, High Blood Pressure, Low Immunity, Dental Problems, Diabetes, Low Bone Density, HIV, Cancer, Premature Aging, Pancreatitis.

Date (*Phoenix dactylifera*)

Wonderfully delicious, dates are one of the most popular fruits with a fabulous list of essential nutrients. Dates are even rich in several vitamins and minerals. These natural products contain calcium, sulphur, iron, potassium, phosphorous, manganese, copper and magnesium which are advantageous for health. It is said that consumption of one date daily is necessary for a balanced and healthy diet. You can have dates just like that. Or you can add it to those mouth-watering desserts. Or you can use dates to make chutney with either tomatoes or tamarind. You can store it in the refrigerator. It is delicious to have it with snacks. Fresh date is made of soft, easily digestible flesh with simple sugar like fructose and dextrose that when eaten replenishes energy and revitalizes the body instantly. Precisely that's why it is used to break fast. The fruit is rich in dietary fibre, which prevents dietary LDL cholesterol absorption. It is also a good bulk laxative. Dates are a good source of Vitamin-A which is known to have antioxidant properties and is essential for vision. Vitamin A also required maintaining healthy mucus membranes and skin. Consumption of natural fruits rich in vitamin A is known to help protect from lung and oral cavity cancers. The fruit is very rich in antioxidant flavonoids such as beta-carotene, lutein, and zeaxanthin. These antioxidants have the ability to help protect cells and other structures in the body from oxygen free radicals. Dates are an excellent source of iron.

Health Benefits

- ☆ Fresh date is made of soft, easily digestible flesh with simple sugar like fructose and dextrose that when eaten replenishes energy and revitalizes the body instantly. Precisely that's why it is used to break fast.
- ☆ The fruit is rich in dietary fibre, which prevents dietary LDL cholesterol absorption. It is also a good bulk laxative.
- ☆ Dates are a good source of Vitamin-A which is known to have antioxidant properties and is essential for vision.
- ☆ Dates very rich in antioxidant flavonoids such as beta-carotene, lutein, and zeaxanthin. These antioxidants have the ability to help protect cells and other structures in the body from oxygen free radicals. Dates are an excellent source of iron.
- ☆ Delicious dates are good source of essential nutrients *viz.* vitamins, and minerals, required for normal growth, development and overall well-being.
- ☆ The fibre content helps to protect the colon mucous membrane by decreasing exposure time and as well as binding to cancer causing chemicals in the colon.
- ☆ They contain many health benefitting flavonoid polyphenolic antioxidants known as tannins.
- ☆ Zeaxanthin, an important dietary carotenoid selectively absorbed into the retinal macula lutea where it is thought to provide antioxidant and protective light-filtering functions; thus it offers protection against age related macular degeneration, especially in elderly populations.
- ☆ Dates are an excellent source of iron, contains 0.90 mg/100 g of fruits. Iron, being a component of hemoglobin inside the red blood cells, determines the oxygen carrying capacity of the blood.
- ☆ They are good in potassium. Potassium is an important component of cell and body fluids that help controlling heart rate and blood pressure; thus offers protection against stroke and coronary heart diseases.
- ☆ They are also rich in minerals like calcium, manganese, copper, and magnesium.
- ☆ Further, the fruit has adequate levels of B-complex group of vitamins as well as vitamin K. It contains very good amounts of pyridoxine (vitamin B-6), niacin, pantothenic acid, and riboflavin. These vitamins are acting as cofactors help body metabolize carbohydrates, protein, and fats. Vitamin K is essential for many coagulant factors in the blood as well as in bone metabolism.

Fig (*Ficus carica*)

Fig is naturally rich in many health benefiting phyto-nutrients, anti-oxidants and vitamins. Dried figs in fact are concentrated source of minerals and vitamins. The fully ripe fig has bell or pear shape with succulent flesh. Figs are very good in preventing constipation. Figs promote bowel function as every three grams of this fruit is loaded with five grams of fibre. It also prevents stomach ache and indigestion. For people who suffer from weight related issues figs can be their answer. The fibre in the figs helps in reducing weight when taken regularly. A type of fibre that is found in fig is **pectin.** Pectin is very good in absorbing cholesterol from the digestive system and carrying them out of the body. So, a regular diet of figs can help you cut down on your cholesterol. Figs are also rich in phenol, **omega-3 fatty acids** and **omega-6 fatty acids.** These compounds are natural heart boosters and so figs can contribute towards reducing the risk of coronary heart diseases.

It is also believed that the fibre in figs absorb cancer causing substances, thus reducing the risk of various types of cancer. Figs are especially good in preventing colon and post-menopausal breast cancer. People suffering from diabetes should adopt a diet of high fibre. For this figs are especially good. This fruit is rich in potassium that controls which helps in controlling the blood sugar. Fig leaves are also known to have diabetic properties. The leaves effectively reduce the amount of insulin that is required by diabetic patients through injections. Hypertension is caused because of low levels of sodium and high levels of potassium in the body. Since figs are high in potassium and low in sodium it can avoid hypertension. Figs also have a high concentration of calcium and so it strengthens the bones. However, a high calcium diet can lead to urinary calcium loss and this is avoided by the potassium content of the figs.

Age results in macular degeneration which leads to vision loss. A diet of figs prevents this. Sore throats can be effectively healed with figs as it also has high mucilage content. Figs are also known to boost the health of the liver. Figs are also high in alkaline and so they keep the body balanced because of the acidic diet by regulating the pH of the body. Figs can also heal various respiratory diseases like asthma and whooping cough. They are also good for earache, boils, venereal diseases and boils. Figs are excellent sources of amino acids that increase libido. They can also improve sexual stamina. **Figs** are perfect fertile booster foods which are full of soluble and insoluble fibre. This juicy taste fruit is a treat for your senses. Good for heart health, figs are a great way to rev up your sexualstamina. Figs are excellent sources of amino acids that increase libido. They can also improve sexual stamina.

Health Benefits of Figs

This fruit is very nutritious and has many health benefits (Anonymous,2013).

- ☆ Pectin, a soluble fibre found in figs is beneficial for the digestive system. This soluble fibre helps in clearing out the accumulating cholesterol in the body.
- ☆ Dried figs are said to contain **omega-3** and **omega-6 fatty acids** along with phenol. These are useful in preventing coronary heart diseases.
- ☆ Figs are also a source of calcium, which is good for strengthening of the bones.
- ☆ Because figs are rich in potassium, it helps in regulating blood sugar. This is beneficial for people who are diabetics. Also, people whose diet is sodium may suffer from urinary calcium loss. The high potassium content in figs, helps in prevention of this.
- ☆ Eating figs are said to be an effective cure for treating sore throat because of their high mucilage content.
- ☆ Traditionally, figs have also been used for treating sexual weakness. The remedy requires one to soak about two to three figs in milk overnight and eat the same next morning.
- ☆ Because figs contain iron, they are useful in treating anemia.
- ☆ Mash about fresh figs and them on your face and leave for 10 to 15 minutes. This is useful for treating acne and pimples.
- ☆ It is said that eating figs can relieve fatigue and boost brain power too.
- ☆ As one grows older, one tends to suffer from macular degeneration *i.e.* weakness of the vision. Eating figs can help prevent this.

- ☆ Hypertension may arise when there's more in take of sodium and low potassium levels in the body. Figs however, are low in sodium and high in potassium and hence, helps in prevention of hypertension.
- ☆ Fig fruit is low in calories. 100 g fresh fruits provide only 74 calories. However they contain health benefiting soluble dietary fibre, minerals, vitamins and pigment anti-oxidants that contribute immensely for optimum health and wellness.
- ☆ Dried figs are excellent source of vitamins and anti-oxidants and minerals, like **calcium**, copper, potassium, manganese, **iron**, selenium and zinc. 100 g of dried figs contain 640 mg of potassium, 162 mg of calcium, and 2.03 mg of iron. In fact dried fruits are concentrated sources of energy.
- ☆ Fresh figs, especially black mission, are good in poly-phenolic flavonoid anti-oxidants such as carotenes, lutein, tannins, chlorgenic acid. etc. Their anti-oxidant value is comparable to that of apples at 3200 umol/100 g.
- ☆ Figs contain adequate levels of some of anti-oxidant vitamins such as vitamin A, E, and K. Altogether these phyto-chemical compounds in fig fruit help scavenge harmful oxygen derived free radicals from the body and thereby protect us from cancers, diabetes, degenerative diseases and infections.
- ☆ The **chlorogenic acid** in these berries help lower blood sugar levels and control blood glucose levels in type-II diabetes mellitus (Adult onset) condition.
- ☆ Fresh as well as dried figs contain good levels of B-complex group of vitamins such as niacin, pyridoxine, folates and pantothenic acid. These vitamins function as co-factors for metabolism of carbohydrates, proteins and fats.
- ☆ If you suffer from osteoporosis, include figs in your diet. Its high calcium content promotes bone health.

Caution

- ☆ Fig leaves and un-ripened fruit produce white latex which can penetrate the skin causing burning discomfort. Fig latex contains several compounds like furocoumarins, 5-methoxypsoralen (5-MOP) etc which can elicit cell-mediated allergic reactions. If left untreated, there may occur severe allergic eruptions all over the exposed parts. Eating fig fruit may also elicit allergic reactions ranging from vomiting, diarroea, and itching of skin and mucus membranes in some sensitized individuals. It is therefore people with history of allergy to figs may be advised to avoid eating them.

Guava (*Psidium guajava*)

Guava is another tropical fruit rich in nutrition. With its unique flavour, taste, and health-promoting qualities, the fruit easily fits in new functional foods category, often called "super fruits."A new study led by an Indian origin researcher has found that guavas are the 'ultimate superfood' with the highest concentration of antioxidants that protect against cell damage which age's skin and can cause cancer. A series of tests conducted on Indian fruits, including Himalayan apples and pomegranates, bananas from the south and grapes from Maharashtra, found that the guava, the poor man's fruit in India, has the highest concentration of antioxidants as compared to all the other fruits. According to scientists from India's National Institute in Hyderabad, the Indian plum, the custard apple and India's beloved mangoes, come after guavas in antioxidant richness. The study found that while there is a presence of antioxidant concentrations of just under 500 milligrams per 100 grams in guavas, 330mg in plums and 135mg in pomegranates, apples have a quarter of the antioxidants in guavas and bananas merely have a tiny fraction with 30 mg per 100 grams. Watermelons and pineapples were found to offer the least protection for the body's fight against free radicals, which can cause cell damage, whereas mangoes, despite a high fructose content, have 170 mg of antioxidants, which is more than three times that of papaya, and grapes were found to be three times more beneficial to the body than oranges.

Guava is a rich source of antioxidants, a rich source of fibre. It's a poor mans fruit because they're quite cheap. A guava a day keeps a doctor away. The fruit is one of the richest sources of vitamin C. It contains 4 to 10 times more vitamin C than do some citrus fruits. The guava contains very little vitamin A or carotene. However, it is fairly rich in most other mineral nutrients.

Guava is one of the best fruits you can eat if you are trying to lose weight. A medium-sized guava fruit has 8 g of dietary fibre that will help support your weight loss diet. Fibre not only keeps your appetite under control, but also keeps your blood sugar stable by slowing the rate that glucose is absorbed and metabolized. Keeping your blood sugar stable is vital to achieving your weight-loss goals. When you raise your blood-sugar levels, your body releases insulin, which promotes fat storage. You want to prevent insulin release as much as possible if you are trying to lose weight. Its fruit is tasty, appetizing, quenches thirst, enhances semen formstion and strengthens the heart. It is anthelminticrich and anti vomiting. It is purgative and cures anthelmin cough. Immature fruits are highly astrigent and not edible. The fruit is best when it is just ripe- that is when the skin has acquired a yellow tinge. Fruits are often plucked right off the tree and eaten just as they are.

Health Benefits

- ☆ Guavas are beneficial in regulating **blood pressure**. It is said that one **guava** contains almost a similar amount of potassium that's present in bananas. Potassium reverses the effects of sodium, thereby regulating the balance of blood pressure. Also, it reduces **cholesterol** levels in the blood by preventing it from thickening.
- ☆ Consumption of guavas helps slow down the absorption of sugar in the blood. It is rich in fibre and is helpful for diabetic individuals. Also, studies have shown that a diet that is high in fibre (5.4 gm per 100 gm of fruit) is linked to a lower risk of developing Type 2 diabetes. Individuals suffering from constipation problems, too, can benefit from the high fibre content.
- ☆ Did you know that guavas contain four times more Vitamin C than oranges? Vitamin C contains antioxidant properties that protect cells from the damage of free radicals and is useful in lowering the risk of cancer.
- ☆ Even though guavas don't contain iodine, they are still beneficial in promoting healthy thyroid function, because it contains copper, which aids the production and absorption of hormones.
- ☆ Guavas are a good source of manganese that acts as an enzyme activator utilising nutrients like thiamine, biotin and ascorbic acid.
- ☆ If you want to optimise your brian function, turn to this fruit. Guavas are rich in the B group of vitamins. Niacin, better known as Vitamin B3, promotes blood circulation, thereby stimulating brain function. Vitamin B6, that is pyridoxine, helps in brain and nerve function.

- ☆ Women with fertility problems can eat guavas as they contain a good amounts of folate, which contain fertility-promoting properties.
- ☆ Eye problems can be kept at bay as guavas contain an abundance of Vitamin A that helps in improving vision.
- ☆ Guava is good for the skin, too. Because of its Vitamin E content, astringent properties and antioxidants, the skin is nourished. Skin ailments like scurvy can be dealt with due to the high Vitamin C content in guavas.
- ☆ Pink guavas are said to contain twice the amount of lycopene present in tomatoes. Lycopene protects the skin from being damaged by UV rays and also works against prostate cancer.
- ☆ Even the leaves of guava have medicinal properties. The juice of the leaves is said to provide relief from cold and cough by reducing the formation of mucus, disinfecting the respiratory tract and preventing bacterial activity in the throat due to its astringent properties.
- ☆ Guavas are low in calories and fats but contain several vital vitamins, minerals, and antioxidant poly-phenolic and flavonoid compounds that play pivotal role in prevention of cancers, anti-aging, immune-booster etc.
- ☆ The fruit is very rich source of soluble dietary fibre), which makes it a good bulk laxative. The fibre content helps protect the colon mucous membrane by decreasing exposure time to toxins as well as binding to cancer causing chemicals in the colon.
- ☆ Fresh guava-fruit is an excellent source of antioxidant **vitamin-C;** provides more than **three times the DRI** (daily-recommended intake). Outer thick rind contains exceptionally higher levels of vitamin C than central pulp.
- ☆ Regular consumption of fruits rich in vitamin C helps body develop resistance against infectious agents and scavenge cancer causing harmful free radicals from the body. Further, the vitamin is required for collagen synthesis in the body. Collagen is the main structural protein in the body required for maintaining the integrity of blood vessels, skin, organs, and bones.
- ☆ The fruit is very good source of **Vitamin-A** and flavonoids like beta-carotene, **lycopene**, lutein and cryptoxanthin. The compounds are known to have antioxidant properties and are essential for optimum health. Further, vitamin-A is also required for maintaining healthy mucus membranes and skin. Consumption of natural fruits rich in carotene is known to protect from lung and oral cavity cancers.
- ☆ **Lycopene** in pink guavas prevents skin damage from UV rays and offers protection from prostate cancer.
- ☆ Fresh fruit is a very rich source of potassium; contains more potassium than banana per 100 g of fruit weight.
- ☆ Fig is also a moderate source of B-complex vitamins such as pantothenic acid, niacin, vitamin-B6 (pyridoxine), vitamin E and K, and minerals like magnesium, copper, and manganese.

Citrus Fruits

Citrus fruits, as such, have long been valued for their wholesome nutritious and antioxidant properties. Citrus fruits like oranges, tangerines, lemons and grapefruit contain anti- inflammatory flavonoids, which help reduce tumour inflammation. It is scientifically established that citrus fruits, especially oranges, by virtue of their richness in vitamins and minerals, have many proven health benefits. Moreover, it is now beginning to be appreciated that the other biologically active, non-nutrient compounds found in citrus fruits such as phyto-chemical antioxidants, soluble and insoluble dietary fibre have been found to be helpful in reduction in the risk for cancers, many chronic diseases like arthritis, and from obesity and coronary heart diseases. These fruits should be consumed in their natural form as salads and fruit chaat etc.

ORANGES (*Citrus reticulata*)

Oranges are delicious and juicy fruit, contains an impressive list of essential nutrients, vitamins, minerals for normal growth and development and overall well-being. If you want to avoid all the extra calorie, it's better to have oranges just like that. So, now you can carry an orange in your bag. Rich in Beta-carotene, oranges can protect our cells. Orange peels are very good for skin too. The fibres in oranges are good for the digestive system and relieving Fibres also reduce cholesterol which is helpful in keeping you in good health and fibres are also good in keeping the blood sugar levels in control.

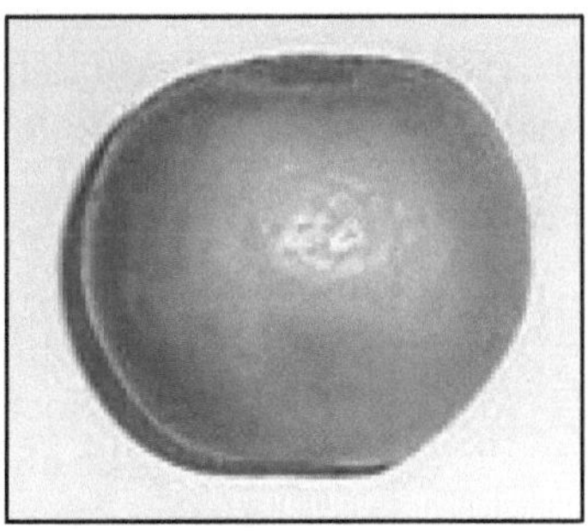

Nagpur Orange

Oranges are also rich in calcium, associated with a healthy bone structure and teeth. Vitamin C stimulates the production of white cells in our body, thus improving the immune system. So, eating oranges regularly will keep you away from viral infections. Regular consumption of oranges significantly reduce the risk of developing kidney stones. They're mainly available in two categories - sweet and bitter, with the former being the type most commonly consumed. Generally an orange should have smoothly textured skin and be firm and heavy for its size. These will have higher juice content than those that are either spongy or lighter in weight. Oranges are extremely high in antioxidants. They help brighten the skin. Orang, a citrus fruit, contains vitamin C, that helps prevent conditions like skin inflammation. One thing known about vitamin C is that, it helps improve the absorption of Vitamin E and iron in the body, and this in turn gives you a radiant complexion. You might have noticed, there are many skin care products that have orange extracts in them. They are an excellent source of vitamin C. One orange supplies116.2 per cent of the daily value for vitamin C. Good intake of vitamin C is associated with a reduced risk of colon cancer as it helps to get of free radicals that cause damage to our DNA. Vitamin C, which is also vital for the proper function of a healthy immune system, is good for preventing colds and preventing recurrent ear infections. One glass of orange juice a day will give you a good dose of vitamin D.

Anti-oxidants in oranges help protect skin from free radical damage known to cause signs of aging. An orange a day can help you look young even at 50! Oranges, being rich in Vitamins B6, help support the production of haemoglobin and also help keep blood pressure under check due to the presence of magnesium. According to a study by US and Canadian researchers, a class of compounds found in citrus fruit peels called Polymethoxylated Flavones (PMFs) have the potential to lower cholesterol more effectively than some prescription drugs without side effects. Fibre in oranges help by keeping blood sugar levels under control thereby making oranges a healthy snack for people with diabetes. The natural fruit sugar in oranges, fructose, can help keep blood sugar levels from rising too high after eating. Consuming foods rich in beta-cryptoxanthin, an orange-red carotenoid found in highest amounts in oranges, corn and pumpkin may significantly lower one's risk of developing lung cancer. Oranges like most citrus fruits, produce more juice when warmer - juice them when they are at room temperature. Rolling the orange under the palm of your hand on a flat surface will also help to extract more juice. Vitamin C gets destroyed fast when exposed to air, so eat an orange quickly once cut up.

Health Benefits

☆ Nutrients in oranges are plentiful and diverse. The fruit is low in calories, contains no saturated fats or cholesterol, but is rich in dietary fibre, **pectin**, which is very effective in persons with excess body weight. Pectin, by its action as bulk laxative, helps to protect the mucous membrane of the colon by decreasing its exposure time to toxic substances as well as by binding to cancer causing chemicals in the colon. Pectin has also been shown to reduce blood cholesterol levels by decreasing its re-absorption in the colon by binding to **bile acids** in the colon.

- ☆ Orange, like other citrus fruits, is an excellent source of **vitamin C.** Vitamin C is a powerful natural antioxidant. Consumption of foods rich in vitamin C helps body develop resistance against infectious agents and scavenge harmful, pro-inflammatory free radicals from the blood. One orange supplies 116.2 per cent of the daily value for vitamin C. Vitamin C, which is also vital for the proper function of a healthy immune system, is good for preventing colds and preventing recurrent ear infections.
- ☆ Orange fruit contains a variety of phytochemicals. **Hesperetin** and **Narigenin** are flavonoids found in citrus fruits. Naringenin is found to have a bio-active effect on human health as antioxidant, free radical scavenger, anti-inflammatory, and immune system modulator.
- ☆ Oranges also contain very good levels of vitamin A, and other flavonoid antioxidants such as alpha and beta-carotenes, beta-cryptoxanthin, zea-xanthin and lutein. These compounds are known to have antioxidant properties.
- ☆ It is also a very good source of B-complex vitamins such as thiamin, pyridoxine, and folates. These vitamins are essential in the sense that body requires them from external sources to replenish.
- ☆ Orange fruit also contains a very good amount of minerals like potassium and calcium. Potassium is an important component of cell and body fluids that helps control heart rate and blood pressure through countering sodium actions.
- ☆ Anti-oxidants in oranges help protect skin from free radical damage known to cause signs of aging. An orange a day can help you look young even at 50!
- ☆ Oranges, being rich in Vitamins B6, help support the production of haemoglobin and also help keep blood pressure under check due to the presence of magnesium.
- ☆ According to a study by US and Canadian researchers, a class of compounds found in citrus fruit peels called Polymethoxylated Flavones (PMFs) have the potential to lower cholesterol more effectively than some prescription drugs without side effects.
- ☆ Fibre in oranges help by keeping blood sugar levels under control thereby making oranges a healthy snack for people with diabetes. The natural fruit sugar in oranges, fructose, can help keep blood sugar levels from rising too high after eating.
- ☆ Consuming foods rich in beta-cryptoxanthin, an orange-red carotenoid found in highest amounts in oranges, corn and pumpkin may significantly lower one's risk of developing lung cancer.
- ☆ Oranges like most citrus fruits, produce more juice when warmer - juice them when they are at room temperature. Rolling the orange under the palm of your hand on a flat surface will also help to extract more juice. Vitamin C gets destroyed fast when exposed to air, so eat an orange quickly once cut up.

Caution

Insecticide sprays are widely applied over orange crops. Therefore, it is recommended to wash the fruits in cold running water before use. Organic orange fruits are devoid of these chemicals and are best suited for zest preparation.

GRAPE FRUIT (*Citrus paradisi*)

Refreshing and delicious grapefruit is rich in phytonutrients such as vitamin A, beta-carotene, and lycopene. It is revered as fruit of "paradise" for its unique health-promoting as well as disease-healing properties, especially among health-conscious, fitness freaks. Grapefruit is called so because it grows in clusters like grapes. In Latin, grapefruits are called citrus paradisi meaning the citrus from paradise. One of the most popular weight loss diets is the Grapefruit diet. This diet involves drinking of at least three glasses of grape fruit juice everyday, as it helps burn calories. The grapefruit looks similar to an orange but is larger in size and belongs to the citrus family of fruits. However, the flesh of the grapefruit is usually red or pink in colour. Grapefruits taste sour and bitter but have a lot of health benefits. Though grapefruits are sour and acidic in taste, they are known to produce an alkaline reaction after digestion. This helps treat acidity which can affect the digestive system. Grapefruit is rich **in** vitamin C which helps maintain the elasticity of the arteries. It also strengthens them. Pectin, a compound, is known to help reduce the accumulation of arterial deposits. Consuming grapefruit on a regular basis will help you boost your immune system. Grapefruits are highly beneficial for diabetics as they help in bringing down blood sugar levels in the body.

Eating grapefruit before bedtime gives good sleep thus, reducing symptoms of insomnia. Pregnant women can reduce water retention in their bodies and swelling in their legs by consuming grapefruit. Grapefruits contain fat-burning enzymes that can promote weightloss. That's because they prevents sugars and starch getting stored in the body. Drinking fresh grapefruit juice can help relieve a sore throat. Grapefruit contains anti-oxidant properties that helps prevent cancer.

Health Benefits of Grapefruit

- ☆ Delicious, grapefruit is very low in calories, consists of just 42 calories per 100 g. Nonetheless, it is rich in dietary insoluble fibre **pectin**, which by acting as bulk laxative helps to protect the colon mucous membrane by decreasing exposure time to toxic substances in the colon as well as binding to cancer causing chemicals in the colon.

- ☆ Pectin has also been shown to reduce blood cholesterol levels by decreasing re-absorption of cholesterol binding bile acids in the colon.
- ☆ The fruit contains very good levels of vitamin-A and flavonoid antioxidants such as **naringenin**, beta-carotene, xanthin and lutein.
- ☆ Vitamin A also required maintaining healthy mucus membranes and skin. Consumption of natural fruits rich in vitamin-A and flavonoids helps to protect from lung and oral cavity cancers.
- ☆ It is a good source of antioxidant vitamin-C; Vitamin-C is a powerful natural anti-oxidant and helps body develop resistance against infectious agents and scavenge harmful free radicals; also is required for the maintenance of normal connective tissue as well for wound healing. It also facilitates dietary iron absorption from the intestine.
- ☆ Fresh fruit is very rich in potassium. Potassium is an important component of cell and body fluids, helps controlling heart rate and blood pressure through countering sodium effects.
- ☆ Red varieties of grapefruits are especially rich in the most powerful flavonoid antioxidant, **lycopene**.
- ☆ It contains moderate levels of B-complex group of vitamins such as folates, riboflavin, pyridoxine, and thiamin in addition to some resourceful minerals such as iron, calcium, copper, and phosphorus.

LEMON (*Citrus lemon*)

Lemon juice is an important ingredient in most Indian recipes. This is also a fruit of all seasons and almost always finds place on our kitchen shelf or refrigerator. With its vitamin C content, its juice will keep your skin beautiful. A glass of warm water with a tsp of honey and a dash of lemon juice on an empty stomach every morning is a great skin cleanser. With its astringent properties, it can be used to lighten the skin tone and also diminish acne scars. Rub the inside of a lemon peel on your elbow remove dark spots. Mix lemon and honey and use it as a natural bleach on your skin. Lemons are an excellent source of vitamin C. Having a lemon squeezed in warm water first thing every morning will cleanse your digestive system. It also increases the levels of immunity thus cutting the chances of infections. Because it has antibacterial and anti-viral properties, you can battle flu and colds by downing a glass of lemon juice. It is also known to purify blood, hence getting rid of toxins from the body. Lemon is also used in aromatherapy. Since it smells so 'clean' it induces the feeling of freshness and purges the feelings of inadequacy thus building confidence.

The essential oil is known to increase the level of concentration. To get rid of a tan apply lemon on hands and feet. Section hair and rub a lemon on hair and voila, you get natural-looking highlights like you just returned from a day at the beach. Lemon juice has more benefits that you'd ever imagine. Lemons have five per cent of citric acid, which gives it its unique taste. Rich in vitamin C, it also has vitamin B, calcium, phosphorus, magnesium, proteins and carbohydrates. Citrus fruits, are valued for their wholesome nutritious and antioxidant properties. Citrus fruits, especially lemons and oranges, by virtue of their richness in vitamins and minerals, have many proven health benefits. Other biologically active, non-nutrient compounds found in citrus fruits such as phyto-chemical antioxidants, soluble and insoluble dietary fibre have been found to be helpful in reduction in the risk for cancers, many chronic diseases like arthritis, and from obesity and coronary heart diseases.

Lemon juice is an important ingredient in most Indian recipes. This is also a fruit of all seasons and almost always finds place on your kitchen shelf or refrigerator. With its vitamin C content, its juice will keep your skin beautiful. A glass of warm water with a tsp of honey and a dash of lemon juice on an empty stomach every morning is a great skin cleanser. With its astringent properties, it can be used to lighten the skin tone and also diminish acne scars. Rub the inside of a lemon peel on your elbow remove dark spots. Mix lemon and honey and use it as a natural bleach on your skin (Anonymous, 2013).

Lemons are known for their 'skin and hair' cleansing properties. Dandruff can be prevented by rubbing lemon on the scalp. Adding two teaspoons of lemon juice in your bath not only deodourises the body but also keeps it fresh. Lemon rinds, when rubbed on scars, acne or dark spots, helps heal them. Rub lemon slices on rough areas of heels and elbows to soften them. Lemons are an excellent source of vitamin C. It also increases the levels of immunity thus cutting the chances of infections. Because it has antibacterial and anti-viral properties, you can battle flu and colds by downing a glass of lemon juice. It is also known to purify blood, hence getting rid of toxins from the body. Since it smells so 'clean' it induces the feeling of freshness and purges the feelings of inadequacy thus building confidence. The essential oil is known to increase the level of concentration. To get rid of a tan apply lemon on hands and feet. Section hair and rub a lemon on hair and voila, you get natural-looking highlights like you just returned from a day at the beach. It is really soothing. Daily consumption of lemon water has a number of health benefits.

- ☆ **Good for Stomach**: Lemon is really helpful when it comes to digestion problems. Due to the digestive qualities of lemon juice, symptoms of indigestion such as heartburn, bloating and belching are relieved. Traditionally grandmoms always recommended a glass of lemon juice when somebody suffered from fever or **stomach** upset.
- ☆ **Glowing skin**: Lemon, being a natural antiseptic medicine, can participate to cure problems related to skin. Lemon is a **vitamin C** rich citrus fruit that rejuvenates skin and results in a glowing skin. Lemon also acts as an anti-aging remedy.
- ☆ **Lose the flab**: In India, many people drink lemon water with a dash of honey early in the morning to lose weight.

- **Cures throat infections**: Lemon is an excellent fruit that aids in fighting problems related to throat infections, sore throat as it has an antibacterial property.
- **Controls high blood pressure**: Lemon water works wonders for people having heart problem owing to its high potassium content. It controls high blood pressure, dizziness, nausea as well as provides relaxation to mind and body.

Health Benefits of Lemon

- Lemons are packed with numerous health benefiting nutrients. The fruit is low in calories, 29 calories per 100 g, one of the lowest among citrus group.
- It contains no saturated fats or cholesterol, but is rich in dietary fibre.
- Its acidic taste is due to **citric acid**. Citric acid is present up to 8 per cent in its juice. Citric acid is a natural preservative, aids digestion. Studies found that citric acid help dissolve kidney stones.
- Lemons, like other citrus fruits, are excellent source of **ascorbic acid** Ascorbic acid or vitamin-C is a powerful water soluble natural anti-oxidant. This vitamin is helpful in preventing scurvy. Besides, consumption of foods rich in vitamin-C helps body develop resistance against infectious agents and scavenge harmful, pro-inflammatory free radicals from the blood.
- Lemons like oranges contain a variety of phytochemicals. **Hesperetin** and **naringenin** are flavonoid glycosides commonly found in citrus fruits. Naringenin is found to have a bio-active effect on human health as antioxidant, free radical scavenger, anti-inflammatory, and immune system modulator. This substance has also been shown to reduce oxidant injury to DNA in the cells in-vitro studies.
- They also a good source of B-complex vitamins such as pantothenic acid, pyridoxine, and folates. These vitamins are essential in the sense that body requires them from external sources to replenish.
- They contain healthy amount of minerals like iron, copper, potassium, and calcium. Potassium in an important component of cell and body fluids helps control heart rate and blood pressure.
- Suffer from indigestion? Mix a few drops of lemon juice with warm water and sip on it. This is useful for treating nausea, heartburn, diarrhoea, bloating and burping.
- Since lemon is a natural antiseptic, it is great to cure skin problems. It clears your skin and also acts as an anti-ageing agent by eliminating wrinkles and blackheads.
- Have a toothache? Apply fresh lemon juice where it hurts. If you have bleeding gums, applying lemon juice can curb the bleeding and stop bad breath.
- A sore throat can be cured by gargling with lemon juice and water regularly.

- ☆ Nimbu paani is high in potassium, which controls high blood pressure, dizziness, nausea and reduces stress. It also cures respiratory disorders like breathing problems and asthma.

Caution

Lemon juice is very low in PH, about 2.0. Its sour taste sometimes causes burning sensation if encounters with mouth, tongue, and lip ulcers. In addition, if taken large amounts may exacerbate acid-peptic disease and stomach ulcer conditions.

Papaya (*Carica papaya*)

The fruit is one of the favorite of fruit lovers for its nutritional, digestive, and medicinal properties. It is a very wholesome fruit. As a source of vitamin A it is unrivalled by any other fruit except perhaps the mango. It is a cheap fruit, available in all seasons, throughout the country. With its deep yellow colour, the ripe papaya fruit contains large quantities of a substance called carotene. This pigment is similar to that found in carrots, beetroot, dark green leafy vegetables like drumstick leaves, palak, curry leaves etc. Carotene in food is converted into vitamin A in our body. Therfore, regular consumption of papaya will ensure a good supply of vitamin A and C, which are both essential for good health. Thus eating ruit will help prevent blindness caused by vitamin A deficiency. Papaya has more carotene compared to other fruits such as apples, guavas, sitaphal, plantains etc. Hence it gives the maximum nutritional returns for the money.

It is cultivated for its fruits and latex papain, an enzyme that is used in food industry. Sensual, earthy and almost wildly exotic in an olfactory sense, the humble papaya is one of earth's finest gifts to mankind. Here are a few reasons why you should include this fruit in your diet: Low in calories, high on nutrition, it's an excellent choice for those on a diet. The peel of the papaya helps in healing skin wounds and injuries. The fruit has anti-inflammatory and anti-cancerous properties. People affected with osteoporosis, arthritis and oedema should consume it regularly. The fruit is good for hair growth and is known to check the growth of dandruff. Papaya is one of the most trusted detox solutions. If taken in moderation, the fruit can help get rid of toxins from the body. Those suffering from constipation and indigestion can rely on the papaya for relief. The benefits of this fruit on skin have perhaps been talked about since the time of our ancestors. Papaya is rich in antioxidants and contain a special enzyme called **papain** that can kill dead cells and cure skin impurities. Papain is present in all parts of the tree and fruit. It is excellent aid to digestion. It is an enzyme which helps to digest the protein in food. Hence it is used in various medicinal preparations. A glass of papaya milk or just applying the flesh of papaya on your skin can do wonders to your skin. Papaya is considered a wonderful wholesome fruit as it meets with some of our most essential vitamins, minerals and protein needs. Both ripe and raw papaya can be used to rejuvenate the skin. Ripe papaya is excellent for exfoliation of the skin and brings about newer looking skin and generally suits all skin types.

Papayas are good cleansers. It has Enzyme Papain which removes blemishes and rejuvenates an undernourished skin. Mash papayas and mix some honey to make a pack which you can smear all over your hands, neck and face. Wash after 15-20 minutes for a blemish free skin. Rich in vitamin A and enzymes, papayas act as a great exfoliator. Mashed papaya, when applied to the face, helps get rid of dead skin and gives your face that glow.

Beauty Uses of Papaya

- ✫ Since the fruit is a good source of Vitamin A and Papain which is a kind of protein, it helps in removing dead skin cells, along with breaking down the inactive proteins.
- ✫ If you apply a finely grounded paste of raw papaya on your face, it will help reduce pimples and blemishes!
- ✫ Mashed papaya can be used for treating the sore and cracked heels.
- ✫ The peel (skin) of the papaya can not only be used on the face, but it should also be used for whitening the skin on your legs.
- ✫ Using papaya on your face regularly will reduce the signs of ageing.
- ✫ If you have rough and dry skin, mash a papaya with honey and apply this mixture for hydration of the skin.
- ✫ Papaya also helps in controlling dandruff. Frequent use of its paste on your hair will improve its texture (Purvaja Sawant, 2013).

Papaya Fruit Health Benefits

- ✫ Papaya is a very diverse fruit that supplements different aspects of health. This fruitful fruit improves immunity, builds energy and has anti-inflammatory benefits.
- ✫ The humble papaya is one of earth's finest gifts to mankind. Here are a few reasons why you should include this fruit in your diet:
- ✫ Low in calories, high on nutrition, it's an excellent choice for those on a diet.
- ✫ The peel of the papaya helps in healing skin wounds and injuries.

- ☆ The fruit has anti-inflammatory and anti-cancerous properties.
- ☆ People affected with osteoporosis, arthritis and oedema should consume it regularly.
- ☆ The fruit is good for hair growth and is known to check the growth of dandruff.
- ☆ Papaya is one of the most trusted detox solutions. If taken in moderation, the fruit can help get rid of toxins from the body.
- ☆ Those suffering from constipation and indigestion can rely on the papaya for relief.
- ☆ The fruit is very low in calories (just 39 cal/100 g) and contains no cholesterol; but is a rich source of phyto-nutrients, minerals, and vitamins.
- ☆ Papayas contain soft, easily digestible pulp/flesh with good amount of soluble dietary fibre that helps to have normal bowel movements; thereby reducing constipation.
- ☆ Fresh, ripe fruit is one of the fruit with highest **vitamin-C** content.
- ☆ It is also an excellent source of **Vitamin-A** and flavonoids like beta-carotenes, lutein, zeaxanthin and cryptoxanthins.
- ☆ Papaya fruit is also rich in many essential B-complex vitamins such as Folic acid, pyridoxine, riboflavin, and thiamin. These vitamins are essential in the sense that body requires them from external sources to replenish and play vital role in metabolism.
- ☆ Fresh papaya also contains good amount of potassium and calcium. Potassium is an important component of cell and body fluids and helps controlling heart rate and blood pressure countering effects of sodium.
- ☆ Papaya has been proven natural remedy for many ailments. In traditional medicine, papaya seeds are anti-inflammatory, anti-parasitic, and analgesic, and they are used to treat stomachache and ringworm infections.
- ☆ Pieces of papaya laid on wounds and surgical incisions are reported to speed up their healing.
- ☆ Papaya is best muscle building food, as it Contains papain which breaks dietary protein into easily absorbable compounds.

Caution

Papayas contain white milk like latex substance, which can cause irritation to skin and provoke allergic reaction in some sensitized persons. Ripe papaya fruit can be safely used by pregnant women. Unripe green papaya should be avoided in pregnant women as it contains lot of papain, a proteolytic enzyme that used commercially to tenderize meat. Unripe papaya seeds (in very small quantities), latex, and leaves also contain **carpaine,** an alkaloid which could be dangerous when eaten in high doses. Unripe papaya, however, used safely as cooked vegetable. Carpaine in large quantities, is said to lower the pulse rate and depress the nervous system. Fortunately, the fleshy part of the fruit is completely free from this toxic substance. Hence, thee delicious fruit can be safely eaten, once the seeds are removed.

Grape (*Vitis vinifera*)

Grapes, "queen of the fruits" are storehouse of numerous health promoting phyto-nutrients such as poly-phenolic antioxidants, vitamins and minerals. So, include them in your regular diet, be it in the form of fresh fruits, juice or in salads!

Grapes Green Red and Black Varieties

Health Benefits

- ☆ Grapes are rich in polyphenolic phytochemical compound **resveratrol**. Resveratrol is one of powerful anti-oxidant, which has been found to play protective function against cancers of colon and prostate, coronary heart disease (CHD), degenerative nerve disease, Alzheimer's disease and viral/ fungal infections.
- ☆ **Resveratrol** reduces stroke risk by altering the molecular mechanisms in the blood vessels. It does so firstly by reducing susceptibility of blood vessels damage through decreased activity of angiotensin (a systemic hormone causing blood vessel constriction that would otherwise elevate blood pressure) and secondly, through increased production of the vasodilator substance, nitric oxide (a beneficial compound that causes relaxation of blood vessels).
- ☆ Anthocyanins are another class of polyphenolic anti-oxidants present abundantly in the red grapes. These phyto-chemicals have been found to have anti-allergic, anti-inflammatory, anti-microbial, as well as anti-cancer activities.

- **Catechins**, a type of flavonoid tannin group of anti-oxidants found in white/ green varieties has also shown to have these health protecting functions.
- The berries are very low in calories. 100 g fresh grapes just provide 69 calories but zero cholesterol levels.
- Grapes are rich source of micronutrient minerals like copper, iron and manganese.
- They are also good source of **vitamin-C**, vitamin A, vitamin K, carotenes, B-complex vitamins such as pyridoxine, riboflavin, and thiamin.

Pears (*Prunus communis*)

Asian varieties feature crispy texture and firm consistency that do not change after harvesting or storage makes them ready-to-eat. Whereas, European types generally becomes soft and juicy when allowed to ripen. In structure, pear fruit has bell or "pyriform" shape; around 5-6 inches long and weigh about 200 gm. Fresh fruit is firm in consistency with mild 'apple' flavour. Externally, its skin is very thin and depending upon the cultivar type, the fruit may have green, red-orange or yellow-orange colours. Internally, it's off white colour pulp is crunchy and juicy. However in fully matured ones, the pulp turns to grainy texture with gritty sensation while cutting with knife. Center of the fruit is more or less similar to apple with centrally located tiny inedible seeds. Pears come in different colours — green, red, yellow and brown.

Health Benefits of Pears

- ☆ Packed with phytonutrients known as flavonols, pears provide antioxidant protection against several diseases.
- ☆ The skin of a pear comprises even more phenolic phytonutrients as its flesh. These include antioxidants and anti-inflammatory flavonoids, apart from anti-cancer phytonutrients such as cinnamic acids. Also, rich with fibre, pear skins should always be consumed.

- ☆ Some researchers say that flavonoids, which have three groups — flavonols, **flavan-3-ols**, and anthocyanins — are abundantly found in pears. This is especially beneficial for those who have type 2 diabetes (Zeenia Baria,2013).
- ☆ Pears contain phytonutrients, which are known to have anti-inflammatory benefits. This results in a decreased risk of common chronic diseases that start with chronic inflammation and oxidative stress like type 2 diabetes and even heart disease.
- ☆ The low acidic nature of pears makes it easy to digest.
- ☆ Pears are also known as being a hypoallergenic food, which means that the chances of one getting an allergy from a pear are very rare.
- ☆ The dietary fibre in pears makes them a good source of immune-supportive vitamin C and bone-building vitamin K.
- ☆ Pear fruit is packed with health benefiting nutrients such as dietary fibre, anti-oxidants, minerals and vitamins which are necessary for optimum health.
- ☆ Pears provide about 3.1 g of dietary fiber per 100g. Regular eating of pears may offer protection against colon cancer. Most of the fiber is **non soluble polysaccharide** (NSP), making them a good bulk laxative. Also, the gritty fiber content binds to cancer causing toxins and chemicals in the colon, protecting its mucous membrane from contact with these compounds.
- ☆ Pear fruit is one of the very low calorie fruits, just provides 58 cal per 100g. Just a few sections a day in the diet can bring significant reduction in weight and blood LDL cholesterol levels.
- ☆ It contains good quantities of vitamin C. Fresh fruits provide about 7 per cent of RDA per 100 g.
- ☆ They are rich in antioxidant flavonoids phyto-nutrients such as beta carotene, lutein and zea-xanthins. These compounds along with vitamin C and A help body protect from harmful free radicals.
- ☆ The fruit is a good source of minerals such as **copper**, iron, potassium, manganese and magnesium as well as B-complex vitamins such as folates, riboflavin and pyridoxine (vitamin B-6).
- ☆ Pears are among the least allergenic of all fruits and are therefore recommended by health practitioners as safe alternative in the preparation of food products for allergy sufferers.
- ☆ Pears have suggested being useful in treating colitis, chronic gallbladder disorders, arthritis and gout.

Peaches (*Prunus persica*)

This fruit is best for dry skin. A combination of peach paste and yogurt, when applied to the skin and rinsed off with lukewarm water not only moisturises the skin but also leaves it supple and soft. Gorging on these yummy fruits is not enough, applying them on the skin and hair helps a lot as well. So, the next time you lunge at that favourite fruit of yours, look beyond its culinary value, for you'd be surprised at its cosmetic benefits.

Health Benefits of Peaches

- ✰ Peaches are low in calories and contain no saturated fats. Nonetheless, they are packed with numerous health promoting compounds, minerals and vitamins.
- ✰ The total measured anti-oxidant strength of 100 g peach fruit is 1814 TE (Trolex equivalents).
- ✰ Fresh peaches are a very good source of antioxidant vitamin C.
- ✰ Fresh peaches are also moderate source of **vitamin A** and **ß-carotene**. ß-carotene is a pro-vitamin, which converts into vitamin A in the body. Vitamin A is essential for vision. It is also required for maintaining healthy mucus membranes and skin.
- ✰ They are rich in many vital minerals such as potassium, fluoride and iron. Iron is required for red blood cell formation. Fluoride is a component of

bones and teeth and is important for prevention of dental caries. Potassium is an important component of cell and body fluids that help regulate heart rate and blood pressure.

☆ Peaches contain many health promoting flavonoid poly phenolic antioxidants such as lutein, zeaxanthin and ß-cryptoxanthin.

Plum (*Prunus domestica*)

Plum is a soft round smooth-skinned sweet fruit, it is also known as a stone fruit tree in the genus Prunus, and the ripe fruit is usually dusty-white outer layer that gives them the appearance of glaze and easily erased. Each fruit is about the size of medium sized tomato, measuring about 5-6 cm in diameter and weigh about 70 g. It has central depression at the stem side. Internally, the pulp is juicy and vary widely from creamy yellow, crimson, light blue or light green in colour depending upon the cultivar type. The main benefit contained in Plum fruit is much beneficial antioxidants for the body, some studies also reported that the plums have anti-cancer substances. There is centrally placed single, smooth but hard stone like seed. Seeds are inedible. It has sweet and tarty taste pulp with pleasant aroma. Some common cultivars of plums are: cherry plum, damson, blackthorn plum. Plums are a good source of iron and potassium. They are low in fat, sodium and cholesterol free. They are also good source of vitamin A and vitamin E. A typical 66 gram plum contains about 36 calories, 0.5 grams of protein and about 9 grams of carbohydrates. Plums are also strong anti-oxidants, which means that they are able to fight off free radicals. Although pruns are better known for thrir laxative effects, plums are just as effective and both forms of the fruit are great sources of dietary fibre.

Health Benefits of Plums

- ☆ The, fleshy, succulent plums are low in calories and contain no saturated fats; but contain numerous health promoting compounds, minerals and vitamins.
- ☆ Plum fruits contain, dietary fibre such as, **sorbitol,** and **isatin** are known to help regulate the functioning of the digestive system and thereby used in constipation conditions.
- ☆ Fresh plums are an excellent source of **vitamin C**, which is also a powerful natural antioxidant.
- ☆ Fresh plums, especially yellow Mirabelle type, are very good source of **vitamin A** and **beta-carotene**. Vitamin A is essential for vision. It is also required for maintaining healthy mucus membranes and skin.
- ☆ The fruit is also good in health promoting flavonoid poly phenolic antioxidants such as lutein, cryptoxanthin and zeaxanthin in significant amounts. **zeaxanthin,** an important dietary carotenoid selectively absorbed into the retinal macula lutea where it is thought to provide antioxidant and protective light-filtering functions.
- ☆ Plums are rich in minerals like potassium, fluoride and iron. Iron is required for red blood cell formation.
- ☆ Rich in B-complex group of vitamins such as niacin, vitamin B-6 and pantothenic acid. These vitamins are acting as cofactors help body metabolize carbohydrates, proteins and fats. They also provide about 5 per cent RDA levels of vitamin K. Vitamin K is essential for many clotting factors function in the blood as well as in bone metabolism and helps reduce Alzheimer's disease in the elderly.

Caution

Plums contain oxalic acid, a naturally occurring substance found in some fruits and vegetables which may crystallise as oxalate stones in the urinary tract in some people. Therefore, people with known oxalate urinary tract stones are advised to avoid eating plums. Adequate intake of water is therefore advised to maintain normal urine output even if these individuals want to eat them.

Litchi (*Litchi chinensis*)

Its outer surface is covered with rough leathery rind or peel featuring pink colour. The peel can be easily removable in the ripe fruits. Inside; consists of edible portion or aril that is white, translucent, sweet and juicy. The fruit has sweet, fragrant flavour and delicious to savor. The pulp has single, glossy brown nut-like seed, 2 cm long, and 1–1.5 cm in diameter. The seeds, like in sapodilla, are not poisonous but should not be eaten. Fresh lychees are readily available in the markets from July to October, about 120-140 days after flowering. The litchi fruit contains vitamins and minerals that promote a healthy diet. Studies have shown that litchi fruit can fight the growth of cancer cells as it contains flavonoids in its pulp, which is effective against breast cancer. It is also a great source of vitamin C that helps the body to fight heart disease apart from cancer. Vitamin C is also good for bones, skin and tissue. Due to high vitamin C content, the fruit benefits those suffering from cold, fever and sore throat. Litchi also helps the body to digest food properly for the best nutrition and is said to have the ability to relieve pain too (Manu Vipin, 2012). Litchi contains a high amount of soluble fibre which protects from colon cancer. The polyphenols present in litchi promote heart health, normalize blood pressure and heart rate thereby protecting against strokes and coronary heart diseases. The fruit is a great source of vitamin C, complex vitamins, folates, phosphorus, calcium, magnesium and protein. It also has oligonol, a polyphenol which has anti-influenza virus actions. It also helps to improve blood circulation, reduces weight and protects the skin from UV A rays. It decreases deep fat and raises sideline blood circulation (Parmer, 2013).

Health Benefits

- ☆ Litche fruits are low in calories, contains no saturated fats or cholesterol, but rich in dietary fibre which, can be very important for individuals who are concerned about their excess body weight.
- ☆ Lychee fruit contains oligonol, a low molecular weight polyphenol. Oligonol has been found to have several anti-oxidant, anti-influenza virus actions. In addition, it helps improve blood flow in organs, reduce weight, and protect skin from harmful UV rays (Takuya Sakurai (Kyorin University, Japan), Biosci. Biotechnol. Biochem., 72(2), 463-476, 2008).
- ☆ Litchi, like citrus fruits, is an excellent source of vitamin C. Consumption of fruits rich in vitamin C helps body develop resistance against infectious agents and scavenge harmful, pro-inflammatory free radicals.
- ☆ It is a very good source of B-complex vitamins such as thiamin, niacin and folates. These vitamins are essential since they function by acting as co-factors to help body metabolize carbohydrates, protein and fats.
- ☆ Litchi also contains a very good amount of minerals like potassium and copper. Potassium is an important component of cell and body fluids help control heart rate and blood pressure; thus offers protection against stroke and coronary heart diseases. Copper is required in the production of red blood cells.

Mango (*Mangifera indica*)

Mango fruit is one of the most popular, nutritionally rich fruit with unique flavour, fragrance, taste, and heath promoting qualities making it a common ingredient in new functional foods often called "super fruits". Mango is one of the delicious tropical seasonal fruit and believed to be originated in the sub-Himalayan plains of Indian subcontinent. **Mangoes**, rightly called the king of fruits for not just its taste but also for health benefits. The soft pulpy fruit has an amazing effect on skin too. Rich in vitamin-A and antioxidants, it fights skin aging, regenerates skin cells and restores the elasticity of skin.

The mango is the national fruit of India, Pakistan and the Philippines. It is also the national tree of Bangladesh. In Hinduism, the perfectly ripe mango is often held by Lord Ganesha as a symbol of attainment, regarding the devotees potential perfection. Mango blossoms are also used in the worship of the goddess Saraswati. No Telugu/Kannada new year's day called Ugadi passes without eating ugadi pacchadi made with mango pieces as one of the ingredients. In Tamil Brahmin homes mango is an ingredient in making vadai paruppu on Sri Rama Navami day (Lord Ram's Birth Day) and also in preparation of pachchadi on Tamil new year's day.

The **Jain** goddess **Ambika** is traditionally represented as sitting under a mango tree. Mango leaves are used to decorate archways and doors in Indian houses and during weddings and celebrations like Ganesh Chaturthi. Mango motifs and paisleys are widely used in different Indian embroidery styles, and are found in Kashmiri shawls, Kanchipuram silk sarees, etc. Paisleys are also common to Iranian art, because of its pre-Islamic Zoroastrian past.

In Tamil Nadu, the mango is considered, along with banana and jackfruit, as one of the three royal fruits (Mukkani-occupying first place in terms of sweetness and flavour. Ma-pala-vazhai).

As the saying goes, 'an apple a day keeps the doctor away'. This old adage could well be used for mangoes because, like apples, they're simply brimming with goodness. Mangoes are high in energy, low in fat, and are a great source of calcium and vitamins essential for good health. Mangoes have a high content of beta-carotene and rank among the top providers of beta-carotene. Beta-carotene is a powerful anti-oxidant that can help protect the body against disease and also fight the signs of ageing by assisting with the growth and repair of cells, tissues and skin.

Not only do they taste great, but mangoes are also loaded with several qualities that are excellent for your health. They are rich in powerful antioxidants that are known to neutralise free radicals that cause damage to cells and lead to health problems like heart disease, premature aging and cancer among other things.

However they are also a great source of nutrients in the concentrated form. In order to reduce the consumption of calories, mangoes are a very good choice, said an expert. Beta-carotene is also known to reduce the risk of certain forms of cancer. To get the right amount of nutrition not too many mangoes need to be consumed. A single fruit is capable of providing of almost a day's supply of Vitamin C to the body. It also helps in making bones stronger since it contains minerals, calcium and magnesium as well as B vitamins. They are also rich in a carotenoid called lycopene, which is an effective antioxidant. In order to lose weight, mangoes can form an important part of the diet however it has to be coupled with other fruits and a complete meal, only after consultation with a nutritionist. Not only losing weight, daily consumption of mangoes make the skin complexion brighter and makes skin softer. A 200g serve of ripe mango (the equivalent of less than one mango) provides you with up to three times your recommended daily intake of Vitamin A and Vitamin C. Vitamin C, an antioxidant important in protecting the body from infection, is required in the formation of collagen, a protein that gives structure to bones, cartilage, muscle and blood vessels. Vitamin C also aids the absorption of iron, necessary for transporting oxygen from the lungs to cells all around the body. **It also promotes healthy immune function.** Mangoes also provide more beta-carotene than any other fruit and are a rich source of fibre and potassium.

In some countries mangoes are used to treat a variety of diseases and conditions. It is believed that immature mango, in which the seed has not fully formed, can be combined with salt and honey to treat diarrhea, dysentery, piles, morning sickness, chronic dyspepsia, indigestion and constipation. Treatments for heat stroke, bilious disorders and scurvy have also been recorded. It's wonderful when a fruit that tastes so delicious has so many wonderful nutrients in it. They are high in Vitamin C, Vitamin B6, Potassium, Copper and Vitamin A. Mangoes also contain several important phytochemicals including: Cryptoxanthin, Lutein, Gallic Acid and Anacardic acid. They are an incredibly healthy snack. You can eat the entire fruit for just over a hundred calories. High in fibre, virtually fat-free, and mangoes contain numerous vitamins. Mangoes also contain beta-carotene which may help slow the aging process, reduce the risk of certain forms of cancer, improve lung function, and reduce complications associated with diabetes.

One cup of mangos is just 100 calories, so it's a sweet treat that won't weigh you down. One serving of mangos provides 100 per cent of your daily vitamin C and 35 per cent of your vitamin A, both important antioxidant nutrients. Vitamin A is important for vision and bone growth. Mangos provide 12 per cent of your daily dietary fibre. Mangos contain over 20 different vitamins and minerals. Each serving of mango is fat free, sodium free and cholesterol free. If your skin frequently feels parched, just eat a mango! Mangoes are bursting with vitamins and minerals, especially vitamin A. Vitamin A normalises the production and life cycle of skin cells. Mangoes also improve your skin if you suffer from acne. In skin with acne, there is an overproduction of cells in the stratum corneum, which is the outermost layer of the skin and is composed of biologically 'dead' cells. These excess dead cells combine with sebum (the skin's own natural oil) to form comedones - the pore plugs that are the defining element of acne. Taken as a dietary supplement, vitamin A helps to prevent overproduction of skin cells in the stratum corneum. You can add 3 to 4 slices of mango to every meal as a refreshing and hydrating naturally sweet treat. Mango fruit is rich in pre-biotic dietary fibre, vitamins, minerals, and has antioxidant compounds. It is beneficial for digestion too.

According to a new research, mangoes have been found to protect against colon, breast and prostate cancers. Fresh mango is a rich source of potassium, which is an important component of cell and body fluids that helps to control heart rate and blood pressure. Vitamin E, which is abundantly present in mangoes, helps to regulate sex hormones and boosts sex drive. Mango helps to clear clogged pores that cause acne. Just slice a mango into thin pieces and keep them on your face for 10 to 15 minutes and then take bath or wash your face and see the results. It's a known fact that mangoes are rich in Iron. People who suffer from anaemia can take mangoes regularly along with their dinner. It is especially good for women after menopause. High level of soluble dietary fibre, Pectin and Vitamin C present in mangoes helps to lower serum cholesterol levels. A recent study shows that not only the flesh of a mango but the leaves can fight diabetes. Mango, rightly called the king of fruits for not just its taste but also for health benefits.

Health Benefits of Mangoes

- ☆ With its high iron content, mangoes are excellent for pregnant women and those who suffer from anemia.
- ☆ Constantly complaining about clogged pores? Place mango slices on your skin and then wash off after 10 minutes.
- ☆ If you suffer from indigestion problems, nothing will help you as much as a mango. They're known to give relief from acidity and aid proper digestion since they contain digestive enzymes that help break down proteins.

- Rich in potassium, mangoes reduce high blood pressure. They also contain pectin, a soluble dietary fibre that is known to lower blood cholesterol levels.
- Trying to put on weight? Include mangoes in your diet. Since it is rich in calories as well as carbohydrates, it could be the perfect fruit to have.
- Some studies say that eating mangoes reduces the risk of kidney stone formation.
- In Chinese medicine, mangoes are considered sweet and sour with a cooling energy. They are useful for those suffering from anaemia, bleeding gums, cough, fever, nausea and even sea sickness.
- This fruit is rich in **glutamine acid** - an important protein for concentration and memory. Instead of snacking on unhealthy chips and cookies, why not feast on slices of mangoes instead.
- Though they are traditionally not considered as aphrodisiacs, mangoes contain Vitamin E which helps boost one's sex life. The vitamin works to regulate the body's sex hormones.
- Mango fruit is rich in pre-biotic dietary fibre, vitamins, minerals, and poly-phenolic flavonoid antioxidant compounds. It is beneficial for digestion too.
- Mango fruit has been found to protect against colon, breast, leukemia and prostate cancers. Studies suggest that *polyphenolic anti-oxidant* compounds in mango are known to offer protection against breast and colon cancers.
- It is an excellent source of **Vitamin-A** and flavonoids like **beta-carotene, alpha-carotene,** and **beta-cryptoxanthin**. Together; these compounds are known to have antioxidant properties and are essential for vision. Vitamin A is also required for maintaining healthy mucus membranes and skin.
- Fresh mango is a very rich source of potassium which is an important component of cell and body fluids that helps controlling heart rate and blood pressure.
- It is also a very good source of vitamin-B6 (pyridoxine), **vitamin-C** and vitamin-E. Consumption of foods rich in vitamin C helps body develop resistance against infectious agents and scavenge harmful oxygen free radicals.
- Copper is a co-factor for many vital enzymes, including cytochrome c-oxidase and superoxide dismutase (other minerals function as co-factors for this enzyme are manganese and zinc). Copper is also required for the production of red blood cells.
- Mango peels are also rich in phytonutrients, such as the pigment antioxidants like carotenoids and polyphenols.
- Fresh mango is a rich source of potassium, which is an important component of cell and body fluids that helps to control heart rate and blood pressure.
- Vitamin E, which is abundantly present in mangoes, helps to regulate sex hormones and boosts sex drive.

- Mango helps to clear clogged pores that cause acne. Just slice a mango into thin pieces and keep them on your face for 10 to 15 minutes and then take bath or wash your face and see the results.
- It's a known fact that mangoes are rich in Iron. People who suffer from anaemia can take mangoes regularly along with their dinner. It is especially good for women after menopause.
- High level of soluble dietary fibre, Pectin and Vitamin C present in mangoes helps to lower serum cholesterol levels.
- A recent study shows that not only the flesh of a mango but the leaves can fight diabetes.
- The soft pulpy fruit has an amazing effect on skin too. Rich in vitamin-A and rich antioxidants, it fights against skin aging, regenerates skin cells and restores the elasticity of skin

Health Tip

Before going to bed put some 10 or 15 mango leaves in warm water and close it with lid. The next day morning filter the water and drink it in empty stomach. Do this regularly.

Caution

Mango latex allergy especially with raw, unripe mangoes is common in some sensitized individuals. Immediate reactions may include itchiness at the angle of the mouth, lips and tip of the tongue. In some people, the reactions can be severe, with manifestations like swelling of the lips, ulceration at the mouth angles, respiratory difficulty, vomiting, and diarrhea. This reaction develops because of **anacardic acid** present in raw, unripe mangoes. Cross-allergic reactions with other anacardiaceae family fruits like "cashew apples" are quite common. It is quite rare with fully ripen fruits; however, people with known case of mango fruit allergy may have to avoid them.

Loquat (*Eriobotry japonica*)

Succulent, tangy and sweet, wonderfully delicious loquat fruits are rich in vitamins, minerals, and anti-oxidants. This unique fruit is originated in the mountainous, evergreen rain forest of Southeastern China, from where it spread all across the world including India. Loquat is a good source of Vitamin A, which is crucial for the visual and dental health of an individual. Extract from loquat leaves is an important ingredient for lung ailments and has been used by Chinese, since ancient times. Loquats are appropriate for maintaining optimum health, as they are low in saturated fat and cholesterol. People with a deficiency of Vitamin A should have loquat in large number, because of its rich content. Loquats are rich in fibre, making them suitable for those who wish to lose weight. The loquat leaf is said to alleviate coughing and nausea. It even dissolves phlegm and is an expectorant. Loquat paste helps in soothing the digestive and respiratory systems of a person. The loquat leaf is known to shorten the recovery time from respiratory illness. Loquat contains malic acid, tartaric acid, citric acid, vitamins A, B and C, and B17. Vitamin B17 is known as amygdaline, laetrile or the anti-cancer vitamin, as it helps prevent cancer. Loquat is very very low in saturated fat, very low in sodium but has no cholesterol or sugar, but very high in vitamin A, and high in vitamin 6. It is high in diatery fibre, manganese and potassium. It has same properties as that of a mango but comparatively low in sugar and calories.

Health Benefits of Loquat Fruit

- Delicious, loquats are very low in calories; provide just 47 cal per 100 g, however, rich in insoluble dietary fibre, pectin. Pectin retains moisture in the colon and thus functions as bulk laxative and by this way, it helps to protect the colon mucous membrane by decreasing exposure time to toxic substances as well as binding to cancer causing chemicals in the colon.

- Pectin has also been shown to reduce blood cholesterol levels by decreasing its re-absorption in the colon by binding bile acids resulting in its excretion from the body.
- Loquat fruit is an excellent source of vitamin-A (provides about 1528 IU per 100g), and phenolic flavonoid antioxidants such as chlorogenic acid, neo-chlorogenic acid, hydroxybenzoic acid, feruloylquinic acid, protocatechuic acid, epicatechin, coumaric acids and ferulic acid. Ripen fruits have more chlorogenic acid concentrations.
- Vitamin A maintains integrity of mucus membranes and skin. Lab studies have shown that consumption of natural fruits rich in vitamin-A and flavonoids helps to protect from lung and oral cavity cancers.
- Fresh fruit is very rich in potassium and some B-complex vitamins such as folates, vitamin B-6 and niacin and contain small amounts of vitamin-C. Potassium is an important component of cell and body fluids, helps controlling heart rate and blood pressure.
- It is also a good source of iron, copper, calcium, manganese, and other minerals. Manganese is used by the body as a co-factor for the antioxidant enzyme, superoxide dismutase. Copper is required in the production of red blood cells. Iron is required for as a cofactor in cellular oxidation as well for red blood cell formation.

Caution

The seeds of loquat fruit contain many toxic alkaloids like cyanogenic glycosides which when consumed can cause serious life threatening symptoms like vomiting, breathlessness, and death. Therefore, especially children are advised to avoid chewing seeds and should be supervised by adults while eating loquat fruits.

Pomegranate (*Punica granatum*)

Pomegranate is among the most popular, nutritionally rich fruit with unique flavour, taste, and heath promoting characteristics. Along with berries, and some tropical exotics such as mango, it too has novel qualities of functional foods often called as "super fruits."The impressive health values of this fruit had already been known since ancient days. The size of a ripe pomegranate can be as small as an orange or as big as a grapefruit, approximately 7-12 cm in diameter, depending on its variety. It has a rounded hexagonal shape, with thick yellowish to reddish outer layer. Inside a pomegranate is about 700-800 tightly packed seed casings called **arils** that are deep red in colour when nicely ripe. The taste of the juice differs depending on the variety and its state of ripeness. But basically, it can be sweet, sour or tangy. The potent anti-oxidant in pomegranate juice is beneficial in preventing the harderning of arteries. This food is very low in Saturated Fat, Cholesterol and Sodium. It is also a good source of Dietary Fibre and Folate, and a very good source of Vitamin C and Vitamin K but a a large portion of the calories in this food come from sugars. Pomegranates are a wonderfully hydrating source of antioxidants. They may be the world's most prolific source of polyphenols. The unique combination of elements in pomegranates increases the protective abilities of sunscreens, which can help prevent sun damage. Add pomegranates to your morning cereal or yogurt, fresh green or fruit salads for a refreshing snack or breakfast.

Pomegranates have had a reputation as a super food for some time, with its high levels of antioxidants thought to reduce the risk of heart disease and cancer It has high levels of iron and vitamin A, C and E. Iron helps maintain a good supply of oxygen to the body, vitamin E can reduce aeging in the brain and might help prevent Alzheimer's while vitamin C boosts skin, teeth and bones. Pomegranate polyphenolics, tannins and anthocyanins could also have beneficial effects. Polyphenols help the body rid itself of cancer causing agents. With pomegranates thought to be particularly good at warding off prostate cancer and possibly breast cancer. Tannins- have been shown to lower blood pressure and stimulate the immune system, while anthocyanins help protect blood vessels and reduce inflammation. It also has effective anti-bacterial properties and can ease the symptoms of osteoarthritis. Pomegranate juice increases sexual urges and strenghtens bones, muscles. And increases testosterone which in turn help improve a person's mood and memory and even relieve stress such as pre-match nerves. In men higher levels of testosterone increases facial hair, deepens the voice and stimulate the sex drive. While the men have higher levels of the hormone, testoterone is also found in women, produced inthe advenal glands and ovaries. (Annamous, 2012). The dried bark of both root and stem, has long been used in the treatment of tapewoms. Extrate of different parts of the tree exhibit anti biotic activity. The seeds of pomegranates yield a drying oil which contain punicic acid forming up to 72 per cent of the fatty acids. This oil possesses antibacterial properties. Pomegranate, is highly recommended for preventing breast cancer. It contains polyphenol- an ellagic acid with anti-oxidant properties that prevent cancer growth. Include this delicious fruit in your diet and discover effective health benefits.

Health Benefits

According to the study, the superfruit contains antioxidants, called polyphenols, that keep arteries free of fat. Polyphenols also help arteries expand and contract to maintain blood flow and keep them from hardening, which is the leading cause of heart attacks. The leaf, fruit rind, seeds, dried bark, stem, root and the fruit itself have a host of benefits. Often called Nature's Power Fruit because of its usefulness, pomegranates are great for your heart since they are an excellent antioxidant. Other benefits include:

- **Fights breast cancer**: Studies show that pomegranate juice destroys breast cancer cells and leaves the healthy cells alone.
- **Stabalises PSA levels**: In a study of men who had undergone treatment for prostate cancer, eight ounces of pomegranate juice every day kept their PSA levels stable thereby reducing the need for more treatment such as chemotherapy or hormone therapy.
- **Protects your arteries**: Pomegranates are known to help prevent plaque from building up in your arteries and reverse previous plaque buildup.
- **Lowers cholesterol:** This fruit is known to lower LDL (bad cholesterol) and raise HDL (good cholesterol).

Other benefits include lowering blood pressure, protecting your teeth (drinking pomegranate juice is a natural way to prevent dental plaque) and even prevent cartilage deterioration.

- Pomegranates have very high content of punicalagins, a potent anti-oxidant component found to be responsible for its superior health benefits.

- ☆ The capacity of anti-oxidant in this fruit is two or three times higher than that of red wine and green tea.
- ☆ The level of anti-oxidant is even higher than those of other fruits known to have high-levels of anti-oxidant, including blueberries, cranberries and oranges. This was attributed to the very high polyphenol content in the fruit.
- ☆ They are also a good source of vitamin B (riboflavin, thiamin and niacin), vitamin C, calcium and phosphorus. These combination and other minerals in pomegranates cause a powerful synergy that prevents and reverses many diseases.
- ☆ Drinking pomegranate juice frequently is extremely beneficial in fighting the hardening of arteries (atherosclerosis). It reduces the oxidation of bad LDL cholesterol which contributes to artery clogging and hardening.
- ☆ Not only does the juice significantly reduce the blood vessel damage, it is found to actually **reverse** the progression of this disease.
- ☆ Pomegranates contain a powerful agent against cancer, particularly prostate cancer.
- ☆ The dried bark of both the root and stem, is used in the treatment of tapeworms. The extracts of different parts of tree exhibit antibiotic activity.
- ☆ The seeds of pomegranates yield a drying oil which contains punicic acid forming up to 72 per cent of the fatty acids. This oil possesses antibacterial properties.
- ☆ This is highly recommended for preventing breast cancer. It contains polyphenol- an ellagic acid with anti-oxidant properties that prevent cancer growth. Include this delicious fruit in your diet and discover effective health benefits.
- ☆ Some studies say that pomegranate juice is known to kill breast cancer cells while leaving healthy cells alone.
- ☆ The fruit is also known to prevent plaque from building up in the arteries as well reversing previous plaque buildup.
- ☆ Pomegranate can significantly lower LDL (bad cholesterol) and increase HDL (good cholesterol).
- ☆ It is beneficial for those suffering from high blood pressure because it is known to lower blood pressure levels.
- ☆ Pomegranate juice is good for your teeth because it prevents dental plaque.
- ☆ The fruit also prevents cartilage deterioration (Anonymous, 2013).

Here are some common ailments that are known to react positively with the use of pomegranate or its juice:

- ☆ **Anemia:** Add a teaspoon of ground cinnamon with a little honey to a cup of pomegranate juice. Especially beneficial for women after monthly loss of blood due to menstruation.

- **Anal Itch:** Itching in the anal region is often caused by parasites in the intestines that go to the anal area to lay their eggs. Roast some pomegranate skin until it is brown and brittle. Then crush it to a fine powder form and mix with a little olive oil. Apply this concoction to the anus to kill the worms.
- **Anti-aging:** We all know that anti-oxidant is highly effective in helping to protect the skin from free radical damage known to cause signs of aging.
- **Asthma:** The high content of ascorbic acid (vitamin C) in this fruit is a powerful anti-inflammatory agent. It can greatly reduce wheezing in young children with asthma.
- **Atherosclerosis:** The highly cleansing power of this miracle juice scrubs away the old build-up of arterial deposits, reducing the risks of heart diseases and stroke.
- **Bleeding Piles:** Pound the (clean) skin of one fruit from the sour variety. Boil the pound pulp in about two cups of water. Sweeten with honey and drink twice a day until healed.
- **Cancer prevention:** The high anti-oxidant content protects cells from damages by free radicals. Regularly drinking juices high in anti-oxidants keep cancer at bay.
- **Cholesterol:** Drinking juices high in anti-oxidant has been proven to fight the oxidative stress that is the main culprit in oxidizing the LDLs in the blood.
- **Dysentery:** Drinking fresh pomegranate juice is an excellent remedy to soothe the pain and inflammation caused by severe diarrhea with blood and mucus in stools.
- **Immune booster:** The anti-oxidant nutrients in pomegranates are critical in building up your immune system. Drink juice high in anti-oxidant when you feel a cold coming.
- **Loss of Appetite:** If you can't eat, at least you can drink! Pomegranate juice can help increase your appetite.
- **Morning Sickness/nausea:** Mix and drink an equal amount of honey with pomegranate juice for relief.
- **Sore Throat:** The anti-inflammatory agent in pomegranate juice significantly reduces the soreness and redness in the throat.
- Pomegranate is touted as a wonder fruit since it shows promise in improving depression and bone mineral desity.

Caution

If you are on any medication, consult with your doctor before you start consuming pomegranate juice regularly. There is some concern that the juice may affect the metabolism of some prescribed medications.

Pineapple (*Ananas comosus*)

It is both juicy and fleshy with the stem serving as the fibrous core. The rough, tough, scaly rind may be dark green, yellow, orange-yellow or reddish when the fruit is ripe. Juicy flesh ranges from creamy white to yellow in colour and has mix of sweet and tart taste with rich flavour. Each fruit measures in size up to 30cm long and weigh 1/2 to 4kg or more. A great skin emollient, pineapples help rejuvenate and cleanse the skin. Rubbing a slice of pineapple on spots like knees, elbows and heels helps soften the skin. It can also act as a loofah or sponge while taking a shower. Though only 60 per cent of it is edible, pineapple is a good source of thiamine (vitamin B1) compared to other fruits and is very rich in vitamin C. Fresh pineapple juice contains an enzyme called Bromelin which aids digestion. Pineapple is highly rich in vitamin C and so acts as a great antioxidant, which prevents colds and flu. It is also a good source of thiamine, B12, folate, copper and dietary fibres. A few slices of pineapple a day in your meal can work wonders for your body and it is a good choice for weight watchers due to its low calorie content (100gm pineapple= 50 KCal.). Besides helping you lose weight, pineapple has an important nutrient that reduces arthritic pain. Bromelain is a magical component which is present in the fruit that has many health benefits. Although it is present in the entire fruit, the stem of a pinappple has the most amount of bromelain. Bromelain extracts aid in digestion and act as an anti-inflammatory agent thus in turn it helps with arthritis pain. Being a proteolytic enzyme, bromelin helps to tenderise your meats by breaking down the protein in it, but remember bromelain can coagulate the milk and prevent the jelly to set hence remember if you want to make a pineapple milkshake you need to cook it to destroy the bromelain content (Trina Remedios, 2013).

Health Benefits

- ☆ Fresh pineapple is storehouse of many health promoting compounds, minerals and vitamins that are essential for optimum health.
- ☆ The fruit is low in calories, contains no saturated fats or cholesterol; but rich source of soluble and insoluble dietary fibre like pectin.
- ☆ Pineapple fruit contains a proteolytic enzyme **bromelain** that digests food by breaking down protein. Bromelain also has anti-inflammatory, anti-clotting and anti-cancer properties. Consumption of pineapple regularly helps fight against arthritis, indigestion and worm infestation. Bromelain (enzyme) is known as a treatment for inflammation and swelling of the nose, ear and ninus.
- ☆ Fresh ananas is an excellent source of antioxidant vitamin; **vitamin C**. It is required for the collagen synthesis in the body. Collagen is the main structural protein in the body required for maintaining the integrity of blood vessels, skin, organs, and bones.
- ☆ It also contains good amount Vitamin A and beta-carotene levels. These compounds are known to have antioxidant properties.
- ☆ In addition, this fruit is rich in B-complex group of vitamins like **folates**, thiamin, pyridoxine, riboflavin and minerals like **copper**, **manganese** and potassium.

Caution

- ☆ Pineapple fruit contains a proteolytic enzyme **bromelain** that may cause excessive uterine bleeding if consumed in large quantities during pregnancy.
- ☆ Juice of ripe fruit is considered to be a diuretic (helps to pass urine easily)while juice fromthe unripe fruit acts as purgative.

Jackfruit (*Artocarpus heterophyllus*)

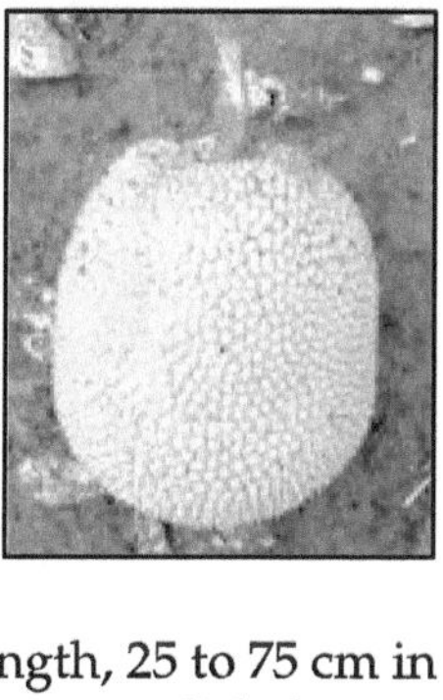

Jackfruit tree is believed to be indigenous to the Southwestern rain forests of India. It is widely cultivated in the tropical regions of Indian subcontinent, Thailand, Malaysia, Indonesia and Brazil for its fruit, seeds, and wood. The tree grows best under tropical humid and rainy climates but rarely survives cold and frosty conditions. Jackfruit is a huge tree that grows to as high as 30 meters, higher than the mango tree. During the season, each tree bears as many as 250 large fruits, supposed to be the largest tree-borne fruits in the world. The fruit varies widely in size, weigh from 3 to 30 kg, and has oblong or round shape measuring 10 cm to 60 cm in length, 25 to 75 cm in diameter. The unripe fruits are green in colour; When ripen, might turn to light brown colour and gives strong sweet, fruity smell. The fruit is very low in saturated fat and sodium. There is no sugar or cholesterol but is high in vitamin C. It is a good source of fibre, vitamin B2, copper, magnesium and potassium. The aroma of ripe fruit is so strong that either you love it or can not stand it. Research studies have revealed that about 16 esters, half a dozen alcohols, volatile vapours and many other compounds combine to present its typical aroma. It also provides appreciable amounts of carotene, thiamine, riboflavin, niacin, vitamin C. In addition it has crude fibre, calcium, iron, potassium and sodium. The seeds of Jack fruit are also nutritious. They provide 7grams of protein, but contain a powerful enzyme inhibitor- particularly carbonic anhydrase. This accounts for the occasional abdominal (stomach upsets) that jackfruit seeds cause.

Health Benefits of Jackfruit

☆ The fruit is made of soft, easily digestible flesh (bulbs) with simple sugars like fructose and sucrose that when eaten replenishes energy and revitalizes the body instantly.

- Jackfruit is rich in dietary fibre, which makes it a good bulk laxative. The fibre content helps to protect the colon mucous membrane by decreasing exposure time and as well as binding to cancer causing chemicals in the colon.
- Fresh fruit has small amounts of vitamin-A and flavonoid pigments such as carotene-ß, xanthin, lutein and cryproxanthin-ß. Together, these compounds play vital roles in antioxidant and vision functions.
- It is one of the rare fruits that is rich in B-complex group of vitamins. It contains very good amounts of vitamin B-6 (pyridoxine), niacin, riboflavin, and folic acid.
- Fresh fruit is a good source of potassium, magnesium, manganese, and iron.
- Being rich in potassium, jackfruit has been found to be helpful in the lowering of blood pressure.
- The extract of Jackfruit root is believed to help cure fever as well as diarrhea.
- Jackfruit contains phytonutrients, with health benefits ranging from anti-cancer to antihypertensive.
- The root of this fruit has been found to be beneficial for those suffering from asthma.
- Jackfruit proves to be a very good source of vitamin C, which is known for its high antioxidant properties.
- The fruit contains isoflavones, antioxidants, and phytonutrients, all of which are credited for their cancer-fighting properties.
- Jackfruit is known to contain anti-ulcer properties and is also good for those suffering from indigestion.
- Boasting of anti-ageing properties, the fruit can help slow down the degeneration of cells and make the skin look young and supple.
- Jackfruit serves as a good supply of proteins, carbohydrates and vitamins, for the human body.
- It is believed that the fruit can help prevent and treat tension and nervousness.
- Since it contains few calories and a very small amount of fat, jackfruit is good for those trying to lose weight.
- If you are suffering from constipation, regular consumption of the fruit will surely prove beneficial.
- The root of jackfruit is said to be good for the treatment of a number of skin problems.
- Jackfruit is known to be a good source of vitamin C and anti-oxidants that help strengthening of the immune system and enhancing the functioning of the white blood cells in the body.

- ✰ Being a good source of potassium, jackfruit is good for regulating blood pressure and thus, reducing one's risk of getting strokes and heart diseases. Potassium helps to maintain the electrolyte balance in the body.
- ✰ Due to the high fibre content present in jackfruit, it is beneficial in reducing constipation and aiding digestion. The fibre content of jackfruit is said to do away with the carcinogenic chemicals in the colon.
- ✰ Besides vitamin C, phytonutrients like saponins, isoflavones and lignans present in jackfruit is said to slow down degeneration of cells in the body due to formation of free radicals that can cause cancer.
- ✰ To maintain healthy skin and vision, vitamin A present and anti-oxidants present in jackfruit are beneficial.
- ✰ Natural sugars like sucrose, fructose present in jackfruit, make it a good source for energy and makes it easily digestible as well.
- ✰ Jackfruit also contains other minerals like manganese, iron, vitamin B6, niacin, folic acid, etc. whose presence in small quantities is required for the optimum functioning of the body.
- ✰ The seeds of the jackfruit are quite nutritious and a good source of protein.

Caution

Jackfruit seeds are indeed very rich in protein and nutritious. In general, the seeds are gathered from the ripe fruit, sun-dried and stored for use in rainy season in many parts of South Indian states. Different variety of recipes prepared in Southern India where they are eaten either by roasting as a snack or added to curries in place of lentils.

Quince (*Cydonia oblonga*)

Quince belonging to the same family as apples and pears, is regarded as their distant relative. The fruit is native to the warm-temperate areas of southwest Asia, falling in the Caucasus region. Quince is very much like pear in shape and gets a golden-yellowish outer-layer, when mature. The Asian variety is softer and much juicier. Apart from being eaten raw, quinces are commonly made into preserves and jellies as well. When you hear the word "quince," you probably think of jams and jellies, or maybe desserts. Unlike apples, pears and other pome fruits, quinces are rarely eaten raw because the raw fruit has an unpleasant and astringent taste. When cooked, quinces get soft and sweet. They can be poached, baked or braised. Quinces can be round, oval or somewhat pear shaped. Their appearance resembles a golden apple or pear. Choose those that are firm with a pale yellow skin. The yellow skin is often somewhat mottled with brown spots that don't affect the flavour or quality. Quinces that are shriveled, soft, or brown all over are no longer fresh. In Medieval times, Europeans thought quinces aided the digestion and prepared them frequently along with meats. The English called the combination chardeqynce meaning flesh of quince.

The Good

This food is very low in Saturated Fat, Cholesterol and Sodium. It is also a good source of Dietary Fibre and Copper, and a very good source of Vitamin C.

The nutritional value and health benefits of the quince makes it ideal for:

- ✰ Maintaining optimum health
- ✰ Weight loss
- ✰ Avoid including quince in your diet if you're interested in:
- ✰ Weight gain

Health and Nutrition Benefits of Eating Quince

- ✰ Being rich in dietary fibre, quince is good for those people who are trying to lose weight and maintain a healthy body.
- ✰ Quince boasts of antioxidant properties, which helps the body fight against free radicals and reduces the risk of cancer.
- ✰ Researches have revealed that quinces might be rich in various anti-viral properties.
- ✰ Consumption of quince has been found to be beneficial for people suffering from gastric ulcer.
- ✰ Quince juice is known to have tonic, antiseptic, analeptic, astringent and diuretic properties.
- ✰ It is believed that eating quince is good for maintaining the optimum health of an individual.
- ✰ Regular consumption of quince not only aids in digestion, but also helps lower cholesterol.
- ✰ The presence of potassium in quince helps the body keep high blood pressure in check.
- ✰ The vitamin C present in quince helps reduce the risk of heart disease in individuals.
- ✰ If consumed on a regular basis, quince proves beneficial for those afflicted with tuberculosis, hepatic insufficiency, diarrhea and dysentery.
- ✰ Those suffering from liver diseases and eye diseases would surely benefit from regular quince consumption.
- ✰ Being rich in antioxidants, quince is believed to be helpful in relieving stress and attaining calm.
- ✰ Quince is low in saturated fat, cholesterol and sodium, while having lots of vitamin C, dietary fibre and copper.
- ✰ Quince juice is good for those suffering from anemia, cardiovascular diseases, respiratory illnesses, diseases of the gastrointestinal tract and even asthma.
- ✰ The juice as well as pulp of boiled or baked quince fruit serves as a good anti-emetic remedy.

Caution

The seeds contain **nitriles** (RCN), which are common in seeds of **the rose family**. In the stomach, enzymes or stomach acid or both cause some of the nitriles to be hydrolyzed and produce HCN **(hydrogen cyanide)**, which is a volatile gas. The seeds are only likely to be toxic if a large quantity is eaten.

Berries

Strawberries, Cran berries, Blueberries, etc., contain antioxidant phytonutrients that get rid of free radicals in the blood and guard your collagen, thereby boosting skin repair. Collagen makes your skin supple, smooth and plump. "Antioxidants that are present in berries such as blueberries, and cranberries and in strawberry, can help keep you skin vibrant and glowing." The canned cranberry juice works for your skin. Eat one to two cups of blueberries a day for your daily 'dose' of anthocyanin, an anti-ageing antioxidant that increases the potency of Vitamin C. Berries atleast a week could cut their risk of developing the disease (parkinson`s) by quarter compared with those who never eate Anonymous, 2012). Flavonoids can protect neurons against disease of brain such as parkinson`s. Cranberries have five times the antioxidant content of broccoli, which means they may protect against *cancer,* stroke and heart disease. But limit the cran berries to about 8 tbsp per pint of juice. Berries have abundant antioxidant properties that support brain function and blood circulation. Raspberries are said to contain the most amount of ellagic acid that improves liver functions, regulating cholesterol levels and eliminating toxins. Since berries are sweet, they are the healthiest choices to satisfy your sugar cravings.

STRAWBERRY (*Fragaria xananassa*)

Delicious and nutrition-rich red coloured strawberries are among the most popular berries. These berries have the taste that varies from cultivar to cultivar and ranges from quite sweet to acidic. The berry features red pulp with tiny yellow colour seeds piercing through its surface from inside. Top end has small, green leafy cap and stem that is adorning its crown. Each berry features conical shape, weighs about 25 g and measures about 3 cm in diameter. Strawberries have higher levels of vitamin C, fibre, folate and potassium than most other fruits like bananas, apples and even

oranges. Apart from the obvious health benefits, it has been shown that eating one serving (about 8-10 strawberries) a day can significantly decrease blood pressure, which may reduce the risk of heart disease. Other studies showed additional nutrition benefits: Strawberries are found to reduce risk of cancer, enhance memory function and rheumatoid arthritis. Strawberries also contain **Ellagic acid** which is also found in raspberries, blackberries, cranberries, grapes, cherries, walnuts, peacans and Brazilnuts, acts as a scavenger to "bind" cancer causing chemicals, making them inactive. It inhibits the ability of other chemicals to cause mutation in bacteria. Strawberries are the only known fruit to carry their seeds outside, with an average of 200 seeds per fruit. The leaves and roots of strawberries are claimed for their medicinal benefits in terms of easing diarrhea, digestive upsets and gout. You'll find strawberries leaves in blended herbal teas. The fruit juice is also used to treat sunburn, skin blemishes and discoloured teeth. Nutrients found in strawberries include vitamin A, C and B6, fibre, potassium, folate and various antioxidants and flavonoids. Your immune system needs these fighters to protect you against diseases related to heart and cancers.

Strawberries are known to be brilliant as it benefits the skin in more than one way. This accessory fruit is known to be rich in coenzyme Q10, a substance that helps fight free radical damage, and lessens the pace of aging, increases both the adeptness and rate of cellular energy production. Quite high in alpha-hydroxyl acid, this fruit helps get rid of dead skin cells and helps the body build collagen. Another great thing about strawberries, it's highly concentrated with vitamin C, an essential antioxidant that helps prevent skin blemishes, fights the aging process and the harmful effects of the sun.

Compared to fruits like apples, oranges or bananas, they score highest in most of the nutrients. For example, comparing the Vitamin C level in 100g of strawberries and 100g of oranges, strawberries exceed oranges by 10mg. Best of all, they're low in calories and absolutely fat-free! The good thing is they are very low in Saturated Fat, Cholesterol and Sodium. It is also a good source of Folate and Potassium, and a very good source of Dietary Fibre, Vitamin C and Manganese. whereas, a large portion of the calories in this food come from sugars. One serving of strawberries - about eight medium strawberries - is high in fibre and an excellent source of vitamin C. The star player in strawberries is vitamin C, an antioxidant that helps repair the body's tissues, boosts immunity, and fights excess free radical damage and with powerful pain-reducing properties, according to research. Some studies suggest vitamin C may help people experience less pain after breaking a bone or having orthopedic surgery. This red, wet drippy fruit makes for an excellent foreplay fruit. Drip the juices over your partner's body and you will do just fine. Strawberries are an excellent source of Vitamin C, a powerful antioxidant. These pretty red berries are low in calories, but have health-promoting phyto-nutrients, minerals and vitamins that are essential for optimum health. They have high amounts of phenolic flavonoid phyto-chemicals called anthocyanin and ellagic acid and studies claim they may help against ageing, inflammation and neurological diseases.

Best Eaten

Tastes best raw with fresh cream. Nutritional facts per cup Vitamin C: 149 per cent Manganese: 29 per cent Vitamin K: 4 per cent Iron: 3 per cent Calcium 2 per cent

Health Benefits

- ✩ Strawberry is low in calories and fats but rich source of many health promoting phyto-nutrients, minerals and vitamins that are essential for optimum health.
- ✩ Strawberries have significantly high amounts of phenolic flavonoid phyto-chemicals called **anthocyanins** and **ellagic acid**. Consumption of straw berries have potential health benefits against cancer, aging, inflammation and neurological diseases.
- ✩ Fresh berries are an excellent source of **vitamin-C**, which is also a powerful natural antioxidant. Consumption of fruits rich in vitamin C helps body develop resistance against infectious agents, counter inflammation and scavenge harmful free radicals.
- ✩ The fruit is rich in B-complex group of vitamins. It contains very good amounts of vitamin B-6, niacin, riboflavin, pantothenic acid and folic acid. These vitamins are acting as co-factors help body metabolize carbohydrates, proteins and fats.
- ✩ Strawberries contain vitamin A, vitamin E and many health promoting flavonoid poly phenolic antioxidants such as lutein, zeaxanthin, beta-carotene in small amounts.
- ✩ They contain good amount of minerals like potassium, manganese, fluorine, copper, iron and iodine. Potassium in an important component of cell and body fluids that helps controlling heart rate and blood pressure. **Manganese** is used by the body as a co-factor for the antioxidant enzyme superoxide dismutase. Copper is required in the production of red blood cells. Iron is required for red blood cell formation. Fluoride is a component of bones and teeth and is important for prevention of dental caries.

Caution

Strawberries may cause serious allergic reactions in some sensitized individuals. Some of the most common symptoms of strawberry allergy include swelling and redness of mouth, lips and tongue, eczema, hives, skin rash, headache, runny nose, itchy eyes, wheezing, gastrointestinal disturbances, depression, hyperactivity and insomnia. Individuals who suspect allergy to these fruits may want to avoid.

RASPBERRIES (*Rubus idaeus*)

Are of various colours *viz.* black, purple and even golden. They are *w*onderfully delicious and rich red coloured raspberry among the most popular berries grown all over the world. They are rich source of many health promoting plant-derived nutrients, minerals, and vitamins that are essential for optimum health. Raspberries are an excellent source of fibre, manganese and vitamin C. They are a very good source of vitamin K and a good source of magnesium, folate, omega-3 fatty acids, copper, vitamin E and potassium. In addition, they contain significant amounts of the anti-cancer phytochemical **ellagic** acid. Rasberries as a are very low in Saturated Fat, Cholesterol and Sodium. They ares also a good source of Vitamin K and Magnesium, and a very good source of Dietary Fibre, Vitamin C, and manganese and a large portion of the calories in this food come from sugars. The leaves of raspberry plant are also commonly used for their medicinal teas and also make delicious herbal teas. The leaves are also reputed to be effective in regulating the menstrual cycle in women. Rasberries rank in the top 10 antioxidant-high fruits and vegetables! They possess almost 50 per cent higher antioxidant activity than strawberries. **Antioxidants** are believed to help prevent and repair oxidative stress, a process that damages body cells and has been linked to the development of diseases such as cancer, heart disease and Alzheimer's disease. Raspberries are low in calories, but are a very rich source of dietary fibre. The large amount ofVitamin C present in them makes these berries very good antioxidants. High levels of a compound known as ellagic acid, a polyphenol prized for its anticarcinogenic properties is present in them.

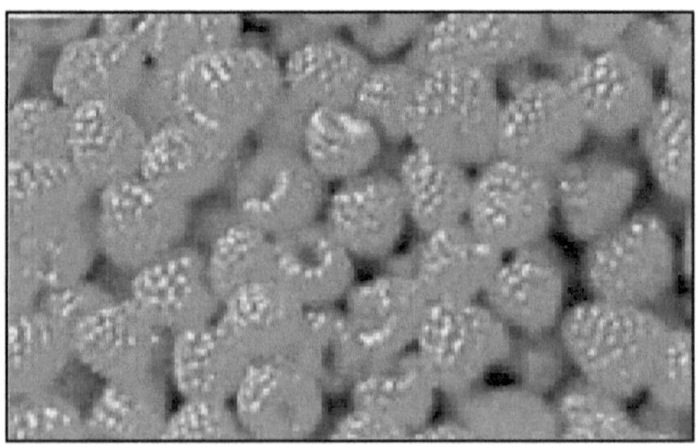

Best Eaten

Can be eaten raw, in fruit salads, as a puree or jam. Make a smoothie in the blender with milk or yogurt. Nutritional facts per cup Vitamin C: 54 per cent Manganese: 41 per cent Vitamin K: 12 per cent Calcium: 3 per cent Iron: 5 per cent

Health Benefits of Raspberries

- ☆ Raspberries are low in calories and saturated fats but are rich source of dietary fibre.
- ☆ Raspberries have high levels of phenolic flavonoid phytochemicals such as anthocyanins, ellagic acid (a tannin), quercetin, gallic acid, cyanidins, pelargonidins, catechins, kaempferol and salicylic acid.

- ☆ **Xylitol**, a low-calorie sugar substitute extracted from raspberries. Xylitol contains just 9.6 calories, as compared to one teaspoon of sugar, which has 15 calories. Xylitol absorbs more slowly than sugar and does not contribute to high blood sugar levels; can be helpful in diabetics.
- ☆ Fresh raspberries are excellent source of **vitamin-C**, which is also a powerful natural anti-oxidant and may contain potential anti cancer agents.
- ☆ Raspberry contains anti-oxidant vitamins like vitamin A, and vitamin E. In addition to the above mentioned antioxidants, is also rich in many other health promoting flavonoid poly phenolic antioxidants such as *lutein, zeaxanthin,* and *beta-carotene* in small amounts.
- ☆ Raspberry has an **ORAC value** (oxygen radical absorbance capacity) of about 4900 per 100 grams, crediting it among the top-ranked ORAC fruits.
- ☆ Raspberries contain good amount of minerals like potassium, manganese, copper, iron and magnesium.
- ☆ They are rich in B-complex group of vitamins and vitamin K. Contain very good amounts of vitamin B-6, niacin, riboflavin, and folic acid.
- ☆ Rasberries have also been linked to a positive promotion in cardiovascular health and cancer.
- ☆ Raspberries provide important anti-inflammatories, including anthocyanins (the pigments in red, purple and blue fruits), which are believed to help reduce cardiovascular disease and diabetes, help to improve eyesight and memory functions.
- ☆ **Ellagic acids**, which may slow down the growth of some cancer cells, is found in raspberries (including the seeds!), in fact raspberries are a leading source of the acid. Research has also shown that ellagic acid from raspberries is easily absorbed by the body. The seeds of raspberries also serve as beneficial "roughage" in the digestive tract.
- ☆ Raspberries, especially the seeds, may become more and more important in the booming cosmeceuticals market **(skin care products** with health benefits). The oil in the seeds of the raspberries is rich in vitamin E, **omega-3 fatty acids** and has a natural SPF (sun protection factor) of 25 to 50.
- ☆ Raspberry Leaf Tea has been praised by many herbalists and midwives for it's medicinal benefits in women's reproductive health. For centuries, leaves of the red raspberry plants have been used to help during pregnancy, in childbirth, with breastfeeding, with irregular menstruation, infertility, and other conditions in connection with women's health.

Caution

Raspberry may cause serious allergic reactions in some sensitized individuals. Some of the most common symptoms include swelling and redness of mouth, lips and tongue, eczema, hives, skin rash, headache, runny nose, itchy eyes, wheezing, gastrointestinal disturbances, depression, hyperactivity and insomnia. Individuals who suspect allergy to these fruits may want to avoid.

KIWI FRUIT or CHINESE GOOSE BERRY (*Actinidia chinensis*)

Chinese Gooseberry The fruit has a soft texture and a sweet but unique flavour. Kiwi fruit or Chinese gooseberry is one of the delicious fruits with full of promising health promoting phyto-chemicals, vitamins and minerals. Named after a flightless bird of the same name, the kiwi is also known by other names like 'macaque peach' and 'chinese gooseberries'. And, did you know that it is the national fruit of China?

Each fruit measures about the size of large hen's egg and weighs up to 125 g. Internally the flesh is soft, juicy, emerald green in colour with rows of tiny, black, edible seeds. Fruit texture is similar to strawberry or sapodilla and the flavour resembles a blend of strawberry, apple and pineapple fruits. Similarly, consuming blue berries help protect brain from Alzheimer's disease and dementia. Kiwi fruit stands number one in nutrient content compared to 27 other fruits. It is packed with twice the amount of Vitamin C – compared to oranges (per 100 mg) – and has twice the amount of nutrients – compared to apples (per 100 mg). Kiwi fruit is an excellent source of Vitamin E, fibre, potassium, folic acid, carotenoids, antioxidants and trace minerals (Anonymous, 2013).

Vitamin E

It's quite unusual for a low fat food - let alone fruit - to be a good source of vitamin E, which is well-known for its heart health and antioxidant properties. However, kiwifruit is a great low-fat, natural source of vitamin E.

Folic Acid

Kiwifruit are a natural source of folic acid which is needed to prevent cardiovascular disease, help brain and cognitive development and prevent neural defects in babies - both before and during pregnancy.

Potassium

Kiwifruit has about the same levels of potassium as bananas but only half the calories - making kiwifruit an excellent low-sodium, high-potassium fruit which can be beneficial in the maintenance of blood pressure and for heart health.

High in vitamin C, dietary fibre, potassium and magnesium

Low in saturated fat, sodium and cholesterol

Health Benefits of Kiwi fruit

Here are some health benefits of the kiwi bet you didn't know:

- ☆ Perhaps, the biggest plus point of this fruit is that it is an abundant source of vitamin C. In fact, they contain twice the amount of vitamin C than that of oranges. This advantage is useful when it comes to boosting the immune system. Also, the abundance of vitamin C helps in fight against the damage of free radicals in the body, reducing the risk of developing cancer. Also, this reduces the signs of ageing through reducing wrinkles, and dark spots.
- ☆ It is said that kiwis promote respiratory health as well. This helps in reducing instances of coughing, shortness of breath, wheezing, etc.
- ☆ Being a good source for dietary fibre, kiwis are great for reducing problems like constipation and improving digestion.
- ☆ Kiwi fruit also contains a high level of serotonin, which is known as a happy hormone and reduces stress.
- ☆ Who would have thought that kiwis are good for impotence? Arginine, an amino acid is useful for treating impotence in men.
- ☆ Kiwis are great for cardiovascular health. Pectin found in kiwis are known to lower cholesterol. Also, deficiency of magnesium is associated with heart diseases and hypertension. Cartenoids, phenolic compounds, antioxidants found in kiwis help prevent oxidation of good **cholesterol** (HDLs).
- ☆ A naturally occurring sugar, inositol plays a factor in regulating diabetes.
- ☆ Kiwis are great for people who want to lose weight as it contains healthy **fats**, healthy sugars and is low in calories.
- ☆ If you want to improve eyesight, then you can turn to this fruit. The combination of vitamins A, C and E aid in maintaining a healthy eyesight and preventing the macular degeneration of the eyes.
- ☆ Kiwifruit is a very rich source of soluble dietary fibre, which makes it a good bulk laxative. The fibre content helps to protect the colon mucous membrane by decreasing exposure time to toxins as well as binding to cancer causing chemicals in the colon.
- ☆ Kiwi fruit contains very good levels of vitamin-A, vitamin-E, **vitamin-K** and flavonoid anti-oxidants such as beta-carotene, lutein and xanthin. Vitamin K has potential role in the increase of bone mass by promoting osteotrophic activity in the bone. It also has established role in Alzheimer's disease patients by limiting neuronal damage in the brain.
- ☆ Kiwi-fruit functions as blood thinner similar to aspirin; helps to prevent clot formation inside the blood vessels and protects from stroke and heart attacks.
- ☆ Kiwi seeds are an excellent source of **omega-3 fatty acids.** Consumption of foods rich in ω-3 fatty acids may reduce the risk of coronary heart disease, stroke, and help prevent development of ADHD, autism, and other developmental differences in children.

- ☆ Fresh fruit is a very rich source of heart healthy electrolyte "potassium". Potassium is an important component of cell and body fluids that helps controlling heart rate and blood pressure by countering effects of sodium.
- ☆ It also contains good amounts of minerals like manganese, iron and magnesium. Manganese is used in the body as a co-factor for the powerful antioxidant enzyme, *superoxide dismutase*. Magnesium is an important bone-strengthening mineral like calcium.

Other Health Benefits

- ☆ Consumption of kiwi fruit makes metabolism stronger as well as improves the nerve function.
- ☆ Prevents wheezing and coughing, particularly in children.
- ☆ Helps to prevent colon cancer
- ☆ Boosts the immune system of the body so helps to fight cold and flu
- ☆ Provides vitamins and antioxidants in a healthy amount
- ☆ Prevents asthma.
- ☆ Helps people with impotence.
- ☆ Rich source of minerals such as magnesium, potassium and copper.
- ☆ Also acts as a natural blood thinner.
- ☆ Assists people suffering from depression.
- ☆ Great for pregnant women as it contains folate which is necessary for cell development.
- ☆ An anti mutagenic component present in kiwifruit helps to prevent the mutations of DNA that may start the cancer process.

CRANBERRIES (*Vaccinium oxycoccus*)

Cranberries are rich in Vitamin C, but their acidic nature and the unusual nature of their proanthocyanidins make them extremely beneficial for health. The entire fruit, and not just the juice, shows presence of phytonutrients that is good for overall health.

Best Eaten

Cranberry can be used in cooking like we use lemon. These berries not only work well with sweets and tarts, but also go well with spicy, flavoured foods. Cranberry extracts have been shown to inhibit the growth of oral, colon, and prostate cancer cells, in addition to inducing cell death of breast cancer cells. Cranberries are loaded with unique antioxidants and other phytonutrients that may protect against heart disease, cancer and variety of other diseases. Phytonutrients in cranberries may also be involved in inhibiting the spread of cancer cells throughout the body. A urinary tract infection is usually caused by bacteria or germs which get into the urinary tract and multiply, causing redness, swelling and pain. But there's an easy way to promote good urinary tract health, and it starts by taking the cranberry glass-a-day challenge. Cranberries contain powerful antioxidants called proanthocyanidins or condensed

tannins, and they prevent certain bacteria from sticking to the urinary tract," says registered dietitian Lois Ferguson. "And when bacteria can't stick, bacteria cannot cause infection.

Nutritional facts per cup Vitamin C: 24 per cent Manganese: 20 per cent Vitamin K: 7 per cent Iron: 2 per cent Vitamin A, Calcium: **1 per cent**

Benefits of the Cranberry

- ☆ Cranberries are a rich source of the flavonoid quercetin which can inhibit the development of both breast and colon cancers.
- ☆ Preliminary studies show that drinking cranberry juice is good for the health of the heart.
- ☆ Research indicates that cranberries are an excellent source of antioxidants which may protect against cancer, heart disease and other diseases.
- ☆ Found to decrease production of cavity and plaque producing bacteria in your mouth.
- ☆ Also found to reduce the bacteria associated with peptic stomach ulcers.
- ☆ In clinical studies, cranberries have been shown to help maintain a healthy urinary tract.
- ☆ Cranberries are especially beneficial to the eyes (they significantly improve symptoms of cataracts, macular degeneration, and diabetic retinopathy).
- ☆ Evidence on how cranberry juice fights bacteria that cause urinary tract infections.
- ☆ Cranberries are part of a heart-healthy diet. Cranberries are a fat free, cholesterol free and low sodium food. Whole cranberries are a good source

of dietary fibre, and all cranberry products contain flavonoids and polyphenolics, natural compounds that offer a wide range of potential health benefits.

- ✰ Urinary Tract Health: Many people know that cranberries help maintain urinary tract health. Research is now showing just how cranberry juice promotes urinary tract health and why cranberries are the only food with this benefit.
- ✰ Drinking cranberry juice can block urinary infections by binding to bacteria so they can't adhere to cell walls. While women often drink unsweetened cranberry juice to treat an infection, there's no hard evidence that works.
- ✰ A compound Howell discovered in cranberries, proanthocyanidine, prevents plaque formation on teeth; mouth washes containing it are being developed to prevent periodontal disease.
- ✰ In some people, regular cranberry juice consumption for months can kill the H. pylori bacteria, which can cause stomach cancer and ulcers.
- ✰ Preliminary research also shows:
- ✰ Drinking cranberry juice daily may increase levels of HDL, or good cholesterol and reduce levels of LDL, or bad cholesterol.
- ✰ Cranberries may prevent tumors from growing rapidly or starting in the first place.
- ✰ Extracts of chemicals in cranberries prevent breast cancer cells from multiplying in a test tube; whether that would work in women is unknown.

Blue Berries

This little berry has so many health benefits that we don't know where to start. Its rich vitamin Ccontent aids blood circulation and provides minerals and salts, which help the body to fight ageing. Also, as it is high in potassium – found in all fruits and veggies – fluid balance in cells is regulated, thus combating puffiness and fluid retention. Have one of the highest antioxidant capacities among all fruits, vegetables, spices and seasonings.

The highest antioxidant capacity of all fresh fruit: Blue Berries, being very rich in anti oxidants like Anthocyanin, vitamin C, B complex, vitamin E, vitamin A, copper (a very effective immune builder and anti-bacterial), selenium, zinc, iron (promotes immunity by raising haemoglobin and oxygen concentration in blood) etc. boost up your immune system and prevent infections. Once your immunity is strong, you won't catch colds, fever, pox and all such nasty viral and bacterial communicable diseases.

Neutralizes free radicals which can affect disease and aging in the body: Blue Berries are laden with anti oxidants and rank number 1 in the world of anti oxidants. This is mainly due to presence of Anthocyanin, a pigment responsible for the blue colour of the blue berries. The abundance of vitamin-C is also a big factor for this as well.

- ☆ Aid in reducing Belly
- ☆ Helps promote urinary tract health
- ☆ Been proved to preserve vision
- ☆ Constipation and Digestion

Nuts

Eating one ounce of mixed nuts (raw unpeeled walnuts, almonds and hazelnuts) a day may benefit those who are at high risk for heart disease and type 2 diabetes, a new study has suggested. Scientists have for the first time found a link between eating nuts and higher levels of serotonin in the bodies of patients with metabolic syndrome (MetS), who are at high risk for heart disease. Serotonin is a substance that helps transmit nerve signals and decreases feelings of hunger, makes people feel happier and improves heart health. Brazil nuts are a very rich source of selenium, a mineral that works very well with Vitamin E to stop oxidative stress and cell damage caused by free radicals; thus, slowing down the ageing process. But should be eaten in moderation (two nuts a day) because of its high fat content. Dry fruits are rich source of nutrients such as vitamins and minerals. Consumption of dry fruits daily enhances the overall bioavailability of nutrients. In addition, mixing dry fruits with some fresh fruits would help you get vitamin C which in turn favours absorption of iron inside the stomach. Almonds and walnuts are a great source of omega and fatty acids and anti oxidants that help in pain control. Add these nuts to your salad or just munch on a handful as an after noon snack. Walnuts and almonds are the healthiest of all nuts. Nuts are fortified with the B group of vitamins, magnesium, zinc and omega oils that helps to keep cortisol (a stress hormone) levels low. Also, they are a good source of energy, balance out sugar cravings and aids metabolism. The monounsaturated fats present in nuts help in curbing overeating. But there's no need to go overboard with nuts as they have high calories

When eaten in moderation, nuts are not fattening. Besides antioxidants, nuts are rich sources of good fats which tend to lower the bad cholesterol, are good for skin and hair and are highly beneficial for health. These can be sliced and sprinkled over salads, cereal/oatmeal porridge and can be roasted and had as snacks.

ALMONDS (*Prunus dulcis*)

Unshelled Nuts

With Shell

Almonds are among the most nutrient-rich tree nuts and just a handful a day is a great way to ensure you're getting more of the good things your body needs and you want. Almonds are delicious, have long been revered as symbol of wellness and health. The king of all nuts with the highest iron content is the humble almond. One ounce of almonds every day provides 6 per cent of iron. In fact, the good news is that almonds are the cheapest of all the other nuts and dry fruits available to us. The nuts are enriched with many health-benefiting nutrients that are essential for optimum health. Almonds are an excellent source of vitamin E, magnesium, manganese, and a good source of fibre, copper, phosphorous, and riboflavin. Plus, a one-ounce serving has 13 grams of good unsaturated fats, just 1 gram of saturated fat, and is always cholesterol free. And when compared ounce for ounce, almonds are the tree nut highest in protein, fibre, calcium, vitamin E, riboflavin, and niacin. The tiny almond is a storehouse of nutritional goodness. Each ounce contains 6 grams of protein and 3 grams of fibre, not to mention healthy doses of iron, zinc, copper, calcium, magnesium, Vitamin E and array antioxidants. Snacking on almonds is a great way to help lower cholesterol and high blood pressure, and can smooth out spikes in blood sugar in diabetics. Calcium is an important nutrient found in almonds. A single serving of almonds can contain up to a quarter of one's daily calcium needs. Such high levels of calcium make almonds a tooth and bone strengthening food as well.

Our diet needs to provide adequate amounts of all of these 'little helpers'. Almonds are a power packed nut which provide you with all the good variants of nutrition like carbohydrates, fats, minerals, proteins and vitamins. Since time immemorial, almonds have been cherished, not only for their delicate flavour and delectable crunch, but also for their nutritional benefits. They are a simple, tasty choice that can make a big difference in the way you feel every single day, not a small feat. And that means you can feel good about what you're eating while enjoying their delicious crunchy flavour. Almonds promote a healthy heart in many ways. The cholesterol-lowering effects of almonds can also help almond snackers maintain healthy hearts. **Alpha-tocopherol vitamin E** found in Almonds are one of the leading food sources of the important antioxidant, which helps neutralize harmful free radicals in your body and help you feel your best day and night. So what are free radicals? They're unstable molecules that form as your body burns oxygen. Free radicals can damage your cells, tissues, and even your DNA. So make a healthy resolution this nutrition week and keep yourself healthy for the rest of the year.

Eat almonds and make yourself stronger forever.

Here's a piece of advice for men who want a healthy sex life - start taking more almonds.

Sin Chew Daily reported that a study showed that almonds have a high quantity of a type of amino acid called **arginine,** which helps to relax blood vessels and improve blood circulation. However, the study added that if almond was not a person's cup of tea, other food items that are high in arginine were beans, salmon and wheat, reports the *Star Online.*

Since ages almonds have been considered as an anti-ageing food because of the high content of Vitamin E. It is great for healthy skin, hair and nails. The body needs Vitamin E to protect the cells from the onslaught of free radicals generated by air pollution, peroxides and ultra violet rays. Experts say that eating around a dozen almonds daily will help. Almond milk, just like coconut milk is a great alternative for vegans. Almond milk is a good source for proteins. It is also low in calories unlike cow'smilk. This milk contains vitamin C, magnesium, monounsaturated fats, mangaese, copper, riboflavins. It boosts energy and lowers heart diseases.

Almonds are a great source of essential fatty acids. These provide the raw material for a man's healthy production of hormones. Additionally, the smell of almonds arouses passion in females. You can add to the sexual ambience by lighting some almond-scented candles to build up her mood.

Libido Dietary Tip

Eat almonds raw (with no added salt or sweetness). Alternately, crush some fresh almonds and sprinkle them on your salad to get the energy you need. They are very low in Cholesterol and Sodium but a good source of Riboflavin, Magnesium and Manganese, and a very good source of Vitamin E (Alpha Tocopherol). All kind of nuts are an excellent source of essential fatty acids, nuts are known to give your passion a hike. Rich in several kinds of minerals, they are pretty good for reproduction and sexual health. Almonds, in particular, are known to arouse enthusiasm in females simply through aroma. **Full of zinc**, nuts of all kinds are known to perk your sexual desire by helping you store energy which will help you last longer. All kinds of nuts provide one with energy that's directly proportional to your stamina while having sex. So, feel free to consume your daily dose of nuts, including almonds, pistachios, walnuts, groundnuts etc. After all, stamina and energy are the driving force behind a satisfying sex life.

One ounce of almonds (about 20 to 24 shelled whole almonds) provides 35 per cent of your daily value for vitamin E. Vitamin E may help promote healthy aging. A study reported in the *Journal of the American Association* (June 26, 2002) suggests a diet rich in foods containing vitamin E may help protect some people against Alzheimer's disease

Nuts also are important for what they DON'T offer:

- ☆ **Cholesterol**: Nuts are cholesterol-free.
- ☆ **Sodium**: Unless salt is added to nuts, they naturually contain, at most, just a trace of sodium.
- ☆ **Good for Brain**: Almond is a source of many nutrients which help in development of the brain. Almond induces high intellectual level and has been considered as an essential food item for growing children. Many mothers give almonds soaked in water to their children daily in the morning (2-3 pieces of soaked almonds are good enough, you can also remove the outer shell if it causes allergy to you).
- ☆ **Regulates Cholesterol**: Regular consumption of almonds helps to increase the level of high density lipoproteins (HDL) and reduce the level of low density lipoproteins (LDL), thereby effectively controlling cholesterol levels. LDL cholesterol is called bad cholesterol.
- ☆ **Good for heart:** Mono-saturated fat, protein and potassium contained in almonds are good for the heart. Vitamin E acts as an antioxidant and reduces the risk of heart diseases. The presence of magnesium in almonds helps to avoid heart attacks. Almonds help reduce C-reactive protein which causes artery-damaging inflammation. Almond is also a source of folic acid. They therefore help to reduce the level of homocystein, which causes fatty plaque buildup in arteries.
- ☆ **Skin care:** The benefits of almond for skin care are well known and hence a massage with almond oil is often recommended for new born babies. Almond milk is also added in some soaps as almonds help in improving the complexion of the skin.
- ☆ **Regulates blood pressure:** Potassium present in almond helps to regulate blood pressure. Almonds are very low in sodium which also helps in containing blood pressure
- ☆ **Prevention of cancer**: Almond improves the movement of food through colon, there by preventing colon cancer.
- ☆ **Protection against diabetes**: Almonds also help in reducing the rise in sugar and insulin levels after meals. This offers protection from diabetes.
- ☆ **Good in pregnancy**: Almond contains folic acid. Folic acid helps to reduce the incidence of birth defects in newborn babes.
- ☆ **Weight loss:** Unsweetened almond milk helps one to reduce weight. The mono-saturated fat contained in almonds satisfies appetite and prevents over eating. Studies have revealed that almond rich low calorie diet is good for obese people to assist in shedding their weight.
- ☆ **Prevention of constipation**: Almonds are rich in fibre. Like most other fibre rich food, almonds also help in preventing constipation. Make sure you drink good amount after eating almonds.
- ☆ **Boosts energy:** The presence of manganese, copper and Riboflavin helps in energy production.

- **Learn more**: Just a hand full of these nuts a day provides enough recommended levels of minerals, vitamins, and protein. Besides, almond oil extracted from the nuts has been used in as base or carrier oil in medicine, aromatherapy and in pharmaceuticals.

Health Benefits of Almonds

- Almond nuts are rich in dietary fibre, vitamins, and minerals and packed with numerous health promoting phyto-chemicals; the kind of well-balanced food ensure protection against diseases and cancers.
- These nuts are rich source of energy and nutrients. They are especially, rich in mono-unsaturated fatty acids like ***oleic*** and ***palmitoleic acids*** that help to lower **LDL** or "bad cholesterol" and increase HDL or "good cholesterol."
- The nuts are an excellent source of **vitamin E**; contain about 25 g per100 g. Vitamin E is a powerful lipid soluble antioxidant, required for maintaining the integrity of cell membrane of mucus membranes and skin by protecting it from harmful oxygen free radicals.
- Almonds are free in gluten and therefore, are a popular ingredient in the preparation of gluten free food formulas. Such formula preparations are in fact healthy alternatives in people with wheat food allergy and celiac disease.
- Nuts are packed with many important B-complex groups of vitamins such as riboflavin, niacin, thiamin, pantothenic acid, vitamin B-6, and folates.
- Almonds are also rich source of minerals like manganese, potassium, calcium, iron, magnesium, zinc, and selenium.
- The sweet almond oil is obtained from the nuts is an excellent emollient; helps to keep skin well protected from dryness. It has also been used in cooking, and as "carrier or base oil" in traditional medicines in aromatherapy, in pharmaceutical and cosmetic industry.
- Almonds are low in saturated fat and contain many other protective nutrients - calcium and magnesium - for strong bones, vitamin E and compounds called phytochemicals, which may help protect against cardiovascular disease and even cancer.
- Alonds are phytochemical powerhouse.
- Reduce heart attack risk and lower cholesterol.
- 20-25 almonds (approximately one ounce) contain as much calcium as 1/4 cup of milk, a valuable tool in preventing osteoporosis.
- Almonds are the best whole food source of vitamin E, in the form of **alpha-tocopherol**, which may help prevent cancer.
- If you're pregnant, or thinking about it, almonds are a great source of the folic acid you need!
- Almonds contain more magnesium than oatmeal or even spinach. Build strong bones and teeth with the phosphorus in almonds. Most of the fat in almonds is monounsaturated, also known as the "good" fat.

☆ Intake of 6 almonds (soaked overnight) is also helpful in keeping a check on diabetes.

Caution

Almond nut allergy, although not so common as other tree nut allergies like cashew, pistachio etc, may cause hypersensitivity cross-reactions in some people to food substances prepared using the nuts. The type and severity of symptoms may vary and may include vomiting, diarrhea, pain abdomen, swelling of lips, and throat leading to breathing difficulty, and chest congestion. Therefore, caution should be exercised in those with nut allergic syndrome while using food preparations that contain nut products.

WALNUTS (*Juglans regia* L)

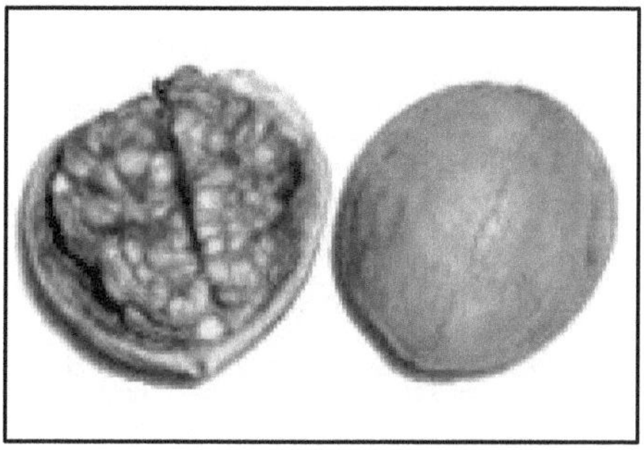

Walnuts are revered since ancient times as symbol of intellectuality since their kernels have convoluted surface inside the shell resembling as that of brain! The nuts are enriched with many health-benefiting nutrients, especially Ω-3 fatty acids that are essential for optimum health. One ounce of walnuts (about 14 shelled walnut halves) is all that is needed to meet the dietary recommendation for omega-3 fatty acids. Each nut is roughly spherical in shape, about the size of medium sized **lemon** weighing about 10-15 g enclosing edible kernel. The walnut kernel consists of two uneven, corrugated lobes, off white in coloUr and covered by a paper thin, light brown skin. The lobes are partially attached to each other. Oil extracted from the nuts, apart from cooking, has also been used in as base or carrier oil in medicine, and aromatherapy. Walnuts are a rich source of Vitamin E and having a handful can help your skin look great. Walnuts are a very good source of manganese and a good source of copper, two minerals that are essential cofactors in a number of enzymes important in antioxidant defenses. For example, the key oxidative enzyme *superoxide dismutase*, which disarms free radicals produced within cell cytoplasm and the mitochondria (the energy production factories within our cells) requires both copper and manganese. Research has shown that having walnuts daily can reduce the risk of heart disease and can help in lowering bad cholesterol (LDL) and the C-reactive protein. When having a meal high in saturated fats, having a handful of raw walnuts can limit the ability of the harmful fat to damage arteries. Walnuts contain less than 4 grams of saturated fat per 50 grams. However, all nuts are high in calories, so, having them in moderation is essential. Walnuts are rich in antioxidants, omega-3-fats and alpha linoleic acid.

Walnuts also contain an antioxidant compound called ellagic acid, which blocks the metabolic pathways that can lead to cancer. Ellagic acid not only helps protect healthy cells from free radical damage, but also helps detoxify potential cancer-causing substances and helps prevent cancer cells from replicating. In a study of over

1,200 elderly people, those who ate the most strawberries (another food that contains ellagic acid) were three times less likely to develop cancer than those who ate few or no strawberries. It reduces vata, aggravates Kapha and pitta, promotes intelligence, promotes digestive functions, oleation, anti -toxin, expectorant. It also increases physical strength, potency and is a bulk promoting herb. Its paste is beneficial for complexion, anti –dermatoses, relieving oedema and is an effective pain reliever (Acharya Balkrishna, 2012). Scientists discovered that these edible seeds cut the risk of dying from cancer by 40 per cent and from cardiovascular disease by at least 55 per cent. In general, nut eaters in the research had a 39 per cent lesser risk of death and walnut eaters in particular a 45 per cent reduced threat (Anonymous, 2013). Research at the Yale-Griffin Prevention Research Centre, Connecticut, found that daily intake of 56 walnuts as a snack or with a meal improves endothelial function in overweight adults with visceral adiposity, and can ward off diabetes and heart disease in at – risk individuals (Anonymous, 2013).

Just a handful of walnuts provides:

2.5 grams of ALA, the plant-based source of omega-3 fatty acid.

4 grams of protein

2 grams of fibre

10 per cent of your daily value of magnesium and phosphorus

Health Benefits of Walnuts

- ☆ **Cures infections:** Walnut oil is known to be a remedy to treat fungal infections. Athlete's foot and Candida are some of the infections which are known to get treated with the application of walnut oil. Apply the oil topically on the affected area. You can also mix this oil with other herbal anti-fungals like garlic to enhance its effectiveness.
- ☆ **Treats psoriasis** : Psoriasis is known to be a painful and persistent skin condition that can be eased with the use of walnut oil. Add this oil to your bath or use it topically on your skin to treat this condition.
- ☆ **Excellent source of Omega-3**: Walnut oil is known to be an excellent plant source of Omega-3. It is also associated with numerous health benefits which includes protection against heart disease, some types of cancerand other inflammatory disorders like eczema etc (Meghna Mukherjee, 2013).

The nuts are rich source of energy and contain many health benefiting nutrients, minerals, antioxidants and vitamins that are essential for optimum health.

They are rich in monounsaturated fatty acids like **oleic acid** and an excellent source of all important **omega-3** essential fatty acids like *linoleic acid, alpha linolenic acid (ALA) and arachidonic acids.* Regular intake of walnuts in the diet help to lower total as well as LDL or "bad cholesterol" and increase HDL or "good cholesterol" levels in the blood. Diet rich in mono-unsaturated fatty acids and omega-3 fatty acids help to prevent coronary artery disease and strokes by favoring healthy blood lipid profile.

Eating 25 g each day provides about 90 per cent of RDI (recommended daily intake) of **omega-3 fatty acids**. Studies have revealed that n-3 fatty acids by their virtue of anti-inflammatory action helps to lower the risk of blood pressure, coronary artery disease, strokes and breast, colon and prostate cancers.

Walnuts are rich source of many phyto-chemical substances that may contribute to their overall anti-oxidant activity, including *melatonin, ellagic acid, vitamin E, carotenoids, and poly-phenolic compounds*. These compounds have potential health effects against cancer, aging, inflammation, and neurological diseases.

Walnuts have highest levels of popyphenolic antioxidants than any other common edible nuts. Eating as few as six to seven average sized nuts a day could help scavange disease causing free radicals from the body.

Walnuts are also excellent source of **vitamin E**, especially rich in *gamma-*tocopherol; contain about 21 g per 100 g. These nuts are packed with many important B-complex groups of vitamins such as riboflavin, niacin, thiamin, pantothenic acid, vitamin B-6, and folates.

They also very are rich source of minerals like **manganese, copper**, potassium, calcium, **iron,** magnesium, zinc, and selenium.

Walnuts oil has flavourful nutty aroma and has an excellent astringent properties; helps to keep skin well protected from dryness. It has also been used in cooking, and as "carrier or base oil" in traditional medicines in massage therapy, aromatherapy, in pharmaceutical and cosmetic industry.

Eating a handful of nuts daily may help prevent weight gain and possibly promote weight loss. The fat, protein and fibre in nuts help you feel full longer, so you may eat less during the day.

Women in a Harvard School of Public Health, reported eating 5 or more 1 ounce servings of nuts/peanuts per week reduced their risk of Type 2 diabetes by almost 30 per cent compared to those who rarely or never ate nuts. Women in the study who ate five tablespoons of peanut butter each week reduced their risk for Type 2 diabetes almost 20 per cent. (*Journal of the American Medical Association*, Nov. 27, 2002.)

Melatonin, a hormone produced by the pineal gland, which is involved in inducing and regulating sleep and is also a powerful antioxidant, has been discovered in walnuts in bio-available form, making them the perfect evening food for a natural good night's sleep.

Walnuts are full of omega-3 fatty acid, which has numerous benefits. They also contain alphalinoleic acid, which strengthens the bones.

Munch a handful of walnuts a day and you will have enough recommended levels of minerals, vitamins, and protein.

Walnuts which are rich in omega-6 fatty acids and arginine help in the production of nitric oxide, thus causing relaxation of arterioles and increasing the blood flow.

Caution

Walnut allergy is a type of hypersensitivity reaction to food substances prepared using these nuts. It is due to prior sensitization of the immune system by allergens in the nuts, which may lead to severe physical symptoms like pain abdomen, vomiting, swelling of lips and throat leading to breathing difficulty, and chest congestion. Therefore, it is advised to avoid any food preparations that contain these nut products in known case of walnut allergic individuals.

AVOCADO (*Persea americana*)

The Aztecs referred to the avocado tree ahuacatl as a "testicle tree." Avocados, like bananas, mature on the tree but ripen only after their harvest. Once ripen, they turn dark green or deep purple and yield to gentle pressure. Inside, cream colour flesh has butter-like consistency with bland taste yet pleasant aroma. The fruit features centrally placed single brown colour seed. On an average, each fruit weighs about 300-700 g, although heavier avocados are quite common in the markets. Avocados contain high levels of folic acid, which helps metabolize proteins, thus giving you more energy. They contain vitamin B6 (a nutrient that increases male hormone production) and potassium (which helps regulate a woman's thyroid gland), two elements that help increase libido in both men and women. Loaded with potassium, avocados boost libido for both sexes. And they're rich in folic acid that provides energy and stamina - both of which are important once your libido is restored. Full of saturated fats, they are rich in folic acid, which helps in metabolizing in the body thus giving more energy. They are indeed excellent for heart health and if they are good for the heart, they have to be good for sex life. This creamy fruit not noly helps lower bad cholesterol and reduce risk of laque build up to hence blood flow and fuel your brain power, Avacoda`s are a great alternative to sea food for those who are not fans of fish. It is an excellent fruit for expecting mothers. It is richy in useful nutrients, particularly folates. Avocado improves various functions of child development taking place in the womb. Research has shown that even a single avocado fruit eaten during a week, balances hormones, sheds unwanted birth weight and prevents cervical cancer in expecting mothers. Avocados receive a lot of flack for being high on fat content. However, these are a great source of good fats. Not a popular fruit in India, avacado is a fruit you can count on for boosting your energy.

Everyone loves avocados, because they are high in vitamin E and are a rich source of skin-protective antioxidants, essential for glowing skin and shiny hair. These wonder fruits also have a high folate content, which is important for skin cell regeneration and will give you a more youthful complexion.

Libido Dietary Tip

Cut an avocado in half, scoop out the insides with your fingers, and then get your woman to lick it off. They are also known as nature's best moisturisers and are loaded with anti-ageing nutrients such as Vitamin C and E, and is also high in **Omega-3 fatty acids.** Though a medium-size avocado has 30 grams of fat, all of it is healthy mono-saturated fats that lower cholesterol and protect the heart. Deepshikha says, "Eating avocados can help stave off wrinkles and keep your skin looking soft and supple, because of the Vitamin E that helps to get your skin glowing. The buttery-fleshed avocado may not seem like a hydrating food, but these healthy gems replenish potassium and contain healthy fats and fibre, which help your body hold on to water. As a bonus, the monosaturated fats in avocados contain **oleic acid**, which has been found to improve fat levels in the body and help control diabetes and cholesterol. Layer your sandwiches with avocado slices, toss diced avocado into your salads or simply scoop avocado out of its peel as a snack. Avocados are full of good energy-rich fats, rich in folic acid which helps in metabolizing proteins in the body thus giving you more energy. They are indeed excellent for your heart health and if they are good for your heart they have to be automatically good for your sex life. Isn't it? Natural conclusions aside, several studies over time have proved that increased overall health leads to higher sexual stimulation in men and women. Avocados are rich in monounsaturated fat, which is easily burned for energy. An avocado has more than twice as much potassium as a banana. For a delicious, creamy salad dressing, mix together avocado and fresh carrot juice.

Health Benefits of Avocado

- ☆ Avocados, like olives, are high in mono-unsaturated fats and calories. However, they are very rich in dietary fibre, vitamins, and minerals and packed with numerous health benefiting plant nutrients.
- ☆ Their creamy pulp is very rich source of mono-unsaturated fatty acids like **oleic** and **palmitoleic acids** as well as omega-6 poly-unsaturated fatty acid **linoleic acid**. They rich in mono-unsaturated fatty acids help lower LDL or bad cholesterol and increase HDL or good cholesterol, thereby, prevent coronary artery disease and strokes by favoring healthy blood lipid profile.
- ☆ They are very good source of soluble and insoluble dietary fibre. 100 g fruit provides 6.7 g or about 18 per cent of recommended daily intake. Dietary fibres help lower blood cholesterol levels and prevent constipation.
- ☆ In addition, the fruit, like persimmons, contain high concentration of tannin. Tannin, a poly-phenolic compound, which was once labeled as anti-nutritional agent is in-fact, has beneficial anti-inflammatory, anti-ulcer and anti-oxidant properties.
- ☆ Its flesh contains many health promoting flavonoid poly-phenolic antioxidants such as *cryptoxanthin, lutein, zeaxanthin, beta and alpha carotenes* in small amounts.
- ☆ They are also a good sourde many vitamins *viz.* Vitamin A, E, and K are especially concentrated in its creamy pulp.

- ☆ Avocados also excellent sources of minerals like iron, copper, magnesium, and manganese.
- ☆ Fresh avocado pear is a very rich source of potassium. 100 g of fruit provides 485 mg or about 10 per cent of daily-required levels.
- ☆ Avocado contains vitamin B6 and folic acid, which help regulate homocysteine levels. High level of homocysteine is associated with an increased risk of heart disease. Besides that, avocado also contains vitamin E, Glutathione and monounsaturated fat which help in maintaining a healthy heart.
- ☆ Avocados are rich in a compound called beta-sitosterol which has been shown to be effective in lowering blood cholesterol levels.
- ☆ Avocados are also a great source of potassium, which helps in controlling blood pressure levels.
- ☆ Phytonutrient compound found in avocados, such as polyphenols and flavonoids have been found to have anti-inflammatory properties, thereby reducing the risk of inflammatory and degenerative disorders.
- ☆ Avocado is an excellent source of carotenoid lutein, which known to help protect against age-related macular degeneration and cataracts.
- ☆ The monounsaturated (good) fats in avocados can reverse insulin resistance, which helping to regulate the blood sugar levels. Avocados also contain more soluble fibre, which keep a steady blood sugar levels.
- ☆ Avocados are rich in folate, a B vitamin commonly known as folic acid. One cup of avocado provides you about 23 per cent of the Daily Value for folate. The high amount of folate in avocado is essential in the prevention of birth defects, such as neural tube defect and spina bifida.
- ☆ The high levels of folate in avocados also protect against stroke.
- ☆ Many studies have shown that avocado can inhibit the growth of prostate cancer.
- ☆ The oleic acid in avocado is also effective in preventing breast cancer.
- ☆ Avocados contain glutathione, a powerful antioxidants to helps fight free radicals in the body.
- ☆ Being rich in antioxidants, avocado is beneficial in preventing aging symptoms. The glutathione in avocado may boosts immune systems, slows the aging process, and encourages a healthy nervous system.
- ☆ Avocados are one of the best natural mouth wash and bad breath remedies. It is cleanses intestine which is the real cause of coated tongue and this unpleasant condition.
- ☆ Avocado intake is linked with increased nutrient absorption. A study suggests that, when participants ate salad included avocados, they absorbed five times the amount of carotenoids (a group of nutrients that includes beta-carotene and lycopene) than those who do not include avocados.

- ☆ The avocado oil is added in many cosmetics because of its ability to nourish the skin and make your skin glow. It also aids in treating psoriasis, a skin disease that causes skin redness and irritation.
- ☆ The avocado has 200 calories for 100 grams. Typically, fruits has approximately 60-80 calories for 100 grams. Due to the high amounts of calories, avocado is a best diet for people who want to gain weight. Avocado is a healthy source of calories, unlike many other calorie-dense foods that may contain excess saturated fats and sugar.

Caution

Raw unripe avocados concentrated with tannins. High tannin content makes them bitter and unappetizing. Very high levels of tannins in the food prevent minerals like iron, calcium and phosphorus and vitamins from absorption in the gut. Although very rare, avocados may result in allergic symptoms in some latex-sensitive persons. The symptoms may include itching in the throat, hives, runny nose, breathlessness etc. Often the symptoms are mild and self-limiting.

JAMUN (*Syzygium cumini*)

The fruit is acidic and astringent in nature, with a sweet taste. Due to its acidic nature, it is usually eaten with a sprinkling of salt. Children are fond of this fruit as it colours the tongue purple due to anthocyanin, a plant pigment. It is a healthy fruit with absolutely no trace of sucrose. It is therefore, the only fruit with minimum calories. Not only the fruit, but the seed and also the leaves and bark of the jamun tree are believed to have medicinal properties. Many scientific researches have shown that dried alcoholic extracts of the seeds of the fruit given to diabetic patients on a regular basis showed a reduction in the level of their blood sugar and glycosuria. Also a concoction made from the mixture of dried seeds and the bark is considered beneficial in the cure of diarrhoea and dysentery. The bark of the tree has high astringent properties and is therefore used for gargles and as a mouthwash. Blackberries are nutritious and are often considered to be the best of the berry family. They are high in antioxidants, anthocyanin, salicylic acid, ellagic acid, and rich in fibre and Vitamins C and K. These **berries** have been shown to help reduce certain cancer risks and are good for cardiovascular health. They are also said to have anti-viral and anti-bacterial properties. The acidic, sour, sweet, and soothing fruit is used to treat diabetes, diarrhoea and ringworm. The fruit also has blood purifying properties. The leaves of the jamun tree possess anti-bacterial properties and is used for making medicines for the strengthening of the teeth and gums. The bark of the tree has antithelmintic properties, and is used to formulate many herbal medicines. The juice of the jamun fruit is extremely soothing and has a colling effect. It helps in the proper functioning of the digestive system. The leaves of the tree also help in controlling the blood pressure and gingivitis.

According to Ayurveda, Jamun cures pitta and skin diseases. Cures burning sensation, vomiting, it is a stabilizer. The fruit promotes digestive functions and body heat. It is digestive in nature, stimulates liver and enhances Vata. When eaten in excess quantities, it hardens stool. The seeds kernel improves digestion. It reduces urine and blood sugar levels. It also reduces amount of urine production. It normalizes heart beat, cures blood disorders and also controls boils, acne and eruptions on skin. It cures diarrhea due to pitta and clears the voice. It relives fatigue, asthma, cough, mouth and throat related disorders and gets rid of worms. Its squash is beneficial in curing vomiting, bleeding diarrhea and piles (Acharya Balkrishna, 2008).

Best Eaten

Pick the fresh, juicy ones and wash them. Do not have the entire fruit. Have small bites of it and avoid the centre.

Nutritional facts per cup Vitamin C: 50 per cent Manganese: 47 per cent Vitamin K: 36 per cent Vitamin A:6 per cent, Calcium: 4 per cent Iron: 5 per cent,

Nutritive Value

70 per cent of the jamun fruit is edible and glucose and furctose are the major sugars found in the ripe fruit. Sucrose is completely absent from the fruit. The fruit is laden with a large number of minerals, and provides fewer calories, as compared to other fruits. The seed of the fruit is also rich in protein and carbohydrates and traces of calcium have also been found.

The juice of ripe fruit or a decoction of it is administered in spleen enlargement, chronic diarrhoea and urine retention.

The seed extract in liquid or powdery form are given to patients with diabetes mellirus or glycosuiria. In many cases blood sugar level drops quickly.

Dried alcoholic extract of Jamun seeds reduces blood sugar and glycosuria. Seeds contain an alkaloid, **jambosine** and **glycoside jambolin** or **antimellin** which halts the conversion of starch into sugar.

Nutrients in Jamun

- ☆ The ripe fruit contains Glucose and Fructose are the major forms of sugar. It also contains Vitamins C and A, riboflavin, nicotinic acid, choline, folic acid, malaic acid, sodium, potassium, calcium, phosphorus, manganese, zinc and iron. Anthocyanins are present in appreciable quantities and are the reason for the antioxidant activity of the fruit.
- ☆ The stem and bark contains tannin, gallic acid, resin, phytosterols.
- ☆ The seed contains the glycoside, jamboline, gallic acid and essential oils.
- ☆ The leaves contain essential oils.
- ☆ The flower contains terpenoids.

Medicinal Properties of Black Berry or Jamun

- ☆ The extracts of the bark, seeds and leaves are used for the treatment of diabetes.
- ☆ The leaves have antibacterial properties and used for strenghthening teeths and gums.
- ☆ Oral administration of dried alcohalic extracts of the seeds to diabetic patients was found to reduce the level of blood sugar and glycosuria in trials conducted at CDRI, Lucknow.
- ☆ The bark of black berry tree is astringent, digestive, diuretic, anthelmintic and is considered useful for throat problems
- ☆ A decoction of the bark and powdered seeds is believed to be very usefil in the treatment of diarrhea, dysentery and dyspepsia
- ☆ The antibiotic activity of black berry extract has been widely studied and found useful against a number of microbial agents.
- ☆ The fruit is also considered to be stomachic, carminative, antiscorbutic and diuretic.
- ☆ Vinegar made from black berry fruit is administered in cases of enlargement of spleen, chronic diarrhea and urine retention
- ☆ For ringworm treatment, water diluted juice is used as lotion
- ☆ A decoction of bark is used in cases of asthma and bronchitis and are gargled or used as mouthwash for the astringent effect on mouth ulcerations, spongy gums, and stomatitis.
- ☆ It is digestive and activates the liver and spleen.
- ☆ It is a good remedy for urinary diseases and diabetes.
- ☆ Regulates heartbeat.
- ☆ Purifies blood, cures anemia and stops skin eruptions.
- ☆ Stops diarrhea and dysentery.
- ☆ Relieves throat affections and other respiratory diseases.
- ☆ Removes worms.

- ☆ Vinegar made from the fruit gives relief in colitis, indigestion, stomach diseases. It relieves gas and improves digestive power.
- ☆ It breaks renal stones.
- ☆ The bark, leaves, fruit and seeds are commonly used for treatments of various disorders.
- ☆ The bark of jamun has astringent, carminative, diuretic, digestive and constipating properties.
- ☆ It is good for sore throat, bronchitis, asthma, thirst and dysentery.

SAPOTA (*Achras sapota*) or CHIKOO

The fruit is sweet with smooth or grainy texture and light musky flavour. It contains 3-10 black, smooth, shiny "biconvex/bean" shaped, inedible seeds located at its center. This delicious tropical fruit is rich in vitamins, minerals and health benefiting anti-oxidant *tannins*. The fruit is soft, composed of easily digestible pulp made up with simple sugars like fructose and sucrose that when eaten replenishes energy and revitalizes the body instantly. t is a round fruit that resembles sometimes like a potato or a kiwi in rounded form. It is sweet in taste and is high calorie fruit. It is famous for its peculiar taste and is very commonly used in making shakes etc.

Health Benefits of Sapodilla

- ☆ Sapote contains a good amount of antioxidant vitamins like **vitamin** and vitamin A. Vitamin A is essential for vision. It is also required for maintaining healthy mucus membranes and skin, lung and oral health even at the old age also helping to enjoy younger happy life.
- ☆ Fresh ripen sapodilla is a good source of minerals like potassium, copper, iron and vitamins like folate, niacin and pantothenic acid. These compounds are essential for optimal health as they involve in various metabolic processes in the body as cofactors for the enzymes.
- ☆ Chiku or sapota contains rich dietary fibre making it a good laxative, Fibres in ripe, sweet and tasty chiku fruits help to prevent constipation.
- ☆ Taking sufficient quantity of ripe chiku in raw form or as chiku shake will protect colon system preventing chances of colon cancer.

- ☆ Chiku contains natural antioxidant properties. Hence eating this seasonal fruits will be the most effective natural health cure medicine to prevent viral, bacterial and parasitic effects in human internal organ system.
- ☆ Vitamin C natural medicinal property in chiku is useful for maintaining healthy and shining skin texture.
- ☆ Eating this delicious and healthy seasonal fruits will develop body resistance to fight against many infectious diseases throughout the year.
- ☆ One of the most important use of this herb is its hemostatic properties. This herb is very beneficial in stopping the blood loss. Helps in reducing bleeding in piles and injury etc.
- ☆ Sapota has been found beneficial in conditions pertaing to infection. It helps in reducing viral as well as bacterial infections in the body. It also reduces inflammation making it very potent in reducing swellings and pain.
- ☆ Sapota is a delicious food and is rich in glucose. This glucose helps in proving instant energy to the body. Hence it is very much recommended to athletes as it provides loads of energy.
- ☆ This wonderful fruit is extremely beneficial in making your skin shiny and elegant thus you require minimum skin care products to enhance skin texture and complexion.

SITAPHAL *(Annona squamosa)*

Useful Part

Fruits, seeds, Roots, Leaves

There are medicinal application of the custard apple tree. The bark and leaves contain annonaine, an alkaloid. A bark decoction is used to stop diarrhea, while the root is used in the treatment of dysentery. A decoction of the leaves is used as a cold remedy and to clarify urine. Annona squamosa Linn., family Annonaceae, is said to show varied medicinal effects, including insecticide, antiovulatory and abortifacient. The fruits of Annona are Haematinic, cooling, sedative, stimulant, expectorant, maturant, tonic. They are useful in anemia, burning sensation. The seeds are abortifacient and insecticidal and are useful in destroying lice in the hair. Leaves are used to overcome hysteria and fainting spells. Fruit is used in making of ice creams and milk beverages.

The fruits are very high in calorific value and are a rich source of minerals and Vitamins. They are used by athletes for their high energy content. The seeds are powerful insecticides and powdered seeds are used for removing head lice. It is also used as an effective pesticide in agriculture and horticulture. Apart from this, various parts of the plant are medicinally useful in a wide variety of diseases from tumor to cough. The leaves are shown to have anti-diabetic properties. It is also known for its hepato-protective powers and scientists have experimentally proven the efficacy of the alcoholic extract of the leaves and stem in malignant tumors.

Its efficacy as anti-depressant and anti oxidant has also been experimentally proven by various scientists. They are also used as anti-depressants, in epilepsy and in spinal cord disorders. Fruits are sweet, haematinic (A medicine which increases the haemoglobin content of the red cells in the blood), cooling, act as a sedative, stimulant and function as expectorant and tonic. Seeds are abortificient and insecticidal. Roots are powerful purgatives and are also used in dysentery. Acetogenins found in the seeds of Sitaphal would actually help fight plant pests. The Sitaphal pesticide is inexpensive, environment-friendly and highly effective in containing a variety of pests on a number of crops. The seed extracts also showed synergistic activity in combination with neem seed extract in pesticidal action.

It is nourishing, increases blood (Hb count), sweet, cooling, heart tonic, provides strength, helps muscle building and alleviates burning sensation, bleeding disorders and disorders of vaata.

Medicinal Uses

In India the crushed leaves are sniffed to overcome hysteria and fainting spells; they are also applied on ulcers and wounds and a leaf decoction is taken in cases of dysentery. Throughout tropical America, a decoction of the leaves alone or with those of other plants is imbibed either as an emmenagogue, febrifuge, tonic, cold remedy, digestive, or to clarify the urine. The leaf decoction is also employed in baths to alleviate rheumatic pain. The green fruit, very astringent, is employed against diarrhea in El Salvador. In India, the crushed ripe fruit, mixed with salt, is applied on tumors. The bark and roots are both highly astringent. The bark decoction is given as a tonic and to halt diarrhea. The root, because of its strong purgative action, is administered as a drastic treatment for dysentery and other ailments.

Health Benefits of Custard Apple

- ☆ The numerous vitamins and nutrients present in custard apple provide protection from various medical conditions such as heart attacks, strokes and cancer.
- ☆ If you have a fluctuating blood pressure then a regular intake of this fruit is effective in keeping it under control.
- ☆ Being a rich source of Vitamin A, custard apples give the consumer a glowing and healthy skin and shiny and lustrous hair.
- ☆ Looking for a natural way to enhance your eye sight? Then the custard apple acts as a storehouse of various vitamins making your vision sharper and brighter.
- ☆ Intake of custard apple helps in curing indigestion problems. Since this is indirectly linked to constipation, these fruits inject copper into the body which ensures the production of stools.

- The high amount of magnesium in custard apples maintains water balance in the body. This promotes the removal of various acids from the joints and lowers the risk of rheumatism and arthritis.
- Custard apples are packed with iron and are the simplest and fastest cure for treating anemia.
- For extremely underweight individuals, custard apples are an ideal way to put on a few kilos. Also, the sugar fibres present in them make the making metabolic system more efficient.
- The skin of custard apple is helpful in fighting against tooth decay and gum pain.
- These fruits are excellent in treating vomiting, gout, lice and vitamin B6 deficiencies.

The custard apple fruit is a sweet fruit that has many benefits. Some of these benefits are:

- It is low in calorie content and therefore good for maintaining optimum health.
- It contains Vitamin A which is good for the skin, hair, eyes.
- It helps in digestion as it is a rich source of dietary fibre.
- It is a coolant, stimulant and expectorant.
- It has healing properties as well – the flesh can be applied on boils, abcessess and ulcers.
- The magnesium in custard apples helps to protect the heart from cardiac diseases. It can relax the muscles as well.
- Custard apples can also be used to treat dysentery and diarrhea.
- It is good for anemic people.
- The potassium present in a custard apple helps fight muscle weakness.
- It is also a rich source of magnesium. It therefore helps to equalize the water balance in the body and this helps in removing acids from the joints. It is thus beneficial for rheumatic and arthritic patients

OLIVE (*Olea europaea*)

Olive contains a compound called **Oleuropein** which can cleanse liver of the toxins. It is the Oleuropein aglycone, the most abundant polyphenol present that is responsible for its characteristic bitter, pungent taste. Oleuropein has anti-oxidant, anti-inflammatory, anti- ageing, anti- cancer, anti-viral and anti- microbial properties. It is a skin protector, as it acts as a free radical scavenger. This polyphenol can also help in the management of Alzheimer's disease. It has healing properties due to its hypotensive; anti-rheumatic, diuretic and anti- pyretic effects. It is helpful in prevention of of diabetic complications associated with oxidative stress. It can help in reduction of coronary heart stress disease and certain cancers. Olive oil is one of the best sources of monounsaturated fats and has the advantage of being less susceptible to oxidation.

In addition, oleic acid, a fatty acid abundant in olive oil, appears to also protect from oxidation of LDL. In addition to bolstering the immune system and helping to protect against viruses, olive oil has also been found to be effective in fighting against diseases.

Health Benefits of Olive

- **Cancer prevention** : Black olives are a great source of vitamin E, which has the brilliant ability to neutralize free radicals in body fat. Especially when working with the stable monounsaturated fats found in olives, vitamin E can make cellular processes safer. When such processes such as **mitochondrial** energy production are not well protected, the free radicals produced can cause oxidation, damaging a cell's mitochondria, and preventing the cell from producing enough energy to supply its needs. If the DNA of a cell is damaged, it may well mutate and become cancerous. Studies have shown that a diet supplemented with olive oil leads to a lower risk of colon cancer, almost as low a risk as a diet rich in fish oil (Anonymous,2013).

  The phyto-nutrient in olive oil, **oleocanthal**, mimics the effect of ibuprofen in reducing inflammation, which can decrease the risk of breast cancer and its recurrence. **Squalene** and **lignans** are among the other olive oil components being studied for their possible effects on cancer.
- **Cardiovascular benefits:** When free radicals oxidize cholesterol, blood vessels are damaged and fat builds up in arteries, possibly leading to a heart attack. The antioxidant nutrients in black olives impede this oxidation of cholesterol, thereby helping to prevent heart disease. Olives do contain fat, but it's the healthy **monounsaturated** kind, which has been found to shrink the risk of **atherosclerosis** and increase good cholesterol. Olive oil helps lower levels of blood cholesterol leading to heart disease.
- **Oxidative Stress**: Olive oil is rich in **antioxidants,** especially vitamin E, long thought to minimize cancer risk. Among plant oils, olive oil is the highest in monounsaturated fat, which doesn't oxidize in the body, and it's low in **polyunsaturated** fat, the kind that does oxidize.
- **Blood Pressure**: Recent studies indicate that regular consumption of olive oil can help decrease both systolic and diastolic blood pressure.
- **Diabetes**: It has been demonstrated that a diet that is rich in olive oil, low in saturated fats, moderately rich in carbohydrates and soluble fiber from fruit, vegetables, pulses and grains is the most effective approach for diabetics. It helps lower "bad" low-density lipoproteins while improving blood sugar control and enhances insulin sensitivity.
- **Obesity**: Although high in calories, olive oil has shown to help reduce levels of obesity.
- **Rheumatoid Arthritis**: Although the reasons are still not fully clear, recent studies have proved that people with diets containing high levels of olive oil are less likely to develop rheumatiod arthritis.

☆ **Osteoporosis**: A high consumption of olive oil appears to improve bone mineralization and calcification. It helps calcium absorption and so plays an important role in aiding sufferers and in preventing the onset of Osteoporosis.

☆ **Bone and connective tissue:** The anti-inflammatory abilities of the monounsaturated fats, vitamin E and polyphenols in black olives may also help dull the asperity of asthma, osteoarthritis, and rheumatoid arthritis. Most of the suffering in having one of these three bone maladies is brought about by high levels of free radicals. Olive oil also contains a chemical called oleocanthal, which acts as a painkiller. Research has found that oleocanthal inhibits inflammation by the same means that drugs like Ibuprofendo.

☆ **Olive oil might help prevent strokes:** According to a new study from France, older individuals who consume olive oil daily may be able to protect themselves from a stroke. The study which is part of the Three-City Study, an ongoing multicenter study of vascular risk factors for dementia, was published in the online issue of Neurology.

☆ **Olive oil diet reduces risk of type 2 diabetes:** Traditionally a low fat diet has been prescribed to prevent various diseases such as heart disease and diabetes. While studies have shown that high fat diets may increase the risk of certain diseases such as cancer and diabetes, it appears that it is the *type* of fat that counts rather than the amount of fat. We now know that a diet rich in monounsaturated fats such as the ones found in olive oil, nuts and seeds actually protects from many of these chronic diseases.

☆ **Olive oil keeps the heart young**: A diet rich in olive oil may be able to slow down the aging of the heart. It is a known fact that as we grow older the heart also goes through a normal aging process. The arteries may not function as well as they did and this can lead to a number of health problems. However, researchers discovered that a diet rich in olive oil or other monounsaturated fats

☆ **Olive oil fights osteoporosis:**. Osteoporosis is a disease characterized by a decrease in bone mass, which in turn causes the architecture of bone tissue to become fragile. This can then increase the possibly of fractures, making even the slightest of knocks potentially fatal for sufferers. The disease is recognized as being particularly prevalent among postmenopausal women for whom a decrease in the production of estrogen then weakens bone structures and most commonly affects the ribs, wrists, and hips. For this study, scientists were particularly interested in how a supplementation of olive oil could be used to help women in this category.

☆ **Digestive tract health:** Frequent consumption of both vitamin E and the monounsaturated fats in black olives is associated with lower rates of colon cancer. These nutrients help prevent colon cancer by neutralizing free radicals. Olive oil's protective function also has a beneficial effect on ulcers and gastritis. Olive oil activates the secretion of bile and pancreatic hormones much more naturally than prescribed drugs, thereby lowering the incidence of gallstone formation. A cup of black olives also contains 17 per cent of the daily allowance of fiber, which promotes digestive tract health by helping to move food through the system at a healthier pace. This keeps any one part of the digestive tract from having to work too hard and supports the ideal balance of chemicals and populations of microorganisms required for a healthy digestive system.

- ✩ **Eye health** :One cup of black olives contains ten per cent of the daily recommended allowance of vitamin A which, when converted into the retinal form, is crucial for healthy eyes. It enables the eye to better distinguish between light and dark, thereby improving night vision. Furthermore, Vitamin A is believed effective against cataracts, macular degeneration, glaucoma and other age-related ocular diseases.
- ✩ **Good source of iron:** Black olives are very high in iron. The ability of red blood cells to carry oxygen throughout the body is due to the presence of iron in the blood. If we suffer from a lack of iron, our tissues don't get enough oxygen, and we may feel cold or weak. Iron also plays a vital role in the production of energy. It is a necessary part of a number of enzymes, including iron catalase, iron peroxidase, and the cytochrome enzymes. It also helps produce carnitine, a nonessential amino acid important for the utilization of fat. To top it all off, the proper function of the immune system is dependent on sufficient iron.
- ✩ **Skin and hair health:** Black olives are rich in fatty acids and antioxidants that nourish, hydrate and protect. Chief among those is vitamin E. Whether applied topically or ingested, vitamin E has been shown to protect skin from ultraviolet radiation, thus guarding against skin cancer and premature aging. You can gain a healthy, glowing complexion by washing your face in warm water, applying a few drops of olive oil to vulnerable spots, and letting it work its magic for 15 minutes before rinsing it off. In fact, you can moisturize with olive oil before any bath, and even condition your hair with it by mixing it with an egg yolk and leaving it before rinsing and washing.
- ✩ **Best muscle building foods (Olive oil)**: The monounsaturated fat in olive oil helps prevent muscle breakdown and protects joints. Olive oil is rich in monounsaturated fats and omega-3 fats. These good fats are heart healthy. Olive oil is anti-inflammatory and helps with muscle pain and inflammation, aiding recovery, making it a perfect muscle building food (Moghul, 2013).

AMLA

Indian Gooseberry (*Phyllanthus emblica*) is rich in vitamin C. Winter brings along its share of colds and cough. To cure this, prepare a mixture of two tablespoons of **amla** with an equal amount of honey. Have this three to four times a day, people suffering from **Vitamin**

C deficiency can greatly benefit by eating amlas, as it is one of its most potent sources. It has 20 times more vitamin C compared to orange. It also contains gallic acid, tannic acid, carbohydrates, albumen, cellulose and calcium. On the other hand, Vitamin C, taken in the form of supplements, is not easily absorbed by the body, unlike when taken in a natural form. Battling with constipation problems, amla contains high fibre content and will ease your problem. Because they are also rich in antioxidants, which protect against the formation of free radicals in the body, amlas, when eaten frequently, help to prevent cancer. To seek relief from painful mouth **ulcers**, simply gargle with a mixture of **water** and amla juice. Are joint pains getting you down? Have amlas, as they contain anti-inflammatory **properties** which will help in reducing the swelling in the joints caused by **arthritis**.

- ☆ The fruit has stool binding propereties and is a blood and urine purifier, It promotes the taste and is beneficial in curing diarrhea, burning sensation, jaundice, bleeding diathesis, gout, piles, constipation, indigestion, breathing disorders, cough etc. It improves eyesight, sexual potency and promotes longitive.
- ☆ Amla is one of the best anti-inflammatory herb. It prevents aging and promotes longevity.
- ☆ It is very useful in skin diseases as it has sheet virya in potency. It inhibits pitta and thus helps in getting relief from all the skin disorders caused by pitta dominance. Regular intake of Amla promotes glow on skin and delays wrinkles or loosening of skin.
- ☆ It improves texture of the hair.
- ☆ It also prevents premature graying of hairs and dandruff.
- ☆ It is very useful in improving eyesight. It also counters diseases like reddening, itching and watering of eyes.
- ☆ Amla is very useful in controlling blood sugar level.
- ☆ It also helps maintain the functioning of the liver.
- ☆ Amla cleanses the mouth and strengthens the teeth.
- ☆ It balances stomach acids because it improves digestion but does not heat the body.
- ☆ It nourishes the brain and mental functioning.
- ☆ Amla has powerful antioxidants, which protects from harmful free radicals that can increase the risk of a wide range of serious diseases including cancer.
- ☆ It is a tridosa healer. Its acid suppresses Vata, mild coolness suppresses Pitta and arid astringence suppresses Kapha.
- ☆ It acts like an anti-dermatoses and anti-pyretic.
- ☆ It is healthy for heart and is a blood purifier.
- ☆ It stimulates sexual desire and stabilizes the womb.
- ☆ It cures cough and cold (kapha, pitta, nashak), cures burning sensation due to dermatoses. It is beneficial for eyes and digestive system and also cures difficult and incurable fever.

- Indian Gooseberry along with harad and pippali cures all types of fevers.
- Indian Gooseberry is bit sweet and sour and pungent in taste. Its juice is good for eyes (Acharaya Balkrishna, 2006).

DRAGON FRUIT (*Hylocereus undatus*)

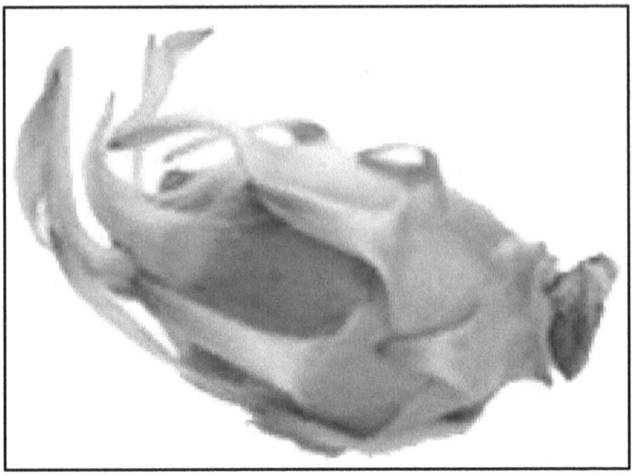

Although the fruit originates in South America, it is now grown in Thailand, Sri Lanka, Bangladesh and the Philippines, where it's loved for its many health benefits. Said to be rich in vitamin C, fibre and calcium, the pitaya - or dragon fruit as it's also known - gives you a bigger bang for your buck than similar super fruits such as acai berries. Already available in Waitrose and Tesco, the bright pink fruit is full of edible black seeds and boasts flesh that tastes like a cross between a kiwifruit and a pear. Along with a mega-dose of vitamin C, pitaya is also thought to help lower cholesterol and is packed with skin-boosting antioxidants. The tiny black seeds are a great source of healthy Omega-3 fat and there's plenty of magnesium and calcium in the flesh. Pitaya is even believed to help regulate blood sugar levels.

Dragon fruits are high in antioxidants, which help to fight carcinogenic free-radicals from forming in the body. In addition, they are a good source of Vitamin C, and are rich in minerals, especially calcium and phosphorus. They are also low in calories and high in fibre, while the seeds are have high polyunsaturated fatty acids. Aside from its nutritional content, the fruit is also said to help excrete heavy metal toxins from the body and lower cholesterol and blood pressure. Dragonfruits are also known to be a natural laxative.

This nutritional information is based on the edible flesh of 1 dragon fruit. The average weight of a dragon fruit is about 3.6 oz. (100 g), although only 2 oz. of this actually edible. The serving information also includes the nutrition from the seeds. This assumes that you'll chew the seeds so that you can digest them.:

Dragonfruits contain phytoalbumins, which have antioxidant properties that help prevent the formation of cancer cells. In addition, dragonfruits are also known to increase the excretion of heavy metal toxins from the body.

Lycopene is also present in dragonfruits, and this is the pigment responsible for their red colour. Lycopene is said to prevent prostate cancer.

They are among the many wonder fruits that are said to provide multiple health benefits. In addition, dragonfruits help protect the environment because they absorb carbon dioxide at nighttime, and then release oxygen to purify the air.

Health Benefits of Dragon Fruit

- ☆ As dragon fruit is a good source of antioxidants, it prevents free radicals and protects from causing cancer and other health detriments from entering your body.
- ☆ This fruit helps in neutralizing the toxic substances such as heavy metals. Besides, high blood pressure and cholesterol levels can be reduced by consuming dragon fruit.
- ☆ Regular consumption of this aids in fighting against cough and asthma.
- ☆ This has high amounts of vitamin C and helps in healing bruises and cuts quickly, and enhancing the immune system.
- ☆ Vitamin B2 present in dragon fruit acts like a multivitamin and helps to improve and recover the loss of appetite.
- ☆ Vitamin B1 in this fruit helps in increasing the energy production and also in metabolizing carbohydrates.
- ☆ Due to the presence of vitamin B3, it helps in lowering bad cholesterol levels. It improves the appearance of your skin by smoothing and moisturizing it.
- ☆ It improves eyesight and prevents hypertension.
- ☆ As it is a good source of phosphorus and calcium, it reinforces bones, helps in tissue formation and forms healthy teeth.
- ☆ Regular consumption of dragon fruit decreases weight, thereby creates a well-balanced body.
- ☆ Dragon fruit is also helpful in reducing blood sugar levels in people suffering from type 2 diabet

Tips

- ☆ Dragon fruits taste best when served chilled.
- ☆ To eat dragon fruit, slice it into half and scoop out the flesh with a spoon. The skin is not eaten.
- ☆ The middle of the fruit is the sweetest part. Hence, save it and eat it at the last.
- ☆ Dragon fruit is eaten either raw or mixed in juices and cocktails.
- ☆ It is also used as preserves and spreads like jam, puree and cordial, salads, sherbets and sorbets, fruit pizza and other beverages.

Vegetables

Vegetables contain various medicinal and therapeutic agents. There are a large array of sedative and soporifics or sleep inducing in the vegetable kingdom. Vegetables like onion, radish and celery exercise a tonic effect and are excellent for nerves. Certain vegetables are highly beneficial in the treatment of various diseases. Carrots are good for blood. White crisp juicy stalks of celery serve a much better medicine in case of rheumatism or nervous dyspepsia than any nerviness that relieves nerve disorder. A dish of spinach or dandelion will be beneficial in the treatment of kidney troubles. Lettuce can be used as a food remedy for insomnia. Onion can be used with adventage in the treatment of cough, cold, influenza, constipation, scurvy and hydrophobia. The leaves of fenugreek are highly valuable in the treatment of indigestion. flatulence and sluggish liver. Garlic can be beneficially used in heart diseases, hypertension, hypoglycemia, diabetes and even in fatal form of meningitis. It has been effectively used in long blood cholesterol and preventing blood clotting. Vegetable are alkaline in nature and are ideal foods to build immunity. They are high in antioxidants, promote good digestive health and help to cleanse and restore the body's pH balance.

Fibres in the vegetables act as the mechanical intestinal expanders, draw more water and protein in them, help easy expulsion of the waste in the form of stool. They prevent habitual constipation and keep the entire intestinal track free from harmful germs. Fibres in the form of cellulose help the elimination of cholesterol. Dietary fibres is very useful for the human body because people who have a diet rich in dietary fibre have low incidence of diseases like coronary heart disease and cancer of intestinal tract, piles, obesity, diabetes, constipation, hiatus hernia, diverticulitis, irritable bowl syndrom, dental caries and gallstones. Beet root, cabbage, carrots, cucumber, green peas and beans are especially valuable in this. They are useful in case of **arteriosclerosis**, high blood pressure and constipation. But when there is inflammation in the intestine, vegetables having less cellulose content such as tomatoes, lettuce, potatoes and vegetable juices should be taken. Pectin food in the vegetables such as brinjal, radish, pumpkin and beet root absorb water, kill certain bacteria and toxins and eliminate them from the body. Garlic, onion, radish and mint contain pectin as well as antimicrobic qualities.

Vegetables supply trace elements which are necessary for human organism. Iodine for instance is essential for throid hormone, which regulates much physical and mental activities, cobalt for increasing the number of blood corpuscles and zinc for proper growth. Parsley, sage, rosemary and thyme have many times more anti-oxidants than typical salad veggies. Also various herbs have been associated with different things like lowering blood pressure and helping to control cholesterol levels. Both fresh and dried herbs taste great and add an anti-oxidant punch. Green leafy vegetables are useful because they are high in anti-oxidants and magnesium. According to scientists, eating one and a half extra servings of green leafy vegetables cut the risk of diabetes by 14 per cent. Emodin present in Chinese Rhubrab *(Rheum palmatum)* and Knot weed *(Polygonum cuspidatum)* inhibits an enzyme called 11-Beta-HSDI which plays a role in resistance to insulin, the hormone that helps clear excess sugar from the blood.

Vegetables can provide widely accessible sources of essential vitamins (particularly A, C, niacin, riboflavin and thiamine), and minerals (calcium and iron), as well as supplementary protein and calories(given with each vegetable). Some vegetables, such as roots, tubers and leafy greens, are capable of producing proteins and calories at rates(per hectare, per day) comparable to those of most efficient stable cereal crops. The greatest advantage with vegetables is that they promote intake of essential nutrients from other foods by making them more palatable. They provide dietary fibre to improve digestion and health, and are essential for properly balanced diet. In many regions around the world, such as India or Southeast Asia, certain cultures survived only on vegetarian meals, either for economic or religious reasons.

India has a long tradition of being predominantly a vegetarian country, as our great sages and saints from millennium have preached vociferously its usefulness. They classified our diet into three categories *viz.* Sattvik, Rajasik, and Tamasvik. Sattvik food which include leafy vegetables, they stressed that it encourages intellectual pursuit, clarity of mind, purer thought, forbearance and keep the nervous system calm and quite. They provide not only the energy rich food like minerals and vitamins. They not only adorn the table, but also enrich health from the most nutritive menu and tone up the energy and vigour of man. They are the cheapest sources of natural nutritive foods. They not only benefit us in protecting against some degenerative diseases but also play a key role in neutralizing the acids produced during digestion of proteinous and fatty foods, also provide roughages which promote and help in preventing constipation etc. The only things that brings down cholesterol and homocysteine levels are vegetables, green leaves and citrus by providing B vitamins, folic acid and vitamin B6 and B12 which help convert homocysteine to other amino acids that are not harmful. According to the recent studies, people who adopt a diet free of meat, eggs and dairy products can drop their homocysteine levels by 20 per cent in one week. This is because folic acid, a vitamin -B found in whole grains, green vegetables, beans and other plant- based foods, helps convert homocysteine to another, more useful amino acid(Manek Gandhi, 2005).

Spinach, kale and collards, appear to help keep the brain young and may slow the mental decline Vegetables are rich sources of vitamin A, thiamine, niacin, folic acid and ß- carotene. Vitamin A (ß- carotene) is essential for the normal functioning of the visual process and the structure of the eye, and a prolong deficiency of the vitamin can lead to blindness. They do not contain active vitamin A compound retinol but only carotenoids, which are converted into active retinol in the body. Carrots, sweet potato, amaranth, basella, spinach are important sources of vitamin A. Some vegetables are rich in calcium, phosphorus and iron. The cruciferous vegetables such as broccoli, cabbages(green savoy, red), Chinese cabbage, collard greens, mustard greens, kale, khol rabi and cauliflower are high in vitamins, monoterpenes, thiols and indoles. Cauliflower has high levels iron, boron, thiol, indoles and sulphur compounds. The greatest benefits found in eating cruciferous vegetables may be due to the concentration of potassium, folate, fibre and dietary flavonoids found in these foods. The nutritional composition of vegetables is influenced by genetic as well as environmental factors, such as temperature, light, moisture and the nutritional status of the soil in which the vegetable is grown. It is also influenced by the cultural practices, stage of maturity, post harvest handling, storage conditions, as well as cooking and canning also influence the nutrient content of vegetables (Roy and Chakraborti, 1993). A study published in journal of Neurology by a researcher, Martha Clare Morris, at the Rush Institute for Healthy Aging at Chicago's Rush University Medical Center, has found that eating green leafy vegetables including sometimes associated with growing old. The researchers were of the view, that this may be because they contain healthy amounts of vitamin E, an antioxidant that is believed to help fight chemicals produced by the body that can damage cells. Vegetables generally contain more vitamin E than fruits, which were not linked with slowed mental decline in the study. Vegetables also are often eaten with healthy fats such as salad oils, which help the body absorb vitamin E and other antioxidants(Anonymous, 2006.). Similarly another researcher reported at a meeting of the American Society for Reproductive Medicine, that men hoping to boost their fertility may want to eat more fruits and veggies. The more produce a man consumed, the less sluggish his sperm, the new research suggests(Anonymous, 2006). Vegetable like carrot help produce female hormones needed for ovulation, whereas, men should eat garlic for sperm development, quality and mobility.

Broccoli is a good alkaline vegetable and can be consumed in plenty by just boiling it and adding some salt and lemon to it for flavour. Brinjal though not a green vegetable is still an alkaline vegetable. It can be consumed in a variety of ways. Brinjals with a few potatoes can make a lovely alkaline vegetable dish. The alkaline vegetable list is unending. Cabbage, Lettuce, mushrooms, onions, peas, pepper bells, parsley, radish, cucumber and almost all vegetables are alkaline vegetables. So go out there and skip the meat section and walk into health and vigour. Brocolli and spinach- both the vegetables have high vitamin-C content. It has been found that vitamin-C prevents age spots caused by clumping of pigments and red lines caused by damaged capillaries. Carrots have constituents like carbohydrates, protein, fat, calcium, iron, fibre, vitamins B1, B2, B6, C, K, biotin, potassium and thiamin which together provide the body important enzymes, vitamins and minerals. Carrots contain beta-carotein that slows down the process of ageing (Indian Syndicate, Tribune, 16th April, 2010).

Plant chemicals in vegetables like tomatoes and pumpkins etc seem to be able to inhibit angiogenesis sothat a single cancer cell or claster of cancer cells is never able to grow enough to cause any mischief. Some plants contain tumour suppressor proteins- which help to curb the growth of cancer cells. Studies have shown that men who eat cooked tomato products two to three times a week reduce their prostrate cancer risk by about 50 per cent. Eating of vegetables such as brussels sprouts, cabbage, cauliflower and broccoli significantly cut the risk of developing colourectal cancer, a new research has found. Mean while, the report also found that both total fruit and vegetable intake- especially apples and dark yellow vegetables- lower the risk of distal colon cancer (Anonymous, 2011).

Carrots and sweet potatoes may prove to be a vital weapon against breast cancer in the early stages of the disease, says Scientists. A study has found that a nutrient in carrots and sweet potatoes, called **retinoic acid**, can reverse the early changes in cells that lead to breast cancer in women. The chemical, which affects cell growth and proliferation, can also rejuvenate the skin, and a weak version of it is used in anti – wrinkle face creams. Watermelon, carrots and potatoes are considered to be good for the body since they help in rejuvenating our bodies.

A vegetarian diets are healthful, nutritionally adequate and may provide health benefits in the prevention and treatment of certain diseases. Well- planned vegetarian diets are appropriate for individuals during all stages of the life cycle, including pregnancy, lactation, infancy, childhood and adolescence and for the althetes. Vegetarian also appear to have lower –low density lipo protein cholesterol levels, lower blood pressure and lower rates of hypertension and type 2 diabetes than non-vegetarians. Further more, vegetarians tend to have a lower body mass index and lower overall cancer rates. Features of a vegetarian diet that may reduce risk of chronic disease include lower intake of saturated fat, cholesterol and higher intake of fruits, vegetables, whole grains, nuts, soy products, fibre and phytochemicals. A vegetarian diet containing more potassium, complex carbohydrates, fibre, calcium, magnesium, vitamin C, all of which may have favourable influence on blood pressure. It can be a great benefit to start lowering your blood pressure naturally. A new study has suggested that sticking to a vegetarian diet can help kidney disease patients avoid accumulating toxic levels of phosphorus in their bodies. Kidney disease patients must limit their phosphorus intake, as high levels of the mineral can lead to heart disease and death.

Vegetable diet will lower the risk of developing cardio - vascular disease, osteoporosis, kidney disease, higher blood pressure, obesity and cancer of colon of lungs. Foods like fruits and vegetables, grains and beans have very little fat and no cholesterol but gives us fibre and other nutrients *viz.* magnesium, potassium, vitamin C and vitamin E, folate, carotenoids that our own body need. Many vegetables and beans like spinach, pinto beans, chick peas have a lot of iron and vitamin B12. Vegetarian diet is associated with a lower risk of death from Ischemic heart disease. Many green vegetables contain potassium which greatly helps calm one's nerves when stressed. Also a diet rich in green vegetables helps the body feel lighter and feel great. Vegetables like basil, black pepper, chillies, cumin, ginger and turmeric stimulate harmone (sex) like Dopamine which enhances mood and confidence.

It won't be wrong if we tag spinach as a natural Viagra. This green leafy vegetable is loaded with sexual benefits. It is rich in Vitamin E, which is a major catalyst in the production of sex hormones in the body. It is also rich in manganese, which facilitates the production of the female hormone estrogen. A deficiency of magnesium also affects a woman's fertility levels, informs nutritionist Aishwarya Rajan. Green leafy veggies are also loaded with zinc, which is known for its libido and sperm production qualities men.

Now, celery may not be the food of choice that you dream of when thinking sex, but it is a great source food for sexual stimulation. Celery contains androsterone, an odorless hormone released through male perspiration and turns women on. Celery is best eaten raw. Wash and cut some, and munch away in bed just before you get down and dirty. Spinach, zucchini, broccoli and brussels sprouts contain a caroteinoid called lutein and a pigment called zeaxanthinan. Both defend your cells from free radicals and help keep your eyes lustrous. Avoid overcooking for best benefits. Steam or saute, instead. Orange and other vegetables that are rich in antioxidants and fibre work against under-eye puffiness.

Reds

What to Eat

Tomatoes, beets, strawberries, watermelon and red peppers.

Why

The same fruits and veggies responsible for staining your best dress shirts are also believed to fight prostate cancer, cardiovascular disease, diabetes and — listen up, guys — male infertility. Crimson-coloured crops contain varying amounts of lycopene and anthocyanin, two naturally occurring chemicals in plants that are as rich in antioxidants as they are in difficult-to-pronounce syllables. Antioxidants, of course, are powerful molecules that cruise around your body, bonding to and safely defusing other less stable molecules (called free radicals, man!), which, if left unchecked, could cause you some serious cellular damage.

Oranges and Yellows

What to Eat

Pumpkins, carrots, yellow squash, lemons and sweet corn.

Why

Sure, a tall glass of Tang can deliver your daily dose of vitamin C, which aids in the healing of wounds and the synthesis of collagen. But actual oranges and similarly shaded foods also provide you with the pigments alpha- and beta-carotene. "Beta-carotene is a precursor to vitamin A," "The body converts this compound into vitamin A, which in turn promotes healthy vision, strong bones and smooth skin." Got psoriasis? Eat more oranges.

Greens

What to Eat

Spinach, broccoli, kale, Brussels sprouts and cabbage.

Why

Milk may do the body good, but spinach may do the body even better. Greens are actually packed with higher and more absorbable concentrations of calcium than dairy products. They also contain the phytochemicals lutein and zeaxanthin, which are important for vision. "And that's to speak nothing of the chlorophyll in greens, which is a great detoxifier. Kelp in particular is high in magnesium, an important nutrient in over 325 chemical activities in your body!" As a rule of thumb, the darker the green, the more chock-full of nutrients it is.

Whites

What to Eat

Cauliflower, potatoes, mushrooms, onions and garlic.

Why

When you feel the need to chill out, reach for an onion. The aromatic bulb, like many white-ish veggies, is rich in the compound allicin. This powerful antioxidant is known to combat high blood pressure and high LDL levels. Pale fruits and veggies are also packed with nutrients that are believed to stimulate your body's B and T cells, which in turn boost your overall immune system. And let's not forget about bananas and potatoes, which are high in heart-healthy potassium.

Blues and Purples

What to Eat

Beets, purple cabbage and eggplant.

Why

Once upon a time, blueberries were largely ignored by nutritionists because of the fruit's low level of vitamin C. Now, the same group of experts is tripping over itself to recommend that you eat 1 to 2 cups of the fruit every day. Why? "They're high in anthocyanins, which can reduce the risk of high blood pressure and improve heart health." Blue fruits and veggies are also high in fibre and packed with antioxidants, and have been shown to reduce the risk of some male cancers. What can't they do? They can't make you fat, since they're really low in calories too.

How Much Vegetables Should be in our Daily Diet?

Eat at least 5-7 servings of fresh vegetables every day. Seasonal vegetables should be encouraged. Include variety in vegetable type and colour in your diet. Yellow and orange colour vegetables are rich in Vitamin A, α and β carotenes, zea-xanthins and crypto-xanthins, where as dark green vegetables are very good source of minerals and phenolic flavonoid anti-oxidants.

Selection of Vegetables

Whenever possible, go for organic farm vegetables to get maximum health benefits. They are not very expensive if you can find them from the nearby local farm owners. Organic verities tend to have smaller in size but have rich flavour and feature good amounts of vitamins, minerals and stuffed with numerous health benefiting anti-oxidants.

- ✰ In the markets, however, always buy small quantities so that they should last within a day or two. There is no point in eating unfit greens!
- ✰ Buy that feature freshness, bright in colour and flavour and feel heavy in your hands.
- ✰ Look carefully for blemishes, spots, fungal mold and signs of insecticide spray. Buy whole vegetables instead of section of them (for example, pumpkin).

Health Benefits of Vegetables

Vegetables, like fruits, are low in fat but contain good amounts of vitamins and minerals. All the Green-Yellow-orange vegetables are rich sources of calcium, magnesium, potassium, iron, beta-carotene, vitamin B-complex, vitamin-C, vitamin A, and vitamin K.

As in fruits, vegetables are also home for many **antioxidants** that; firstly, help body protect from oxidant stress, diseases and cancers, and secondly, help body develop capacity to fight against these by boosting immunity.

In addition, vegetables contain soluble as well as insoluble dietary fibre known as non-starch polysaccharides (NSP) like cellulose, gums, mucilage, pectin etc., that absorb excess water in the colon and retain good amount of moisture in the fecal matter, thereby helps its easy passage out of the body. Thus helpful in conditions like hemorrhoids, chronic constipation, rectal fissures etc.

Vegetable nutrition has widely drawn the attention of fitness conscious as well as food scientists alike for their proven health benefits. Many vegetables are very low in calories, such as *celery* which is even as less than 10 cal per 100 g and here is the long list of vegetables whose calorie is less than 20 per 100 g: Bottle gourd, Bitter melon, Cabbage, Chinese cabbage, Bok choy, Eggplant, Endive, Spinach, Summer squash, Swiss chard, etc. Scientific studies have shown that low calorie but nutrient rich foods help body stay fit and disease free.

And, also our body spends considerable amount of energy during digestion of foods which is known as **BMR** or **Basal metabolism rate**. So just imagine.... when you add more vegetable nutrition in the diet, in fact you set to lose more weight than you would gain..Right!. This is the concept behind the **"negative calorie foods"**.

Best Muscle Building Fruits and Vegetables

Naturally available muscle building foods that will ensure that your time spent in the gym isn't a waste of sweat. They are Almonds, Banana, Pineapple, Papaya, olive oil, Quinoa, Spinach, Broccoli, Watercress, Turmeric, Cucumber and Sweet potato etc.

How to Use Vegetables?

First thing- you need to do immediately after shopping, wash them, especially green leafy vegetables. Rinse in salt water for few minutes and gently swish in cool water until you are satisfied with cleanliness. This way you ensure they are free from dust, sand and any residual chemical sprays. Use them early while fresh because firstly, certain vegetables have very short shelf life and secondly, the health benefiting properties of a vegetable declines with time. However, if you need to store them, then place in plastic wrappings or in zip pouches in order to preserve nutrition for short periods until you use them.

Chemical Constituents of Vegetables

Vegetable	*Parts*	*Chemical Compounds*
Asparagus	Sprouts	Lutein, Zeaxathin, Carotenes, Crypto-xanthins
Amaranth		α-spinastevol, Saponins.
Aloe vera	Root	Barbiloine (glucoside).
Artichoke	Bitter Principle	Cyanarin and Sesquiterpene, lactones (Cholesterol reduction action)
	Anti-oxidants	Silymarin, Caffeic acid and Ferulic acid
Arugula	Phyto-chemicals	Indoles, Thiocyanates, Sulforaphane and Isothiocyanates di-indolyl- methane (DIM) Anti-bacterial and Anti viral
Ajwain (seeds)	Aromatic fragrance	Thymol, pinene, cymene, limonene and terpinene
Beets (red)	Pigment	Betanin and Betacyanin
Beet (yellow)		β-xanthin
	Root	Glycine Betaine
Bell pepper	Capiscum	Capsaicin-Aggravates Gastro-esophageal reflux condition (GER).
Black pepper		Piperin, Piperidin, piperetine, Chavicine.
Bitter Gourd		Polypeptide-P (plant insulin known to lower blood sugar levels), Mamoridisine.
	Hypoglycemic agent	Charantin (increases glucose uptake and glycogen synthesis in liver cells etc.
	Seeds	Gourdin
Broccoli	Phyto-nutrients	Thiocyanates, indoles, sulforaphane, isothiocyanates
	Flavonoids	β-carotene, cryptoxanthin, lutein and zeaxanthin
	Flowers head	Omega-3-fatty acids
Brussels Sprouts	i. Flavonoids	Thiocyanates, zeaxanthin, sulforaphane and isothiocyanates
	ii. Anti-bacterial, viral	di-indolyl-methane (DIM)
	iii. Glucoside	Sinigrin (helps protect colon cancers by destroying pre-cancerous cells)
	iv. Carotenoid	Zeaxanthin
	v.	Goitrogens
Butternut squash	Flavonoids	α and β-carotens, cryptoxanthin-β and lutein
	Seed	Tryptophan (converts to health benefiting GABA neuro-chemical in the brain).

Contd...

Table–*Contd...*

Vegetable	*Parts*	*Chemical Compounds*
Basil	Flavonoids Oils	Orientin and Vicenin Eugenol*, citronellol, linalool, citral, limonene, terpineol, Zeaxanthin**
Brinjal	Deep blue/purple	Anthocyanins
Borage	Fatty acid	Gamma-linolenic acid (GLA)***
Burdock	Rot	Inulin, glucoside-lappin, mucilage
Black pepper	Pungent	Peperine#, sabinene, pinene, terpenene, limonene, mercene
Cabbage	Thiocyanates	Indol-3-carbinol, lutein, sulforaphane, isothiocyanate, zeaxanthin, goitrogen
Cauliflower	Phyto-chemical	Indole-3-carbinol##, di-indolyl-methane, goitrogen
Carrot	Root	Falcarinol (destroys pre-cancerous cells in the tumers
Celery	Leaves and seeds	Limonene (75-80 per cent), sesquiterpenes like β- seliene (10 per cent) and humulene
Chives	Thio-sulfides	Diallyl disulfide, diallyl trisulfide and allyal propyl disulfide+
Coriander	Leaves and seeds Leaves/stem tips Seeds	Borneol, linalool, cineole, cymene, terpineol, dipentene, phellandrene, pipene and terpinolene, Coriandol Quercetin, kaempferol, rhamnctin and epigenin Petroselinic acid, linoleic acid, oleic acid, palmitic acid, linalool, α-pinene, geraniol, camphene and terpine
Caraway	Seeds	Carvone, limonene, carveol, pinen, cumuninic-aldehyde, fururol and thujone
Cardamom	Seeds	Pinene, sabinene, myrcene, phellandrene, limonene,1,8-cineole, p-cymene, linalool, ciniol
Cloves	Essential oils	Euginol, acetyl-eugenol, β-caryophyllene, vanillin, crategolic acid, eugenin, kaempferol, eugenitin, oleanolic acid, stigmasterol, campesterol, Emitil, sesquiterpenes
Cumin		Cuminaldehyde (4-isopropylbenzaldehyde), pyrazines, 2-methoxy-3-sec-butyl-pyrazine, 2-ethoxy-3-isopropyl pyrazine and 2-methoxy-3-methyl pyrazine. Thymoquinone-checks proliferaton of cells responsible for prostate cancer.
Cinamon		Eugenol@, cinnamate, linalool, cinnamaldehyde, β-caryophyllene, methyl chavicol
Collards green	Leaves	Di-indolyl-methane (DIM) and Sulforaphane
Cucumber		β-carotene and α-carotene, Zeaxanthin and lutein
Capers	Flower buds	Rutin++(rutoside), Quercetin
Dandelion	Roots-Bitter crystalline compound	Taraxain, Taraxacerin (acrid resin), Inulin and Levulin (therapeutic properties)
Endive		β-carotene, folic acid, bitter glycosides and inulin
Fennel	oils	Anethole!, estragole and fenochone (fenchyl acetate)
Fenugreek	Seeds	4-hydroxy-isoleucine!!

Contd...

Table–*Contd...*

Vegetable	*Parts*	*Chemical Compounds*
Green beans	Phenolic compounds	Zeaxanthin, lutein, β-carotene, folates, Oxalic acid
Garlic	Thio-sulfinites	Diallyl-disulfide, diallyl-trisulfide, allyl propyl disulfide+++
Ginger	Roots	Gingerol^, Zingerone^^, shogaol, farnesene, Gingerine (bitterness).
Horse radish	Pungency	Allyl isothiocyanate, 3-butenyl isothiocyanate, 2-propenyl glucosinlate (sinigrin), 2-pentyl isothiocyanate and phenyl ethyl isothiocyanate
Kale	Leaves	Sulforaphane, indole-3-carbinol, di- indolyl- methane (DIM).
Knol Khol		Isothiocyanates, sulforaphane, indole-3-carbinol, Goitrogens
Leeks		Diallyl disulfide, diallyl trisulfide and allyal propyl disulfide
Lettuce	Leaves	β-carotene, zeaxanthin, folate
Mustard	i. Seed	i. Brassicasterol, campesterol, sitosterol, avenasterol, stigmasterol also sinigrin, myrocin, erucic, eicosenoic, oleic, palmitic acid, Sinigrin, Sinapin
	ii. Green Oil	ii. Indoles, sulforaphane, di-indolyl-methane
Mint		Menthol, menthone, menthol acetate (analgesic).
Okra	Pod	Β-carotenes, xanthin, lutein
Onion		Bulb (Allyl sulfide gas) Allium, allyl-disulfide conver to Allicin, Quercetin, pyridoxine, chromium
Oregano		Phyto-chemical quercetin which restricts growth of malignant cells.
Parsley	Oil	Eugenol
	Antioxidants	Apiin, apigenin, crisoeriol and luteolin zea-xanthin~
Purslane		Omega-3-fatty acid
	Alkaloids pigment	Betalain-β-cyanins (reddish), β-xanthins (yellow)
Pumpkin	Fruit	α and β carotenes, cryptoxanthin, lutein, zeaxanthin
	Seed	Tryptophan (health promoting amino acid) selenium, zinc
Peas	Pods (phytosterols)	β-sitosterol, carotenes, lutein, zeaxanthin
Potato	Toxic alkaloid?	Solanine and Chaconine
Parsnip		Falcocarinol, falcarindiol, panaxydiol and methy-falcarindol
Radish		Goitrogens"
Rhubarb	Leaf blade	Oxalates
Rosemary	Flower tops	Rosmarinic acid
	Volatile oils	Cineol, camphene, borneol, bornyl acetate, α- pipene.+*
Saffron	Flavour	Picrocrocin and Safranal
	Colour	Crocin (dicarboxylic acid) gives saffron its golden yellow hue.
Shallot		Diallyl-disulfide, diallyl-trisulfide, allyl propyl disulfide
Sage		Thujone****, Salvigenin

Contd...

Table–*Contd...*

Vegetable	*Parts*	*Chemical Compounds*
Spinach$		Omega-3-fatty acid, lutein, zeaxanthin, β-carotene, oxalic acid.
i. Savory ii. Sweet corn iii. Sweet potato iv. Swiss chard		i. Thymol and carvacrol<, linalool, camphene, caryophyllene, terpineol, myrcene ii. Ferulic acid iii. oxalic acid, β-carotene iv. Omega -3- fatty acid.
Thyme	Oil Antioxidants	Thymol Zeaxanthin, lutein, pigenin, naringenin, luteolin, thymonin
Turmeric	Pigment	**Curcumin-** Anti tumor, i arthritic, i amyloid, i septic and anti inflammatory properties
Tomato	Fruit	Lycopene-A flavonoid anti oxidant
Turnip	Roots/tops	Oxalic acid
Tarragon		Estragole (methyl chavicol), cineol, ocimene, phellandrene
Water cress	Stems (Pappery flavour)	**Gluconasturtiin-**On hydrolysis provide 2- phenethyl isothiocyanate (PEITC)
Watermelon		Is believed to be cancer preventive inhibition of phase 1 enzymes. Citrulline.
Yam		Vit. B6, Thiamin, riboflavin, folic acid, pantothemic acid, niacin

*: Anti–inflammatory function.

**: Yellow flavonoid carotenoid compound is selectively absorbed into the retinal maculaluteua where it found to filter harmful UV rays from reaching retina.

***: linolenic acid is Omega-6- fatty acid that plays vital role in restoring of joint health, immunity, healthy skin and mucus membranes.

#: Amine alkaloid- which gives strong spicy pungent character to pepper.

##: Acts as a anti-estrogent agent

+: Convert to allicin by enzymatic reaction when its leaves are disrupt (crushing, cutting.

++: Strengthens capillaries and inhibits platelet clump formation in the blood vessels. Both thes actions of rutin help in smooth circulation of blood in very small vessels.

@: A phenylpropanoids class of chemical compound, which gives pleasant, sweet aromatic fragrance.

!: Anethole- antifungal and anti bacterial properties.

!!: Present in the fenugreek seed, has faciliator action on inulin secretion.

+++: That can form allicin by enzymatic reaction, which is activated by disruption of bulb.- Allicin-(volatile compound)has anti bacterial, anti viral and anti fungal activities.

^: Helps improve the intestinal motility and has anti –nflammatory, painkiller (analgesic), nerve soothing, anti –pyretic as well as anti bacterial properties.

^^: Zingerone- Which gives pungent character to the ginger root, has been found to effective against E. coli induced diarrhea, especially in children.

": May cause swelling of the thyroid gland. Should be avoided in individuals with thyroid dysfunction.

Contd...

Table–*Contd...*

?: These toxic substances are destroyed when cooked at high temperature (over 170 ºc).

~: Helps prevent age related macular- degeneration (ARMD) in the retina of the eye in the old age population, through its anti-oxidant and ultra- violet light filtering functions.

+*: Are known to have rubefacient, anti inflammation, anti-allergic, anti- fungal and anti septic properties.

****: It enhances concentration, attention, span and quickens the senses. It helps to deal with grief and depression.

,: Inhibits the growth of several bacteria strains like E. Coli and Bacillus cereus.

$: Reheating of spinach left over may cause conversion of nitrates into nitrites and nitrosamines by certain bacteria that thrive on pre-preparad nitrate rich foods such as spinach and many other green vegetables. These poisonous compounds may be harmful to health, especially to children.

Oxalic acid: A naturaly occuring substance found in some vegetables which may crystallize as oxalate stones in the urinary tract in some people.

Asparagus (*Asparagus officinalis*)

Asparagus officinalis is widely consumed worldwide and has long been used as an herbal medicine due to its anticancer effects. An excellent food for pregnant women, asparagus is full of fibre, folate and Vitamin B6 qualities which help in maximizing heart's health. These long veggie sticks are also pretty low in sodium and high in potassium content, which makes them ideal for the healthy working of our large intestine. The wonder vegetable has very high levels of vitamin E, which plays a major role in skin cell restoration. It is rich in folic acid and contains magnesium that are famed for their mood-lightening properties. Baby asparagus stalks are tender, crunchy and tasty all the same time making them the ferfect for any time snack. Consumption of asparagus helps make your skin more firm and clears blemishes on it. Its shoots contain a white crystalline substance, **asparagines** (an amino acid) which is used in the medicine as diuretic in cardiac dropsy and chronic gout. Among vegetables, asparagus is the leading supplier of **folic acid** for human nutrition, which is necessary for blood cell formation, growth and prevention of liver disease. Folic acid plays a significant role in the prevention of neutral tube defects, such as spina bifida, that cause paralysis and death in infants. Asparagus is low in calories, contains no fat or cholesterol, is very low in sodium and is a good source of potassium, fibre, thiamin, B6 and one of the richest source of rutin, a chemical that strengthens capillary walls(Michigan Asparagus Council, 2005), making asparagus a wise choice for health- conscious consumers. Asparagus rhizomes and root is used ethnomedically to treat urinary tract infections as well as kidney and bladder stones. It has also aphrodisiac properties (this belief is at least partially due to the phallic shape of the shoot). Ingestion of asparagus may bring on an attack of gout in certain individuals, due to the high level of purines. According to modern physicists it inhibits stimulation of uterus and normalizes its functioning. The plant is also a diuretic and can aid the elimination of water through urination, a special boon in cases of water retension (oedema). Asparagus can be usaed to treat many kidney conditions, but it should not be used in cases of inflammations. It is also helpful in cleansing cholesterol from arteries and hence is useful in treating vascular problems such as hypertension and arteriosclerosis. Asparagus is high in the B vitamin folate that helps increase your production of histamine. Histamine is important for increasing the sex drive in men and women.

According to **Bhawa Prakash** it is heavy, cool and bitter and nutritive. It enhances intelligence, promotes digestion, pacifies vata and pitta disorders and cures reduced semen production. It also cures reduction of breast milk. According to **Acharya Sushart** asparagus is beneficial in dryness and promotes physical strength. It purifies semen, enhances intelligence and promotes digestive fire. While according to **Dhanvantari nighantu**, Shatawar can cure even a very weak person. It promotes physical strength in the patients and enables the patient to fight the disease. It enhances the immune system and suppresses vata pitta, produces semen, is cool and sweet. It is a semen enhancing herb, Shatavari, sariva, arand, punnarva and hansraj all these suppress pitta and vayu and cures respiratory disorders.

According to **Acharya Charak**, it is beneficial for physical strength and arresting aging (Acharya Balkrishna, 2008).

Health Benefits

- ☆ Asparagus's biggest talent is its ability to encourage the body to flush out toxins, due to its natural diuretic abilities. Asparagus is both cleansing and anti-inflammatory to the body. It has the antioxidant glutathione, which can lower your risk factor for heart disease and cancer. It's useful for all inflammatory conditions, such as arthritis and irritable bowel syndrome.
- ☆ Asparagus is a very low calorie vegetable. 100 g fresh spears give only 20 calories. More calories will be burnt to digest than gained, the fact, which fits in to the category of low calorie or negative calorie vegetables.
- ☆ In addition, the shoots have good levels of dietary fibre. Dietary fibre helps control constipation conditions, decrease bad, "LDL" cholesterol levels by binding to it in the intestines, and regulates blood sugar levels. In addition, high fibre diet helps prevent colon-rectal cancer risks by preventing toxic compounds in the food from absorption.
- ☆ Its shoots have long been used in many traditional medicines to treat conditions like dropsy and irritable bowel syndrome.
- ☆ Fresh asparagus spears are good source of anti-oxidants such as *lutein, zeaxanthin, carotenes*, and *crypto-xanthins*. Together, these flavonoid compounds help remove harmful oxidant free radicals from the body protect it from possible cancer, neuro-degenerative diseases, and viral infections.

- ☆ Fresh asparagus are rich in folates. 100 g of spears provide about 54 mcg or 14 per cent of RDA of folic acid. Folates are one of the important co-factors for DNA synthesis inside the cell. Scientific studies have shown that adequate consumption of folates in the diet during pre-conception period and during early pregnancy help prevent neural tube defects in the newborn baby.
- ☆ The shoots are also rich in B-complex group of vitamins such as thiamin, riboflavin, niacin, vitamin B-6 (pyridoxine), and pantothenic acid those are essential for optimum cellular enzymatic and metabolic functions.
- ☆ Fresh asparagus also contains fair amounts of anti-oxidant vitamins such as vitamin-C, vitamin-A and vitamin-E. Regular consumption of foods rich in these vitamins helps body develop resistance against infectious agents and scavenge harmful, pro-inflammatory free radicals from the body.
- ☆ Its shoots are also good source of vitamin K; provides about 35 per cent of DRI. Vitamin K has potential role bone health by promoting osteotrophic (bone formation) activity. Adequate vitamin-K levels in the diet helps limiting neuronal damage in the brain; thus, has established role in the treatment of patients suffering from Alzheimer's disease.
- ☆ Asparagus is good in minerals especially copper and iron. In addition, it has small amounts of some other essential minerals and electrolytes such as calcium, potassium, manganese, and phosphorus. Potassium is an important component of cell and body fluids that helps controlling heart rate and blood pressure by countering effects of sodium. Manganese is used by the body as a co-factor for the antioxidant enzyme, *superoxide dismutase*. Copper is required in the production of red blood cells. Iron is required for cellular respiration and red blood cell formation.
- ☆ Asparagus is high in glutathione, an important anticarcinogen.
- ☆ It also contains rutin, which protects small blood vessels from rupturing and may protect against radiation.
- ☆ Asparagus is a good source of vitamins A, C and E, B-complex vitamins, potassium and zinc.
- ☆ It also has antifungal, anti-inflammatory and diuretic properties.
- ☆ The biological functions of asparagus can help alleviate alcohol hangover and protect liver cells (Anonymous, 2012).

Caution

In general asparagus is well tolerated in general population and allergic reactions are quite rare to occur. Ingestion of young shoots may give offensive smell to urine. This is due to the metabolism of *asparagusic acid*, which breaks down into various sulphur-containing degradation products such as methanethiol, sulfides etc. The condition, however, is harmless.

Tuber Crops

Tuber crops species like Dioscorea, Alocasia, Sweet potato, Elephant foot yam and Coleus are known to contain medicinal properties. Betacarotene and atocopherol(vitamin E) are known to have anti- cancer and anti-oxidant properties. Sweet potato varieties with high content of these principles may be developed to enhance their pharmaceutical potential. Many yam and aroid tubers contain high levels of mucilage. Tuber rich in mucilage shall be exploited for extraction of mucilage for industrial as well as medicinal application. Researchers have found that white potatoes are the largest and most affordable source of potassium of any vegetable or fruit.

POTATO (*Solanum tuberosum*)

Research has shown that it is the peel that antioxidants are contained. These antioxidants can help revitalize body cells as well as fight infection. This potato fibre has also parts that lower cholesterol levels. Also the layer of potato lying just below the skin is very rich in protein and mineral salts. There is a "Saying that "cook the peel to eat and use the rest to feed the chicken." Potato fibre has also parts that lower cholesterol level. Potato is perhaps the most alkaline vegetable. It is rich in soda and potash but low in lime. Therefore, it is helpful in maintaining the alkaline reserve of body and is natural antidote for an overdose of acidic foods. They are a highly starchy food and when eaten the with skin or peel a good source of vegetable protein, potassium, vitamin C, iron, phosphorus, niacin and enzymes - although old potatoes are low in vitamin C.

Medicinally, potatoes can help to relieve arthritis and reduce water retension due to their high potassium content. They also neutralize body acids and potato juice can be used to treat stomach and duodenal ulcers. The fresh juice is sometimes used to reduce hypertension and promote intestinal flora.

Nutritionally, potatoes are best known for their carbohydrate content (approximately 26 grams in a medium potato). The predominant form of this carbohydrate is **starch.** A small but significant portion of this starch is resistant to digestion by enzymes in the stomach and small intestine and so reaches the large intestine essentially intact. This resistant starch is considered to have similar physiological effects and health benefits as fibre; it provide bulk, offers protection against colon cancer, improves glucose tolerance and insulin sensitivity, lowers plasma cholesterol and trigylyceride concentrations, increases satiety, and possibly even reduces fat storage (Cummings *et al.*, 1996). The amount of resistant starch in potatoes depends much on preparation methods. Cooking and then cooling potatoes significantly increases resistant starch. For example, cooked potato starch contains about 7 per cent resistant starch, which increases to about 13 per cent upon cooling (Englyst *et al.*, 1992). Individuals who consumed potatoes (baked, boiled and roasted) had higher intakes of potassium and vitamin C and consumed more total vegetables in a day compared to those who did not consume potatoes.

Potatoes contain vitamins and minerals which are vital to human nutrition. Potatoes consist of about 80 per cent water, 20 per cent solid matter and have a high nutritional value. Starch makes up about 85 per cent of this solid mass and the rest is protein. As humans can subsist healthily on a diet of potato and milk, the later supplies vitamin A and vitamin D. A normal sized potato (150g) with skin intact provides 27 mg of vitamin C(45 per cent of the Daily Value DV)), 620mg of potassium (18 per cent of DV), 0.2mg vitamin B6(10 per cent DV) and trace amounts of thiamine, riboflavin, folate, niacin, magnesium, phosphorus, iron and zinc. Potassium is essential to the body because of its role in attaining optimal muscle performance and improving the nerve's response to stimulation. Iron is essential in helping the body convert food to energy as well as resist infection, is also present. Potato contains no fat or cholesterol and mineral sodium. What they do have is natural fibre in the skins, all those vitamins and minerals and great flavour. And 170grams of potato contains 2 grams of highly digestible protein, almost as much as half a glass of milk, making it a great foundation for a whole meal.

Moreover, the fibre content of a potato with skin (2grams) equals that of many whole grain breads, pastas and cereals. Potatoes also contain phytochemicals such as carotenoids and polyphenols. The notion that "all of the potato`s nutrients" are found in the skin is a misnomer. While skin does contain approximately half of total dietary fibre, more than 50 per cent of the nutrients are found within the potato itself. The cooking method used can impact significantly the nutrient availability of the potato. Almost all the protein content of a potato is contained in a thin layer just under its skin. This is evident when the skin of a boiled potato is carefully peeled, it appears as a yellowish film. For maximum utilization of this small, but available dietary source of protein, potatoes should be consumed whole or peeled or peeled after cooking.

One medium sized potato with skin contains 620mg of potassium, followed by one medium stalk of broccoli at 540mg. Foods that are a good source of potassium and low in sodium- such as potatoes- may reduce the risk of high blood pressure and stroke. Potatoes are also an excellent source of vitamin C and bring important B vitamins and fibre. A 5.3 ounce potato has a mere 100 calories and contains no fat, cholesterol or sodium.

Potato	1 medium	620mg
Broccoli	1medium stalk	540mg
Tomato (Plum roma)	1 medium	360mg
Tomato (other)	1 medium	360mg
Celery	2medium stalks	350mg
Cantaloupe	1\4medium	280mg
Sweet pepper	1 medium	270mg
Onion	1 medium	240mg
Sweet corn	1 medium ear	240mg
Water melon	2 coups (diced)	230mg
Cucumber	1\3 medium	170 mg
Lettuce (Iceberg)	1\6 medium head	120mg

Carbo Facts

One of the most popular nutritional myths today concerns the dietary evils of carbohydrates.

Despite the media hype, scientific evidence indicate that when consumed in normal amounts.

- ☆ Carbohydrates DO NOT cause weight gain and obesity.
- ☆ Carbohydrates DO NOT automatically cause blood sugar "spikes" and insulin " surges".
- ☆ Carbohydrates DO NOT cause diabetes, heart disease or cancer.

Despite the popular notion, the majority of nutrients are not found in the skin, but in the potato it self. Nonetheless, leaving the skin on the potato retains all the nutrients, the fibre in the skin and makes potato easier to prepare.

- ☆ **Potatoes and blood sugar level:** Unlike the general perception of high glycemic index food, potatoes have a medium glycemic index and they do not raise the blood sugar levels to crazy. Their glycemic index is a decent 56. However, the forms in which they are consumed make them fattening such as deep fried, shallow fried and so on.
- ☆ **Potatoes and recommended amount**: The moderate amount recommended is 1 medium-sized potato a day.

- **Potatoes and calories:** 1 large potato (260gms) will have 278 calories and 63 grams *i. e.* 21 per cent of carbohydrate. Calories from potatoes result in weight gain only if taken regularly in large amounts. Also, as stated earlier, the form in which they are consumed matters. The best preparations are baked and steamed.
- **Potatoes and top health benefits**: Potential health benefits of potatoes are: Weight gain for the lean: They could result in weight gain as most calories come from a high percentage of carbohydrates in potatoes.
- **Facilitates digestion**: Potatoes contain simple carbohydrates that are easy to digest and thus facilitate digestion. They are a source of instant energy, but they should be consumed only in moderate quantities to avoid other health complications.
- **Helps skin care**: Potatoes contain vitamin-C and B-complex and minerals like potassium, magnesium and zinc which are good for the skin.
- **Prevents scurvy**:The vitamin-C present in potatoes can help prevent scurvy which is caused due to lack of vitamin-C.
- **Relieves inflammation**:Potato is a simple carbohydrate especially in steamed and mashed form and is easily digested and absorbed through our digestive system. It relieves inflammation of the intestines and the digestive system.
- **Keeps brain function active**: Carbohydrates are a major and easiest form of glucose and the good news is that potatoes are high in carbohydrates. By providing instant calories, potatoes maintain instant replenished levels of glucose in the blood which keeps the brain active instantly.
- **Heart disease:** Usually, potatoes should not be given during treatment and immediately after treatment of heart disease. This is because, usually, heart diseases are related with obesity and this would be a time to eat a lean diet devoid of simple carbs. Although, as per your cardiologist's recommendation, potatoes can be taken in a healthy balanced manner.

Health Benefits

- Potatoes are one of the richest sources of starch, vitamins, minerals and dietary fibre. Contains very low fat (just 0.1 g per100 g) and no cholesterol.
- Both soluble and insoluble fibre in them increases the bulk of the stool, thus, it helps prevent constipation, decrease absorption of dietary cholesterol and there by lower plasma LDL cholesterol. Additionally, the rich fibre content also helps protect from colon cancer.
- The fibre content helps slow absorption of starch in the gut and thereby keeping blood sugar levels within normal range. For the same reason, potato is still favoured source of carbohydrates in diabetics.
- The tubers are one of the richest sources of B-complex group of vitamins such as Vitamin B6, Niacin, Pantothenic acid and folates.

- ✰ Fresh potato skin as well as flesh are good source of antioxidant vitamin; vitamin-C. Regular consumption of foods rich in vitamin-C helps body develop resistance against infectious agents and scavenge harmful, pro-inflammatory free radicals.
- ✰ They also contain adequate amounts of many essential minerals like Iron, manganese, copper and potassium.
- ✰ Red and *russet potatoes* contain good amount vitamin A, and flavonoids like carotenes and zeaxanthins.
- ✰ Recent studies at Agricultural research service (by plant genetics scientist Roy Navarre) suggests that flavonoid antioxidant, *quercetin* present in potatoes has anti-cancer and cardio-protective properties.
- ✰ Eat potatoes as they resist the accumulation of calcium in the kidney, thus preventing the formation of kidney stones.

Caution

Potatoes may contain toxic alkaloids, *solanine* and *chaconine.* These alkaloids present in the greatest concentrations just underneath the skin and increase proportionately with age and exposure to sun light. Cooking at high temperatures (over 170 °C) partly destroys these toxic substances. When consumed in sufficient amounts, these compounds may cause headache, weakness, muscle cramps and, in severe cases loss of consciousness and coma; however, poisoning from potatoes occurs very rarely. Exposure to light also causes green discolouration; thus giving a visual clue as areas of the tuber that may have more toxins; however, this does not provide a definitive clue, as greening and solanine accumulation can occur independently to each other. Some varieties contain greater solanine concentrations than others.

SWEET POTATO (*Ipomoea batatas)*

Sweet potato leaves (top greens) are also edible; in fact, the greens contain more nutrients and dietary fibre than some green leafy vegetables like spinach (for example, 100 g sweet potato leaves provide 1028 IU of vitamin A). Grown under earth, sweet potatoes are full of ant-defending qualities as they are rich in manganese, Vitamin A and Vitamin C. Good for our digestive system, they are full of iron and fibre which helps in giving us good amounts of energy. Outstandingly rich in vitamin A (many times more so than potatoes), sweet potatoes also provide vitamin C, calcium, iron, potassium and sodium, but they must be baked or boiled in their skin to retain these nutrients. Due to their exceptional vitamin A content, sweet potatoes can be used to improve night vision.

They are also reported to increase milk secretion in nursing mothers, remove toxins from the body and treat underweight and diarrhoea.

It is a rich source of **lutein**. Its use as a leafy vegetable is novel in most countries and potentially may serve a market niche given its enhanced nutritional status. Lutein is a carotenoid capable of delaying blindness-related muscular degeneration. The content of lutein in sweet potato ranged from 0.38 – 0.58mg. g^{-1} fresh weight. Lutein is not synthesized donovo and must be ingested from outside sources. Dark green vegetables have been identified as major edible source of lutein, by decreasing order, are kale, spinach, lettuce and broccoli (Johnson, 2002). Sweet potatoes are an excellent source of carotenoid antioxidants. They contain calcium, are high in vitamins A and C and contain thiamine.

Sweet potatoes have a slightly lower carbohydrate content than regular potatoes. One medium potato contains 26 grams of carbohydrates, while one medium sweet potato contains 23 grams of carbohydrates.

Sweet potatoes contain almost twice as much fibre as other types of potatoes. Contributing close to 7 grams of fibre per serving. The high fibre content gives them a 'slow burning quality'. Which means their caloric energy is used more slowly and efficiently than a low-fibre carbohydrate.

They are an excellent source of vitamin A and a good source of potassium and vitamin C, B6, riboflavin, copper, pantothetic acid and folic acid. A medium sweet potato has 28 per cent more potassium than a banana. The high amount of water that is found in sweet potatoes helps you lose weight. Like fibre, water takes up a lot of room in your stomach. Therefore, eating foods that contain high amounts of water makes you feel full and will prevent overeating and snacking between meals.

Health Benefits of Sweet Potato

Best Muscle Building Foods

- ☆ Sweet potatoes are an excellent source of vitamin A and a good source of potassium and vitamin C, B6, riboflavin, copper, pantothentic acid and folic acid. A medium sweet potato has 28 per cent more potassium than a banana. They replenish energy stores and fuel the muscle-building process (Moghul,2013).
- ☆ Sweet potatoes are far superior than the run-of-the-mill white potato.The orange variety contains beta carotene, which makes them filled with robust antioxidant, antiviral, and anticancer abilities.They're also full of fibre and the vitamin E they contain is healthy for the skin.

They are High in Vitamin B6

- ☆ Vitamin B6 helps reduce the chemical **homocysteine** in our bodies. Homocysteine has been linked with degenerative diseases, including the prevention of heart attacks.

They are a Good Source of Vitamin C

- ☆ While most people know that vitamin C is important to help ward off cold and flu viruses, few people are aware that this crucial vitamin plays an important role in bone and tooth formation, digestion, and blood cell formation. It helps accelerate wound healing, produces collagen which helps maintain skin's youthful elasticity, and is essential to helping us cope with stress. It even appears to help protect our body against toxins that may be linked to cancer.

They Contain Vitamin D

- ☆ Vitamin D, which is critical for immune system and overall health at this time of year. Both a vitamin and a hormone, vitamin D is primarily made in our bodies as a result of getting adequate sunlight. You may have heard about seasonal affective disorder (or SAD, as it is also called), which is linked to inadequate sunlight and therefore a vitamin D deficiency. Vitamin D plays an important role in our energy levels, moods, and helps to build healthy bones, heart, nerves, skin, and teeth, and it supports the thyroid gland.

Sweet Potatoes contain Iron

- ☆ Most people are aware that we need the mineral iron to have adequate energy, but iron plays other important roles in our body, including red and white blood cell production, resistance to stress, proper immune functioning, and the metabolizing of protein, among other things.
- ☆ Sweet potatoes are a good source of magnesium, which is the relaxation and anti-stress mineral. Magnesium is necessary for healthy artery, blood, bone, heart, muscle, and nerve function, yet experts estimate that approximately 80 per cent of the population in North America may be deficient in this important mineral.
- ☆ Nutritious sweet potatoes are low in calories (provide just 90 cal/100 g, on comparison with starch rich cereals) and contains no saturated fats and cholesterol; but are rich source of dietary fibre, anti-oxidants, vitamins and minerals.
- ☆ They are storehouse of starch, a complex carbohydrate, which raises the blood sugar levels *slowly* on comparison to simple sugars; therefore, recommended as a healthy food supplement even in diabetes.
- ☆ They are excellent source of *flavonoids* like beta-carotene and vitamin A. The value is one of the highest among root vegetables category. These compounds are powerful natural antioxidants. Vitamin A is also required by the body to maintain integrity of healthy mucus membranes and skin. It is also vital nutrient for vision. Consumption of natural vegetables and fruits rich in flavonoids helps to protect from lung and oral cavity cancers.
- ☆ The tubers are packed with many essential vitamins such as pantothenic acid (vitamin B5), pyridoxine (vitamin B-6) and thiamin (vitamin B-1), niacin, and riboflavin. These vitamins are essential in the sense that body requires them from external sources to replenish. These vitamins function as co-factors for various enzymes during metabolism.

- ☆ They also contain good amounts of minerals like iron, calcium, magnesium, manganese, and potassium that are very essential for body metabolism.
- ☆ Be careful: eating too many may cause abdominal swelling and indigestion.
- ☆ Sweet potatoes are also high in sugar and therefore should be used sparingly.
- ☆ Sweet potatoes are not related to the potato nor the yam—they are actually a member of the morning.
- ☆ Sweet potato leaves (top greens) are also edible. In fact, the greens contain more nutrients and dietary fibre than some of the green-leafy vegetables.

Caution

Sweet potatoes contain *oxalic acid,* a naturally occurring substance found in some vegetables that may crystallize as oxalate stones in the urinary tract in some people. It is, therefore, individuals with known history of oxalate urinary tract stones may have to avoid eating them. Adequate intake of water is therefore advised to maintain normal urine output in these individuals to minimize stone risk. Since sweet potatoes can cause indigestion and abdominal swelling, they should be eaten in moderate servings.

BEET (*Beta vulgaris*)

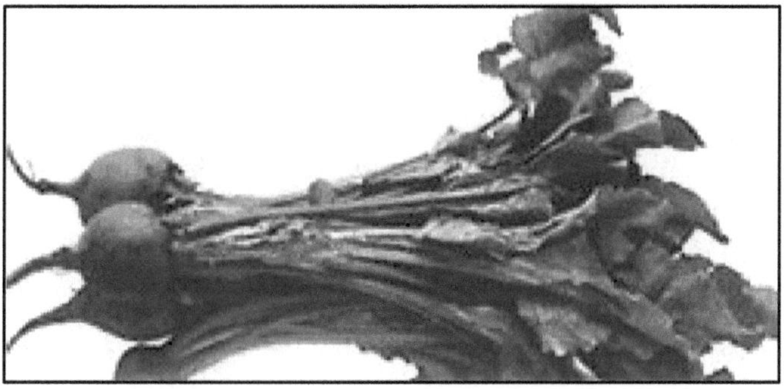

Beets are highly nutritious and "cardiovascular health" friendly root vegetables. Certain unique pigment antioxidants present in root as well as top greens have found to offer protection against coronary artery disease and stroke, lower cholesterol levels in the body and have anti-aging effects. This colourful tuber is a great source of folate also known as vitamin B9. This vitamin promotes functioning of the liver and the nervous system, which usually gets congested under chronic stres

Romans used beet root as a treatment for fevers and constipations, amongst other ailments, considered its juice as an aphrodisiac. The root is a rich source of natural sugar. It also contains potassium, calcium, chlorine, iodine, iron, copper and a lot of vitamin A and B. The iron content in the beet regenerates and reactivates the red blood corpuscles. This blood forming quality helps in cases of anaemia. Although it is normally the root part of the plant that is eaten, in fact its leaves are a very good source of beta-carotene. Its high sodium and potassium content acts as a good solvent for excess calcium in the body thus, preventing deposits of calcium forming in the joints. This complaint is common as the process of ageing begins. The root is also valuable in the treatment of hypertension and heart trouble. The glamorous root performs certain mundane functions too. It relieves the habitually constipated ones, cleanses the liver particularly during jaundice. Being alkaline in nature, the root is useful in combating acidosis. For those who have jumpy nerves, beat it with beet. It is so soothing that schools of thought believe that beet roots actually calm the nerves, particularly at menopause or even during pre-menstrual tensions. The red colour in beet is decided by slightly water soluble xanthine pigments called **betalains**. High ratio of betacyanin to betaxanthine leads to violet, medium to red and orange tuber colours and these ratios are controlled by a triallelic system at the R Locus in the chromosome with incomplete dominance. Beet root is excellent for those in their late middle age. **Betanins-** obtained from the roots are used industrially as red food colourants, *e. g.* to improve the colour of tomato paste, sauces, desserts, jams and jellies, ice creams and break fast cereals. So make beetroot available for a large part of the year a regular part of one's meal. Researchers from the William Harvey Research Institute at Barts and the London School of Medicine have discovered that drinking just 500 ml of beetroot juice a day can significantly reduce BP in one hour. This decrease in BP resulted from the chemical formation of **nitrite** from the dietary nitrate in the juice. However, the human body in itself is incapable of changing nitrate into nitrite. It is the bacteria on the tongue or the saliva that converts nitrate in the juice into nitrite. This nitrite- containing saliva is swallowed and in the acid environment of the stomach is either converted into nitric oxide or re-enters the circulation as nitrite(Sinha, 2008). Beet juice cleanses the blood and strengthens the gall bladder and liver. It is a rich source of the mineral boron, which plays an important role in the production of human sex harmones. Today the beetroot is still championed as a universal panacea. Beetroot not only provides a multitude of health benefits, the vibrant vegetable is also a great source of longlasting energy. First of all, beetroot has a high sugar content which helps to provide an instant energy boost. It is also high in many energy-boosting nutrients including magnesium, iron, vitamin C and nitrate.

Health Benefits

- ☆ Garden beet is very low in calories (contain only 45 kcal/100 g), and contain only small amount of fat. Its nutrition benefits come particularly from fibre, vitamins, minerals, and unique plant derived anti-oxidants.
- ☆ The root is rich source of phytochemical compound ***Glycine betaine.*** Betaine has the property of lowering *homocysteine* levels in the blood. Homocysteine, one of highly toxic metabolite, promotes platelet clot as well as atherosclerotic-plaque formation, which is otherwise can be harmful to blood vessels. High levels of homocystiene in the blood results in the development of coronary heart disease (CHD), stroke and peripheral vascular diseases.
- ☆ Raw beets are an excellent source of **folates**; contains about 109 mcg/100 g (**Provides 27 per cent of RDA**). However, extensive cooking may significantly depletes its level in food. Folates are necessary for DNA

synthesis in the cells. When given during peri-conception period folates can prevent neural tube defects in the baby.

- ☆ It contains significant amounts of **vitamin-C**, one of the powerful natural antioxidant, which helps body scavenge deleterious free radicals one of the reasons for cancers development.
- ☆ Beet greens (tops) are an excellent source of carotenoids, flavonoid anti-oxidants, and vitamin A; contain these compounds several times more than that of in the roots. Vitamin A is required maintaining healthy mucus membranes and skin and is essential for vision. Consumption of natural vegetables rich in flavonoids helps to protect from lung and oral cavity cancers.
- ☆ The root is also rich source of niacin (vit. B-3), pantothenic acid (vit. B-5), pyridoxine (vit. B-6) and carotenoids, and minerals such as iron, manganese, and magnesium.
- ☆ In addition, the root indeed has very good levels of potassium. 100 g fresh root has 325 mg of potassium or 7 per cent of daily requirements. Potassium lowers heart rate and regulates metabolism inside the cells by countering detrimental effects of sodium.
- ☆ It is used for purifying the blood.
- ☆ Improves circulation.
- ☆ Promotes menstruation.
- ☆ Stimulates the bowels.
- ☆ It helps calm nervousness and is of benefit in vascular congestion.
- ☆ Recommended for intestinal cleansing of parasites

Beet Greens/Root

- ☆ Beet greens contain notable amounts of calcium, iron, magnesium and phosphorus
- ☆ They also contain vitamins A, B-complex and C
- ☆ Beet roots are high in carbohydrate levels and should therefore be used sparingly

Caution

Beeturia is a harmless condition of passing red or pink colour urine after eating beets and its top greens. The condition can be found in around 10-15 per cent of the population who are genetically unable to break down betacyanin pigment.

Beet greens contain *oxalic acid*, a naturally occurring substance found in some vegetables, which may crystallize as oxalate stones in the urinary tract in some people. It is therefore; in individuals with known oxalate urinary tract stones are advised to avoid eating excess greens.

Green beets are rich in oxalic acid but, if eaten in excess, they can interfere with calciun metabolism and promote the formation of calcium metabolism and promote the formation of calcium – 0xalate kidney stones.

Cruciferous Crops

Cruciferous vegetables such as cabbage, cauliflower and broccoli are rich in Indole -3- Carbinal and isothiocyanates, which inhibit carcinogenesis in the lungs and esophagus and reduces breast cancer by metabolizing the female hormone **estrogen.** Green leafy vegetables, which are rich in chlorophyll and chlorophyllides are antimutagenic and antitumorigenic in humans. Globble down green vegetable such as spinach, broccoli, brussels sprouts, parsley to keep your mind sharp, thanks to their high amounts of folic acid. Feeding of vegetables induces enzymes of xenobiotic mechanism and thereby accelerates the metabolic disposal of xenobiotics. Glucosinolates (thioglucosides) are sulphur containing compounds in brassica vegetables. The glucosinolates induce physiological changes in humans including carcinogensis inhibition. The greatest benefits found from eating cruciferous and other green leafy vegetables is the concentration of potassium, folate, fibre, and dietary flavonoids found in these foods. The chemical called **Allyl isothio cyanate** (AITC) is created when some vegetables of brassica family are chopped, chewed, cooked, processed and digested, AITC- is a break- down product of **Sinigrin**- chemical compound in cruciferous vegetables. AITC - sabotages the uncontrolled cell division of colon cancer cells. Sinigrin- a chemical compound found in brassicas, is converted by processing or eating in to Ally-isothiocynate(AITC). Brassica crops are a source of natural antioxidants. In ayurvedic medicine, cabbage leaves are prescribed for cough, fever, skin diseases, peptic ulcers, urinary discharges and hermorrhoids. The seeds are diuretic, laxative, stomachic and antihelminthic. The chemoprotective components include ascorbic acid, tocopherols, carotenoids, isothiocyanates, indoles and flavonoids. Fresh cabbage is reported to contain a heat- labile anti –peptic cancer components, vitamin U(98- 100). Cabbage juice has ulcer-healing capabilities, but should be used in conjunction with a doctor's prescribed therapy. However, patients with gas trouble should avoid it.

A new study, conducted on mice, has found that compounds extracted from green vegetables like broccoli and cabbage could be a potent drug against melanoma, a type of skin cancer. The study found that these compounds, when combined with selenium, target tumours more safely and effectively than conventional therapy. "There are currently no drugs to target the proteins that trigger melanoma", said Gavin Robertson, associate professor of pharmacology, pathology and dermatology, Penn State College of Medicine (ANI, 2009). These vegetables are good for memory.

CABBAGE (*Brassica oleracea capitata*)

A family of cruciferous vegetables that include cabbage, broccoli, brussels sprouts, cauliflower and kale. They are rich in vitamins and minerals, particularly sulphur, the 'beauty mineral' which produces collagen and boosts the immune system against germs and viruses. They also contain compounds with anti- cancer properties, such as dithiolthiones and indoles which protect against breast and conon cancer. This family of vegetables also has a beneficial effect on the liver.

Crucifer's well known cancer fighting properties are thought to result from their high level active phytochemicals called **Glucosinolates,** which our bodies metabolize into powerful anti-carcinogens called **isothiocyanates.** Isothiocyanates offer the bladder in particular, significant protection, most likely because the majority of compounds produced by isothiocyanate metabolism travel through the blader enroute to excretion in the urine, suggested the researchers. Those eating the most cruciferous vegetables were found to have a 29 per cent lower risk of blader cancer compared to participants eating the least of this family of vegetables. Cabbage raw or cooked, inhibits the conversion of sugar and other carbohydrates in to fat. Hence, it is of great value in weight reduction.

This leafy vegetable is a delight for people wanting to loose weight. It is packed with vitamin A which protects your eyes and skin, vitamin C, vitamin E, vitamin B which helps in boosting energy metabolism. Cabbage extracts have been proven to kill certain viruses and bacteria. Cabbage boosts the immune system's ability to produce more antibodies. Cabbage provides high levels of iron and sulphur, minerals that work in part as cleansing agents for the digestive system. Cabbage shredded and boiled 150 grams contains 33 calories.

Cabbage is a wise vegetable. It reminds people that youth is health. It is very good cleaner of stomach if taken raw and unsalted. It prevents and cures ulcers. Some cancers are also prevented by the green leafy vegetables. A combination of vitamins that the cabbage contains helps you always feeling young. It keeps your blood vessels stronger and its tartonic acid content inhibits the conversion of sugar and other carbohydrates in to fat. A meal of cabbages gives even the child enough nutrition and roughage, as 100 g of cabbage yields 27 kilo-calories of energy which is a lot more than most other foods of equal measure can give. It is said to be great for the skin and eyes as it is full of vitamin A. The phytonutrients in this vegetable help the body fight against the free radicals. It also helps lungs, stomach and colon prostate cancer.

Those interested in body building may want to eat cabbage as it is full of iodine that is needed for muscle development.

Cabbage contains a lot of vitamin C and keeps the teeth and gums healthy. It is now highly regarded as one of the most nutritious vegetables available today, and is thought to have strong anti-aging and anti-cancer properties. Studies show that it should be brought to the table at least 2 or 3 times a week. It is an excellent source of vitamin C and Beta-carotene. These anti-oxidants not only slow down the process of aging, but also cut down the risk of cataracts. Cabbage also reduces risk of heart disease and stroke, alleviates rheumatisms and skin problems. All this, while being quite low on calories. To top it all, this is truly a miracle vegetable for cancer patients. Cabbage is high in both fibres and chemicals, whose combined action is known to prevent colon cancer. This is because fibres help the intestines stay healthy by promoting transit movements. Similarly, the chemicals present inhibit tumour growth and protects cells against free radicals. Some of these chemicals are also believed to reduce the incidence of breast, uterus and ovarian cancers as well. Moreover, juiced cabbage promotes healing of ulcers(Gita Angara, 2007).

Cabbage is recommended in natural medicine pratice for improving digestion, treating constipation, preventing the common cold and alleviuating depression. It also contains a factor called Vitamin U, which is a remedy for ulcers and raw cabbage juice has been reported to assist in the cure of both peptic and duodenal ulcers. The recommended method is to drink half a cup of freshly made cabbage juice two or three times a day between meals, on empty stomach. Many of the healing properties of cabbage as blood purofer are due to its high sulphur content. Grated cabbage can be made into a poultice to be applied externally in the treatment of wounds, varicose veins and leg ulcers. Cabbage (purple) is a great store of vitamin C and vitamin K. The rich deep colour of this vegetable is due to a high concentration of anthocyanin polyphenols (strong dietary antioxidants, possessing anti-inflammatory properties), making it have even more phytonutrients than a green cabbage. A few studies also show that anthocyanins may help reduce the risk of heart disease, diabetes and certain types of cancer. It is rich in anti-oxidants which makes skin supple and clear.

But there are a few things to keep in mind before rushing to grab a bite of the healing cabbage. Here are a few do's and don'ts when it comes to cooking, storing and eating cabbage:

Do's	*Don'ts*
Add a few seeds of caraway/cumin or fennel (or both) before or while cooking cabbage, as it helps to digest the vegetable as also to avoid flatulence.	Avoid cutting the vegetable in advance, or you could lose vitamins, especially vitamin C. As the leaves in the inside are usually well tight and clean, you should just remove the exterior leaves and rinse the whole cabbage under running water. You can then cook it.
Shredded in small pieces, raw cabbage can be added to any salad, or it can be a salad itself, with a few pieces of apples, raisins and cheese.	Do not cook cabbage in aluminium pot, this discolours the vegetable and also alters its flavour.

Contd...

Table–*Contd...*

Do's	*Don'ts*
When shredded it can be cooked with dal or cereals.	Don't miss out on your iodine intake. Cabbage reduces **iodine absorption**, so if you do eat more than 2 or 3 times a week, ensure your iodine intake through other means.
A whole cabbage can be stored in a refrigerator or cold place for at least two weeks, as it keeps well.	

Health Benefits

- ☆ Fresh, dark green-leafy cabbage is incredibly nutritious and low in fat and calories (Provides just 25 cal/100 g).
- ☆ It is storehouse to many phyto-chemicals like *thiocyanates, indole-3-carbinol, lutein, zeaxanthin, sulforaphane* and *isothiocyanates*. These compounds are powerful anti-oxidants and known to help protect against breast, colon, and prostate cancers and help reduce LDL or "bad cholesterol" levels in the blood.
- ☆ Fresh cabbage is an excellent source of natural antioxidant, vitamin C. Provides about 61 per cent of RDA, more than that of in the oranges. Regular consumption of foods rich in vitamin C helps body develop resistance against infectious agents and scavenge harmful, pro-inflammatory free radicals.
- ☆ It is also rich in many essential vitamins such as pantothenic acid (vitamin B5), pyridoxine (vitamin B-6) and thiamin (vitamin B-1). These vitamins are essential in the sense that our body requires them from external sources to replenish.
- ☆ It also contains good amount of minerals like potassium, manganese, iron, and magnesium. Potassium is an important component of cell and body fluids that helps controlling heart rate and blood pressure. Manganese is used by the body as a co-factor for the antioxidant enzyme, *superoxide dismutase.* Iron is required for the red blood cell formation.
- ☆ Cabbage is very good source of vitamin K, provides about 63 per cent of RDA levels. Vitamin-K has potential role in bone metabolism by promoting osteotrophic activity in them. So enough vitamin K in the diet gives you healthy bones. In addition, vitamin-K also has established role in curing Alzheimer's disease patients by limiting neuronal damage in their brain.
- ☆ Cabbage is rich in the following nutrients:
- ☆ Vitamin A: responsible for the protection of your skin and eyes.
- ☆ Vitamin C: an all important anti-oxidant and helps the mitochondria to burn fat.
- ☆ Vitamin E: a fat soluble anti-oxidant which plays a role in skin integrity.

- ☆ Vitamin B: helps maintain integrity of nerve endings and boosts energy metabolism.
- ☆ Cabbage extracts have been proven to kill certain viruses and bacteria. Cabbage boosts the immune system's ability to produce more antibodies.
- ☆ Cabbage provides high levels of iron and sulphur, minerals that work in part as cleansing agents for the digestive system.
- ☆ Cabbage is an excellent source of vitamins K and C: 1 cup (150 grams) of shredded, boiled cabbage contains 91 per cent of the RDA for vitamin K and 50 per cent for vitamin C.
- ☆ Cabbage is also a very good source of dietary fibre, manganese, vitaminB6 and folate, and a good source of thiamin, riboflavin, calcium, potassium, Vitamin A, tryptophan, protein and magnesium.
- ☆ It also aids in draining out excess fluids from your body, which in turn naturally detoxifies your system.

Caution

Cabbage may contain "goitrogens", a plant based compounds, especially found in cruciferous vegetables like cauliflower, broccoli etc, may cause swelling of thyroid gland and should be avoided in individuals with thyroid dysfunction. However, they may be used liberally in healthy persons.

CAULIFLOWER (*Brassica oleracea Botrytis*)

The allicin in cauliflower is known to promote a healthy heart and reduce the risk of strokes. It contains selenium and Vitamin C, both of which work together to strengthen the immune system. It helps in maintaining healthy cholesterol level. The high amount of fibre in cauliflower improves colon health and helps prevent cancer. Cauliflower even acts as a blood and liver detoxifier. People suffering from ailments like arthritis, asthma, constipation, high blood pressure and kidney bladder disorders can benefit from the consumption of cauliflower. Packed with rich nutrients, cauliflower or cabbage flower is one of the commonly used flower-vegetable. The flower heads contains numerous health benefiting phtyo nutrients such as indole-3-carbinol, sulforaphane etc that help prevent prostate, ovarian and cervical cancers. Botanically, it is a member of the cruciferous or brassicaceae family; has got similar nutritional and phyto-chemistry profile with broccoli and cabbage. Several cultivars exists other than common snow-white variety including green, orange, purple, and romanesco heads. Cauliflower lacks the green chlorophyll found in other vegetables; cabbage, broccoli and kale.

Several Researches on this flowery vegetable have revealed that it has compounds to resist cancer; it has the ability to eliminate cancer enzymes.

- ☆ **Good for detoxification of liver**: It contains glucosinolates and thiocyanates (including sulforaphane and isothiocyanates), which help to increase the liver's ability to neutrolise potentially toxic substances that could lead to cancer if left unattended. The presence of enzymes like glutathione transferase, glucuronosyl transferase and quinine reductase also help in the detoxification.
- ☆ **Prevent cancer**: Researches have shown that cauliflower helps keep off various kinds of cancers viz lung, colon, breast and ovarian, with the most recent addition being bladder cancer.
- ☆ **Shield against rheumatoid arthritis:** Cauliflower, which is rich in vitamin C, is said to provide a shield against inflammatory polyarthritis, a form of rheumatoid arthritis affecting two or more joints (Anonymous, 2013).

Health Benefits

- ☆ Very low in calories. 100 g of fresh cauliflower has only 26 calories. However, it is very low in fat and contains no cholesterol.
- ☆ Its florets contain about 2 g of dietary fibre per 100 g; providing about 5 per cent of recommended value.
- ☆ Cauliflower contains several anti cancer phyto-chemicals like **sulforaphane** and plant sterols such as **indole-3-carbinol** which appears to function as an anti-estrogen agent. Together these compounds have proven benefits against prostate, breast, cervical, colon, ovarian cancers by virtue of their cancer cell growth inhibition, cytotoxic effects on cancer cells.
- ☆ Also, **Di-indolyl-methane** (DIM), a lipid soluble compound present abundantly in brassica group of vegetables has found effective as immune modulator, anti-bacterial and anti-viral compound by potentiating Interferon-Gamma receptors and its production. DIM has currently been found application in the treatment of recurring respiratory papillomatosis caused by the Human Papilloma Virus (HPV) and is in Phase III clinical trials for cervical dysplasia.
- ☆ Fresh cauliflower is excellent source of **vitamin C**; 100 g provides about 48.2 mg or 80 per cent of daily recommended value. Vitamin-C is a proven antioxidant helps fight against harmful free radicals, boosts immunity and prevent from infections and cancers.
- ☆ It contains good amounts of many essential B-complex group of vitamins such as folates, pantothenic acid (vitamin B5), pyridoxine (vitamin B6) and thiamin (vitamin B1), niacin (B3) as well as vitamin K. These vitamins are essential in the sense that body requires them from external sources to replenish and required for fat, protein and carbohydrates metabolism.
- ☆ It is also good source of minerals such as manganese, copper, iron, calcium and potassium. Manganese is used in the body as a co-factor for the

antioxidant enzyme *superoxide dismutase. Potassium* is an important intracellular electrolyte helps counter the hypertension effects of sodium.

- The allicin in cauliflower is known to promote a healthy heart and reduce the risk of strokes.
- It contains selenium and Vitamin C, both of which work together to strengthen the immune system.
- It helps in maintaining healthy cholesterol level. The high amount of fibre in cauliflower improves colon health and helps prevent cancer.
- Cauliflower even acts as a blood and liver detoxifier.
- People suffering from ailments like arthritis, asthma, constipation, high blood pressure and kidney bladder disorders can benefit from the consumption of cauliflower.
- Cauliflower contains 'indole 3-carbinol', a substance that can prevent breast and other female cancers.
- Being rich in nutrient folate, cauliflower helps in improving cell growth and cell replication mechanism.
- The milk, sweet and delicious nutty flavour of cauliflower is at its best when it is in season from December through March

Caution

Like other members of the brassica/cruciferous family, prolong/excessive use of cauliflower may cause swelling of thyroid gland and thyroid hormone deficicency. This is due to the presence of certain plant compounds known as *goitrogens* in these group of vegetables. It is therefore advised to avoid, especially in individuals with thyroid dysfunction. However, these vegetables may be used liberally in healthy person.

BROCCOLI (*Brassica oleracea* Italica)

When one talks about vegetables which can fight diseases effectively, broccoli ranks amongst the highest. Broccoli heads are rich source of phyto-nutrients that help protect from prostate cancer and stroke risks. It is actually a flower vegetable and known for its notable and unique nutrients that are found to have disease prevention and health promoting properties. They are a storehouse of antioxidants which help in fighting rectal, lungs and stomach cancers. They are also rich in folate, Vitamin C and beta-carotene which help in boosting your immunity to fight against colds and flus. Broccoli heads are rich source of phyto-nutrients that help protect

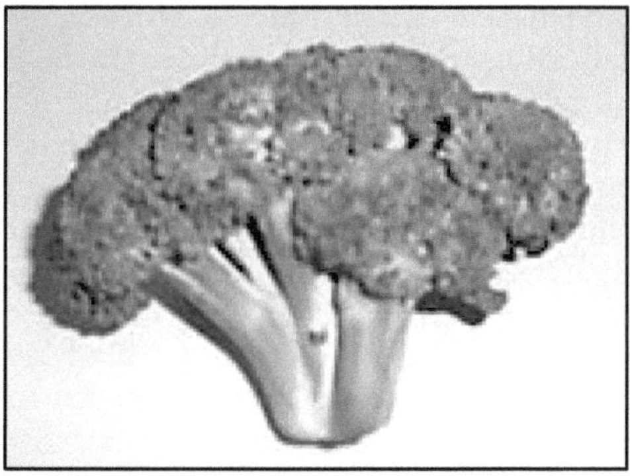

Cauliflower Head

from prostate cancer and stroke risks. It is actually a flower vegetable and known for its notable and unique nutrients that are found to have disease prevention and health promoting properties. It is considered a wonder food for its ability to help prevent cancer. It has the powerful anti cancer component ever detected. Scientists have isolated an ingredient in broccoli that kindles the activity of critical enzyme in the cell known to guard against tumour. The newly discovered chemical in broccoli is **Sulfora phane**- is by far the most powerful inciter. Besides anticarcinogenic properties, broccoli is rich source of vitamins, minerals, proteins, etc. It has about 130 times more Vitamin A content than cauliflower and 22 times more than cabbage. It is a rich source of **Sulphoraphane,** a compound associated with reducing the risk of cancer. Broccoli which is known for its anti cancer properties but it could also boost the immune system in older people and slow down the effects of ageing according to the new research, as **sulforaphane** present in cruciferous vegetables was found to activate a number of antioxidant genes and enzymes in immune cells. These prevent free radicals from damaging cells (Methew, 2008). When one talks about vegetables which can fight diseases effectively, broccoli ranks amongst the highest. They are a storehouse of antioxidants which help in fighting rectal, lungs and stomach cancers. They are also rich in folate that helps in reducing stress and irritability, VitaminC and beta-carotene which help in boosting your immunity to fight against colds and flues. **It** is considered as one of nature's superfoods.

Broccoli is exceptionally nutritious, containing an abundance of sulphur, iron and cholorophyll, which purifies the blood; it is also rich in vitamins A, B-Complex and especially C. In fact, it contains more vitamin C than oranges - one cup provides as much as 70 mg of vitamin C.

It provides stress relief, helps enhance mood with regard to depression, panic, anxiety and other emotional disorders. A team in Britain has found that **sulforaphane**, a compound present in broccoli, boosts the production of enzymes which protect the blood vessels and encourages a reduction in high levels of molecules which cause significant cell damage and may help counter processes linked to the development of vascular disease in diabetes as high glucose levels (hyperglycaemia), are associated with diabetes. Their study found that a 73 per cent reduction of molecules in the body called **Reactive Oxygen Species** (ROS). Hyperglycaemia can cause levels of ROS to increase nearly three fold and such high levels can damage human cells. **Sulforaphane** present in broccoli activates a protein in the body called **nrf2**, which protects cells and tissues from damage by activating protective antioxidant and detoxifying enzymes (Anonymous, 2008). Sulforaphane may offer special protection to those with colon cancer-susceptible genes, suggests a study conducted at Rutgers University and Published online in Journal Carcinogenesis. Sulforaphane (SFN) is the hydrolysis product of **glucoraphanin**, particularly high in young sprouts of broccoli and cauliflower. SFN can also be obtained by eating cruciferous vegetables such as Brussels sprouts, broccoli, cauliflower, cabbage kale, collards, arugula, broccoli raab, turnip, radish, mustard, knolkhol and water cress. SFN- is therefore able to prevent, delay or reverse preneoplastic lesions, as well as to act on cancer cells as a therapeutic agent.

Broccoli contains high levels of calcium, pantothenic acid, vitamin A, thiols and indoles and sulphur compounds that have antimicrobial activity. Direct sulforaphane (glucosinolate), a naturally occurring compound found in cruciferous vegetables and broccoli sprouts may even be more versatile and effective than vitamin C and E in protecting cells from oxidative damage. It plays a key role in increasing the body's natural antioxidant defence systems and may function as a powerful indirect antioxidant with a longer lasting effect than other antioxidants such as vitamin C and vitamin E. Other sources are broccoli, cabbage and cauliflower, kale, knolkhol, collards, radish, rutabaga and turnips. Cruciferous vegetables contain high concentration of carotenoids, which are believed to be chemopreventive and associated with decreased risk for various human cancers in epidemiological studies. Researchers at the University of California, Los Angeles, found that **diindolymethane**(DIM) a compound resulting from digestion of cruciferous vegetable and genistein, an isoflavone in soy, reduce the production of two proteins needed for breast and ovarian cancers to spread(Reuters, 2007).

Researchers put broccoli sprouts to the test and found that they contain high levels of **glucoraphanin** also known as **sulforaphane glucosinolates** (SGS). In addition to being a cancer preventative, SGS boosts production of phase 2 enzymes, which are part of the bodys anti-oxidant system. This extra boost lowers blood pressure and the body's inflammatory responses, leading to better cardiovascular health over all. As just 5 grams of broccoli sprouts contain concentrations of SGS equal to that found in 150grams of mature broccoli. That makes you just one salad a day away from better health. Known for its great cance - combating ability, broccoli was found to contain **Sulphoraphane,** a sulphur -based compound, which helps to kill cancer causing substances in food. When sulphoraphane is released in the gut, it steps up production of powerful enzymes that destroy carcinogenic substances such as those found in heavily barbecued meat. Another naturally occuring compound found in broccoli - **13C-** was found in US study to increase DNA repair proteins, thus preventing genetic information damage and lowering the risk of developing cancer. The vegetable is recommended for those suffering from heart disorders and strokes. Being packed with iron it is ideal for anaemia patients as well.

According to a study in cancer prevention research, a journal of the American Association of Cancer Research, that regular eating of 70g, three-day old broccoli sprouts can suppress a bacteria *(H. pylori)* linked to stomach cancer, gastritis and ulcers. Helicobactor pylori infections are one of the most common stomach infections affecting people globally and are a major cause of stomach cancer.

Broccoli has recently entered the public awareness as a preventive dietary agent. This study supports the emerging evidence that broccoli sprouts may be able to prevent cancer, in humans not just in laboratory animals, said Jed Fahey, a faculty research associate in the department of pharmacology at John Hopkins School of Medicine (Anonymous, 2009).

In this study, researchers enrolled 48 Helicobacter infected Japanese men and women and randomly assigned them to eat 70 grams of fresh broccoli sprouts for eight weeks or even equivalent amount of alfalfa sprouts. Researchers assessed the severity of H. pylori infection at enrolment and again at 4 and 8 weeks using standard breath, serum and stool tests. H. pylori were significantly lower at 8 weeks on all measures among those patients who had eaten broccoli sprouts, while they remained the same for patients who had eaten alfalfa sprouts (Anonymous, 2009).

Health Benefits

- ☆ Broccoli is very low in calories, provides just 34 cal per 100 g. However, it is rich in dietary fibre, minerals, vitamins, and anti-oxidants that have proven health benefits.
- ☆ Fresh Broccoli is a storehouse of many phyto-nutrients such as *thiocyanates, indoles, sulforaphae, isothiocyanates* and flavonoids like beta-carotene cryptoxanthin, lutein, and zeaxanthin. Studies have shown that these compounds by modifying positive signalling at molecular receptor levels help protect from prostate, colon, urinary bladder, pancreatic, and breast cancers.
- ☆ Fresh vegetable is exceptionally rich source of vitamin-C. Provides 89.2 mg or about 150 per cent of RDA per 100 g. Vitamin-C is a powerful natural anti-oxidant and immune modulator, helps fight against flu causing viruses.
- ☆ Furher, it contains very good amounts of another anti-oxidant vitamin, vitamin-A. 100 g fresh head provies 623 IU or 21 per cent of recommended daily levels. Together with other pro-vitamins like beta-carotene, alpha-carotene, and zeaxanthin, vitamin A helps maintain integrity of skin and mucus membranes. Vitamin A is essential for vision and helps prevent from macular degeneration of retina in the elderly population.
- ☆ It is also a good source of minerals like calcium, manganese, iron, magnesium, selenium, zinc and phosphorus.
- ☆ Fresh broccoli heads are an excellent source of folates; contains about 63 mcg/100 g (Provides 16 per cent of RDA). Studies have shown that consumption of fresh vegetables and fruits rich in folates during pre-conception and pregnancy helps prevent neural tube defects in the offspring.
- ☆ Broccoli leaves (green tops) are an excellent source of carotenoids and vitamin A; (provide 16000 IU of vitamin A per 100 g) contain these compounds several times more than in the roots.
- ☆ This flower vegetable is also rich source of other vitamin-K and B-complex group of vitamins like Niacin (vit B-3), pantothenic acid (vit. B-5), pyridoxine (vit. B-6) and riboflavin. The flower heads also have some amount omega-3 fatty acids.
- ☆ Broccoli contains twice the vitamin C of an orange.
- ☆ It has almost as much calcium as whole milk—and the calcium is better absorbed.
- ☆ It contains selenium, a mineral that has been found to have anti-cancer and anti-viral properties.
- ☆ Broccoli is a modest source of vitamin A and alpha-tocopherol vitamin E.
- ☆ It also has antioxidant properties.

Caution

Like other members of the cruciferous family, broccoli contain "goitrogens" which may cause swelling of thyroid gland and therefore, should be avoided in individuals with thyroid dysfunction. However, it may be used liberally in healthy person.

BRUSSELS SPROUTS (*Brassica oleracea* Gemmifera)

Brussels Sprouts Heads

Brussel sprouts are small leafy green buds resemble like miniature cabbages in appearance. The buds are exceptionally rich in protein, dietary fibre, vitamins, minerals and antioxidants which work wonders to get rid of many health troubles. Plant phytonutrients found in brussels sprouts enhance the activity of body's natural defences to protect against diseases, including cancers, Scientists have found that sulforaphane, one of the powerful glucosinolate phytonutrients found in brussels sprouts and other cruciferous vegetables, boosts the bodys detoxification enzymes, potentially by altering gene expression, thus helping to clear potential carcinogenic substances more quickly. Brussels sprouts are excellent for pregnant women, these little veggies are full of B-vitamin and folic acid which keeps women away from uteral tube defects. They are also quite rich in omega 3 fatty acids, potassium, fibre and Vitamin C and K.

Brussels sprouts are similar to cabbage in flavour and nutrient content and provide an excellent source of vitamins C, B1 and β-carotene. Similarly they are also rich in potassium, calcium, and sulphur but unlike cabbage and broccoli, which can be enjoyed both raw and cooked, brussels sprouts are palatable only as a cooked vegetable. Excellent for pregnant women, these little veggies are full of B-vitamin and folic acid which keeps women away from uteral tube defects. They are also quite rich in omega 3 fatty acids, potassium, fibre and Vitamin C and K.

Brussels sprouts stronger protective effects are thought to be due to the fact that this cruciferous vegetable contains 2-3 times the amount of glucosinolates than are found in red cabbage. Glucosinolates increase phase 11 glucuronidation activity, one of the primary pathways through which toxins made even more dangerous by phase 1 are rendered water soluble and ready for elimination from the body.

Cruciferous vegetable like Brussels sprouts, cabbage, cauliflower, broccoli kale help prevent cancer, when these vegetables are cut, chewed or digested, a sulphur-containing compound called singrin is brought into contact with enzyme myrosinase, resulting in the release of glucose and breakdown products, including highly reactive compounds called isothiocyanates. Isothiocyanates are not only potent inducers of the liver's phase 11 enzymes, which detoxify carcinogens but research recently conducted at the Institute for Food Research in the U. K. shows one of these compounds allyl isothiocyanate, also inhibits mitosis (cell division) and stimulates apoptosis (programmed cell death) in human tumour cells.

Brussels sprouts contain goitrogens, naturally-occuring substances in certain foods, can interfere with the functioning of the thyroid gland. Therefore, individually with already existing and untreated thyroid problems should avoid brussel sprouts for this reason.

Health Benefits

- One of the nutritious vegetable that should be considered in weight reduction programs. 100 g Brussel sprouts just provide 45 calories but contain 3.8 g of dietary fibre (10 per cent of RDA) and no cholesterol.
- In fact, brussel sprouts are storehouse of flavonoid anti-oxidants like *thiocyanates, indoles, lutein, zeaxanthin, sulforaphane* and *isothiocyanates*. Together these phytochemicals offers protection from prostate, colon, prostate and endometrial cancers.
- Di-indolyl-methane (DIM), a metabolite of *indole-3-carbinol* has been found to be an effective immune modulator, anti-bacterial and anti-viral agent through its action of potentiating "Interferon-γ" receptors.
- In addition brussel sprouts contain glucoside, sinigrin. Early laboratory studies suggest that *sinigrin* helps protect from colon cancers by destroying pre-cancerous cells.
- Brussel sprouts are an excellent source of vitamin C; 100 g sprouts provide about 142 per cent of RDA. Along with other antioxidant vitamins, vitamin A and E; it protects body by trapping harmful free radicals.
- Zeaxanthin, an important dietary carotenoid in sprouts, is selectively absorbed into the retinal macula lutea in the eyes where it is thought to provide anti-oxidant and protective light-filtering functions from UV rays. Thus, it helps prevent retinal damage, "age related macular degeneration disease" (ARMD), in the elderly.
- Sprouts are good source of another anti-oxidant vitamin A, provides about 754 IU per 100g. Vitamin A is required for maintaining healthy mucus membranes and skin and is also essential for acuity of vision. Foods rich in this vitamin offer protection against lung and oral cavity cancers.
- It is one of the excellent vegetable sources for vitamin-K; 100 g provides about 177 mcg or about 147 per cent of RDA. Vitamin K has potential role bone health by promoting osteotrophic (bone formation and strengthening) activity. Adequate vitamin-K levels in the diet helps limiting neuronal damage in the brain; helps prevnt or at least delay onset of Alzheimer's disease.
- The sprouts are notably good in many B-complex group of vitamins such as niacin, vitamin B-6 (pyridoxine), thiamin, pantothenic acid, etc that are essential for substrate metabolism in the body.
- They are also rich source of minerals like copper, calcium, potassium, iron, manganese and phosphorus. Potassium in an important component of cell and body fluids that helps controlling heart rate and blood pressure by countering effects of sodium. Manganese is used by the body as a co-factor for the antioxidant enzyme *superoxide dismutase*. Iron is required for cellular oxidation and red blood cell formation.

Brussel sprouts are incredibly nutritious vegetable that offers protection from vitamin A deficiency, bone loss, iron deficiency anaemia, and believed to protect from cardiovascular diseases and, colon and prostate cancers.

Caution

Brussel sprouts may contain goitrogens which may cause swelling of thyroid gland and should be avoided in individuals with thyroid dysfunction. However, they may be used liberally in healthy person.

CHINESE CABBAGE

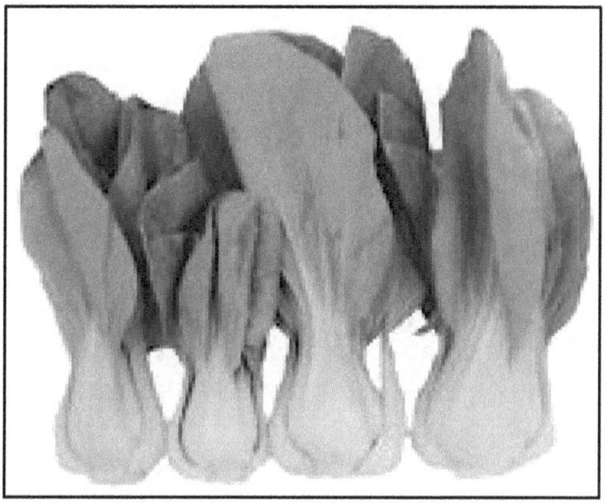

In structure, Chinese Cabbage resembles collards and could be described as a non-heading cabbage (Acephala group). It is basically a small plant which grows upright from the ground with smooth white romaine lettuce like stalks which spread in the end to fine, glossy green oval or round leaves. Fully grown up plant may reach about 12-18 inches in length. leafy Chinese cabbage is one of the popular mainland crop in China, Philippines, Vietnam, India and other regions; nonetheless this humble brassica family vegetable has gained popularity even in the western world for its sweet, succulent nutritious stalks.

Brassica campestris group can be further categorized according to the colour of the petiole in its leaves. White petiole varieties include joi choi, pak-choy white, prize choi, lei choi, taisai, canton pak choi etc. Green petiole types are chinese pak choi green, mei qing choi etc.

Health Benefits

- ☆ It is one of the rare vegetable that is very low in calories. However, it is very rich source of many vital phyto-nutrients, vitamins, minerals and anti-oxidants.
- ☆ 100 g of bok choy provides just 13 cal. It is one of the recommended vegetable in the zero calorie or negative calorie category of foods which when eaten would add no extra weight to the body but in-turn facilitate calorie burns and reduction of weight.
- ☆ Like other brasisca family vegetables, bok choy contains certain anti oxidant plant chemicals like thiocyanates, indole-3-carbinol, lutein, zeaxanthin, sulforaphane and isothiocyanates. Along with dietary fibre, vitamins these compounds help to protect against breast, colon and prostate cancers and help reduce LDL or "bad cholesterol" levels in the blood.

- Fresh pak choi is an excellent source of water soluble antioxidant, **vitamin-C** (ascorbic acid). 100 g provides 45 mg or 75 per cent of daily requirements of vitamin C. Regular consumption of foods rich in vitamin C helps body develop resistance against infectious agents and scavenge harmful, pro-inflammatory free radicals.
- It has more vitamin A, carotenes, and other flavonoid polyphenolic anti-oxidants than cabbage, cauliflower etc. Just 100 g of fresh bok choy provides 4468 IU or 149 per cent of daily required levels vitamin A.
- Pak choi is very good source of **vitamin K,** provides about 38 per cent of RDA levels. Vitamin-K has potential role in bone metabolism by promoting osteotrophic activity in bone cells. Therefore, enough vitamin K in the diet makes your bone stronger, healthier and delay osteoporosis. Further, vitamin-K also has established role in curing Alzheimer's disease patients by limiting neuronal damage in their brain.
- Fresh bok choy has many vital B-complex vitamins such as pyridoxine (vitamin B6), riboflavin, pantothenic acid (vitamin B5), pyridoxine and thiamin (vitamin B-1). These vitamins are essential in the sense that our body requires them from external sources to replenish.
- It also contains good amount of minerals like calcium, phosphorous, potassium, manganese, iron and magnesium. Potassium is an important component of cell and body fluids that helps controlling heart rate and blood pressure. Manganese is used by the body as a co-factor for the antioxidant enzyme *superoxide dismutase.* Iron is required for the red blood cell formation.
- Chinese cabbage has anti-inflammatory properties.
- It is an excellent source of folic acid.
- Chinese cabbage is low in calories and low in sodium.
- It is also high in vitamin A and a good source of potassium.

KALE (*Brassica oleracea acephale*)

Similar to cabbage, Kale *reduces cholesterol and is great for digestion*. Although relatively unheard of by most people, kale is now fast emerging as a kind of super-food. Whether it is consumed as a soup, cooked with food or in a smoothie, this slightly bitter vegetable is chock full of antioxidants and help to protect you from free radical damage, which is said to cause premature ageing and disease. If you like cabbage, then chances are you will like the flavour, texture of kale. It is in fact a type of cabbage though on examination, you will find that its leaves resemble crisp lettuce leaves. Be sure to cook it, however, as consumed raw, the vegetable is hard to digest.

Kale is rich in numerous health benefiting polyphenolic flavonoid compounds such as lutein, zeaxanthin, and beta-carotene, and vitamins than found in any other green leafy vegetables. It is widely cultivated across Europe, Japan, United States and in some regions of India for its "frilly"leaves. It features distinctive very long, curly, blue-green leaves with embossed surface resembling dinosaur skin, giving its name as dinosaur kale. It is considered to be a highly nutritious vegetable, with powerful anti-oxidant properties and is anti-inflammatory. Kale is very high in beta -carotene, vitamin-K, vitamin -C, lutein, zeaxanthin and reasonably rich in calcium. Because of its high vitamin-K content, patients taking anti-coagulants such as Warfarin are encouraged to avoid this food since it increases vitamin -K concentration in the blood, which is what the drugs are often attempting to lower. This effectively raises the effective dose of the drug. Kale, as with broccoli and other brassicas, contain sulforaphane, a chemical believed to have potent anti- cancer properties, particularly when chopped. Kale, as with **broccoli**, it contains sulforaphane (particularly when chopped or minced), a chemical with potent **anti-cancer** properties. Kale is also a source of indole-3-carbinol, a chemical which boosts DNA repair in cells and appears to block the growth of cancer cells.

In addition to its unique organosulphur compounds, kale is well known for its carotenoids, especially **lutein** and **zeaxanthin.** These carotenoids act like sunglass filters and prevent damage to the eyes from excessive exposure to ultra violet light. Studies have shown the protective effect of these nutrients against the risk of cataracts, where increased eye cloudiness leads to blurred vision. In one study, people who had a diet history of eating lutein-rich foods like kale-had a 50 per cent lower risk for new cataracts.

Kale also emerged from our food ranking system as an excellent source of traditional nutrients, including vitamin A, vitamin C, vitamin B6, and manganese. It is also a very good source of dietary fibre, calcium, copper, vitamin B6 and potassium. This combination of vitamins, minerals and phyto-nutrients makes kale a healthy superstar. Both vitamin A and beta-carotenes are important vision nutrients.

Kale being an excellent source of vitamin C as just one cup of this cooked vegetable supplies 88.8 per cent of the daily value of vitamin C. Vitamin C is the primary water soluble antioxidant in the body, disarming free radicals and preventing damage in the aqueous environment both inside and outside cells. Inside cells a potential result of free radical damage to DNA is cancer. Especially in areas of the body where cellular turnover is especially rapid, such as digestive system, preventing DNA mutations translates into preventing cancer. This is why a good intake of vitamin C is associated with a reduced risk of colon cancer. Vitamin C, which is also vital for the proper function of a healthy immune system, is good for preventing colds and may be helpful in preventing recurrent ear infections.

Kale's health benefits continue with its fibre; a cup of kale provides 10.4 per cent of the daily value for fibre, which has been shown to reduce high cholesterol level, thus helping to prevent **atherosclerosis.** Fibre can also help out by keeping blood sugar level under control, so kale is an excellent vegetable for people with diabetes. Kale's fibre binds to cancer causing chemicals, keeping them away from the cells lining the colon, preventing yet another line of protection from colon cancer.

Kale is also very good source of calcium. Calcium is one of the nutrients needed to make healthy bones and dairy products are a heavily promoted source of this nutrient, but unlike dairy products, kale is not a highly allergenic food, nor does it contains any saturated fat-plus. A cup of kale supplies 93.6 mg of calcium (9.4 per cent of the daily value for this mineral) for only 36.4calories, in contrast, a cup of 2 per cent cows milk provides 296.7mg of calcium, but costs is high; 121.2 calories and 14.6 per cent of days suggested limit on saturated fat.

Health Benefits

- ✰ Kale is very versatile and nutritious green leafy vegetable. It is widely recognized as an incredibly nutritious vegetable since ancient Greek and Roman times for its low fat, no cholesterol but health benefiting anti-oxidant properties.
- ✰ Kale, like other members of the brassica family, contains health-promoting phytochemicals, sulforaphane and indole-3-carbinol that are appears to protect against prostate and colon cancers.
- ✰ Di-indolyl-methane (DIM), a metabolite of *indole-3-carbinol* has been found to be an effective immune modulator, anti-bacterial and anti-viral agent through its action of potentiating "Interferon-Gamma" receptors.
- ✰ Borecole is very rich source of ß-carotene, lutein and zeaxanthin. These flavonoids have strong anti-oxidant and anti-cancer activities. Beta-carotene is converted to vitamin A in the body.
- ✰ Zeaxanthin, an important dietary carotenoid, is selectively absorbed into the retinal macula lutea in the eyes where it is thought to provide antioxidant and protective light-filtering functions. Thus, it helps prevent retinal detachment and offers protection against "age related macular degeneration disease" (ARMD) in the elderly.
- ✰ It is very rich in vitamin A, 100 g leaves provide 512 per cent of RDA. Vitamin A is required for maintaining healthy mucus membranes and skin and is essential for vision. Foods rich in this vitamin offer protection against lung and oral cavity cancers.
- ✰ It is one of the excellent vegetable sources for vitamin-K; 100 g provides about 700 per cent of recommended intake. Vitamin K has potential role bone health by promoting osteotrophic (bone formation and strengthening) activity. Adequate vitamin-K levels in the diet helps limiting neuronal damage in the brain; thus, has established role in the treatment of patients suffering from Alzheimer's disease.
- ✰ 100 g of fresh leaves contain 120 mg or 200 per cent of daily-recommended levels of vitamin C. Scottish curly leaf variety yet has more of this vitamin, 130 mg/100g. Vitamin C is a powerful antioxidant, which helps body develop resistance against infectious agents and scavenge harmful oxygen free radicals.

- ☆ This leafy vegetable is notably good in many B-complex group of vitamins such as niacin, vit. B-6 (pyridoxine), thiamin, pantothenic acid, etc that are essential for substrate metabolism in the body.
- ☆ It is also rich source of minerals like copper, calcium, sodium, potassium, iron, manganese, and phosphorus. Potassium is an important component of cell and body fluids that helps controlling heart rate and blood pressure by countering effects of sodium. Manganese is used by the body as a co-factor for the antioxidant enzyme, *superoxide dismutase*. Iron is required for cellular oxidation and red blood cell formation.
- ☆ Kale eases lung congestion and is beneficial to the stomach, liver and immune system
- ☆ It contains lutein and zeaxanthin, which protect the eyes from macular degeneration
- ☆ It also contains indole-3-carbinol, which may protect against colon cancer
- ☆ Kale is an excellent source of calcium, iron, vitamins A and C, and chlorophyll
- ☆ Kale provides rich nutrition ingredients that offer protection from vitamin A deficiency, osteoporosis, iron deficiency anemia, and believed to protect from cardiovascular diseases and, colon and prostate cancers.

Caution

Because of its high vitamin K content, patients taking anti-coagulants such as warfarin are encouraged to avoid kale since it increases the vitamin K concentration in the blood, which is what the drugs are attempting to lower. This effectively raises the effective dose of the drug and causes toxicity.

Its leaves contain 0.2 g/100 g of oxalic acid, a value far less than some other comparable greens such as **spinach** (0.97 g/100) and **purslane** (1.31 g/100 g). It may be used, however with caution, even in individuals with known oxalate urinary tract stones are advised to avoid eating other greens and certain vegetables belonging to amaranthaceae and brassica family because of their high oxalic acid content. Adequate intake of water is therefore advised to maintain normal urine output.

COLLARD (*Brassica oleracea* L. Acephala)

Collard greens are highly nutritious staple green "cabbage-like leaves" vegetable. Widely considered to be healthy food, collards are good source of vitamin C and soluble fibre and contains multiple nutrients with potential anti-cancer properties, such as **diindolylmethane** and **sulforaphane.** Researchers at the University of California at Berkeley have recently discovered that **3,3-diidolylmethane** in brassica vegetables such as collard greens is a potent modulator of the innate immune response system with potent anti-viral, anti-bacterial and anticancer activity.

Health Benefits of Collard (Greens)

- Wonderfully nutritious collard leaves are very low in calories (provide only 30 cal per 100 g) and contain no cholesterol. However, these greeny leaves contain very good amount of soluble and insoluble dietary fibre that helps control LDL cholesterol levels and also; offers protection against hemorrhoids, constipation as well as colon cancer diseases.
- Widely considered to be healthful foods, collards are rich in invaluable sources of phyto-nutrients with potent anti-cancer properties, such as di-indolyl-methane (DIM) and sulforaphane that have proven benefits against *prostate, breast, cervical, colon, ovarian* cancers by virtue of their cancer cell growth inhibition and cytotoxic effects on cancer cells.
- Di-indolyl-methane has also found to be effective immune modulator, anti-bacterial and anti-viral properties by potentiating Interferon-gamma receptors and production.
- The leaves are also an excellent source of folates, provides about 166 mcg *or* 41.5 per cent of RDA. Folates are important in DNA synthesis and when given during peri-conception period can prevent neural tube defects in the baby.
- Fresh collard leaves are also rich in vitamin-C. Provides about 59 per cent of RDA per 100 g. Vitamin-C is a powerful natural anti-oxidant that offers protection against free radical injury and flu-like viral infections.
- Collard greens are also an excellent source of vitamin-A (222 per cent of RDA per 100 g) and flavonoid poly-phenolic anti-oxidants such as *lutein, carotenes, zea-xanthin, crypto-xanthin* etc. These compounds are scientifically found to have antioxidant properties. Vitamin A also required maintaining healthy mucus membranes and skin and is also essential for vision. Consumption of natural fruits rich in flavonoids helps to protect from lung and oral cavity cancers.
- This leafy vegetable contain amazingly high levels of vitamin-K, provides staggering 426 per cent of recommended daily levels per 100 leaves. Vitamin K has potential role in the increase of bone mass by promoting osteotrophic activity in the bone. It also has beneficial effect in Alzheimer's disease patients by limiting neuronal damage in their brain.
- Collards are rich in many vital B-complex groups of minerals such as niacin (vitamin B-3), pantothenic acid (vitamin B-5), pyridoxine (vitamin B-6) and riboflavin.
- The leaves and stems are good in minerals like iron, calcium, copper, manganese, selenium and zinc.

Caution

- ✰ Like other members of the brassica family, collards may contain goitrogens which may cause swelling of thyroid gland and therefore, should be avoided in individuals with thyroid dysfunction. However, it may be used liberally in healthy person.
- ✰ Should be used sparingly with people suffering from oxalate kidney stones.
- ✰ 100 g of raw collard greens provide more than 500 mcg of vitamin K well above daily recommended value; it is therefore, should be used cautiously in people taking anticoagulants like warferin.

KNOL-KHOL (*Brassica oleracea* Gongylodes)

It is grown for its large, edible stem which has a slight bitter –sweet flavour similar to turnip. Knol khol can improve blood circulation, strengthen digestion and help to stabilize **blood sugar** levels, making it beneficial for diabetics and hypoglycaemics. It can also help to relieve painful urination, stop internal bleeding and purify the body from toxins and alcohol. The juice is considered a remedy for nose bleeds. The edible stem attains maturity and ready for harvest in 55-60 days after sowing. Approximate weight is 150 g. The tuber is tolerant to cracking and has good standing ability for up to 30 days after maturity. Knol-knols have similar taste and texture as that of a broccoli stem or cabbage, but milder and sweeter. The younger stems have crispy, pleasant taste and rich flavour.

Health Benefits of Kohlrabi (Knol-khol)

- ✰ Mildly sweet, succulent Kohlrabi is notably rich in vitamins and Dietary fibre; but has only 27 calories per 100 g, negligible amount of fat, and zero cholesterol.
- ✰ Fresh kohlrabi stem is rich source of vitamin-C; provides 62 g per 100g weight that is about 102 per cent of RDA. Vitamin C (ascorbic acid) is a water-soluble vitamin and powerful anti-oxidant. It helps body maintain healthy connective tissue, teeth, and gum. Its anti-oxidant property helps body protect from diseases and cancers by scavenging harmful free radicals from the body.
- ✰ Kohlrabi, like other members of the brassica family, contains health-promoting phytochemicals such as isothiocyanates, sulforaphane and indole-3-carbinol that are appears to protect against prostate and colon cancers.
- ✰ This stem vegetable is especially contain good amounts of many B-complex group of vitamins such as niacin, vitamin B-6 (pyridoxine), thiamin,

pantothenic acid etc that acts as co-factors to enzymes during various metabolism inside the body.

- ☆ Knol-knol notably has good levels of minerals; copper, calcium, potassium, manganese, iron, and phosphorus are especially available in the stem. Potassium is an important component of cell and body fluids that helps controlling heart rate and blood pressure by countering effects of sodium. Manganese is used by the body as a co-factor for the antioxidant enzyme, *superoxide dismutase*.
- ☆ In addition, its creamy colour flesh contains small amounts of vitamin A and carotenes.
- ☆ Knolkhob leaves or tops, like turnip greens, are also very nutritious greens abundant in carotenes, vitamin A, vitamin K, minerals, and B-complex group of vitamins.
- ☆ Kohlrabi, which belongs to the cabbage family, is an excellent source of vitamin C and potassium.
- ☆ It is also high in fibre.
- ☆ Kohlrabi helps to stabilize blood sugar and is therefore useful hypoglycemia and diabetes.
- ☆ It can also be effective against edema, candida and viral conditions.

Caution

Kohlrabi may contain **goitrogens**, a plant based compounds found in cruciferous vegetable like cauliflower, broccoli etc, may cause swelling of thyroid gland and should be avoided in individuals with thyroid dysfunction. However, knol-knols may be eaten liberally in healthy person.

MUSTARD GREENS (*Brassica juncea*)

Spicy, crunchy mustard greens or leafy mustard is indeed one of the most nutritious green leafy vegetables. The greens actually have more vitamin A, carotenes, vitamin K, and flavonoid anti-oxidants than many commonly consumed fruits and vegetables. Actually, its young tender green leaves that is harvested when the plant reaches about 60cm in height and used as green leafy vegetable. Completely grown plant reaches about 120-150cm in height and bears golden yellow coloured flowers. Fresh mustards feature dark green coloured broad leaves with flat surface and may have either toothed, frilled or lacy edges depending up on the cultivar type. Its light green stem branches out with many laterals.

Mustard seeds used as spice are being used extensively in cooking as well as in oil production all over South-Asian region.

Health Benefits

- ☆ Mustard greens like spinach are the storehouse of many phytonutrients that have health promotional and disease prevention properties.
- ☆ Mustards are very low in calories (26 kcal per 100 g raw leaves) and fats, but rich in dietary fibre; recommended in cholesterol controlling and weight reduction programs. However, its dark green leaves contain very good amount of dietary fibre that helps control cholesterol level and also help protect against hemorrhoids, constipation as well as colon cancer diseases.
- ☆ Fresh mustard greens are an excellent source of several vital anti-oxidants and minerals like vitamin C, vitamin A, vitamin E, carotenes as well essential minerals such as calcium, iron, magnesium, potassium, zinc, selenium, and manganese.
- ☆ The greens are supposed to be one of the highest among leafy vegetables which provide **vitamin K**. 100 g of fresh leaves contain about 497 mcg or about 500 per cent of daily requirement of vitamin K 1 (phylloquinone). Vitamin K has found to have potential role in bone mass building function by promoting osteo-trophic activity in the bone. It also has established role in Alzheimer's disease patients by limiting neuronal damage in the brain.
- ☆ Fresh leaves are also very good source of **folic acid**. 100 g provide about 187 mcg (about 47 per cent of RDA) of folic acid. This water-soluble vitamin has an important role in DNA synthesis and when given before and early pregnancy help prevent neural tube defects in the baby.
- ☆ Mustard greens are rich source of anti-oxidants flavonoids, indoles, sulforaphane, carotenes, lutein and zeaxanthin. Indoles, mainly **di-indolyl-methane (DIM)** and **sulforaphane** have proven benefits against prostate, breast, colon and ovarian cancers by virtue of their cancer cell growth inhibition, cytotoxic effects on cancer cells.
- ☆ Fresh mustard leaves are excellent source of **vitamin-C**. Provides 70 mcg or about 117 per cent of RDA per 100 g. Vitamin-C (ascorbic acid) is a powerful natural anti-oxidant that offers protection against free radical injury and flu-like viral infections.
- ☆ The leaves are also an excellent source of **vitamin-A** (provide 10500 IU or 350 per cent of RDA per 100 g). Vitamin A is essential nutrient for maintaining healthy mucus membranes and skin and is also essential for vision. Consumption of natural fruits rich in flavonoids helps to protect from lung and oral cavity cancers.
- ☆ Regular consumption of mustard greens in the diet is known to prevent arthritis, osteoporosis, iron deficiency anemia and believed to protect from cardiovascular diseases, asthma and colon and prostate cancers.
- ☆ Mustard greens are an excellent anticancer vegetable.

- They may also be beneficial for colds, arthritis or depression.
- While mustard greens sold in the United States are relatively mild in flavour, some mustard green varieties, especially those in Asia, can be as hot as a jalapeno pepper depending on their mustard oil content.

Caution

- Like spinach, reheating of mustard green leftovers may cause conversion of nitrates to nitrites and nitrosamines by certain bacteria that thrive on prepared nitrate-rich foods. These poisonous compounds may be harmful to health.
- Phytates and dietary fibre present in the mustard greens may interfere with the bioavailability of iron, calcium and magnesium.
- Because of its high vitamin K content, patients taking anti-coagulants such as warfarin are encouraged to avoid this food since it increases the vitamin K concentration in the blood, which is what the drugs are often attempting to lower. This effectively raises the effective dose of the drug.
- Mustards contain oxalic acid, a naturally occurring substance found in some vegetables, which may crystallise as oxalate stones in the urinary tract in some people. It is therefore, people with known oxalate urinary tract stones are advised to avoid eating vegetables belong to brassica family.
- Mustard greens may also contain goitrogens, which may interfere with thyroid hormone production and can cause thyroxin hormone deficiency in individuals with thyroid dysfunction.

BROCCOLI RABE (*Broccoli rabe*, Rapini)

It is a bitter dark leafy green vegetable that resembles broccoli but with thin stalks. While it appears to have baby-like broccoli heads surrounded by leaves on long thin stems (hence the name), the florets are anything but a toned-down version of broccoli. In fact, they are bitter and pungent and more closely resemble mustard greens with a zesty flavour and intense emerald colour. Rabe is a member of the Brassiceae tribe (or cruciferous vegetables) along with mustards and cabbages, and shares a subspecies with the turnip. Broccoli rabe (pronounced "robb") is a non-heading variety of broccoli with long, thin leafy stalks topped with small florets.

Nutrition

It gives 60 calories (40 from fat), 4.5g total fat, 0.5g saturated fat, 0mg cholesterol, 210mg sodium, 3g total carbohydrate (0g dietary fibre, 0g sugar), 2g protein per serving.

Because of its impressive nutritional profile that includes beta-carotene, vitamin C, calcium, fibres and phytochemicals, specially indoles and aromatic isothiocynates, broccoli and its kin may be responsible for boosting certain enzymes that help to detoxify the body. These enzymes help to prevent cancer, diabetes, heart diseases, osteoporosis and high blood pressure. Broccoli along with onions, carrots and cabbage may also help to lower blood cholesterol. At U. S. Department of Agriculture's Regional Research Center in Philadelphia two researchers, Dr. Peter Hoagland and Dr. Philip Pfeffer, discovered that vegetables contain certain pectin fibre called calcium pectate that binds to bile acids, holding more cholesterol in the liver and releasing less into the blood stream. They found broccoli equally as effective as some cholesterol lowering drugs.

Broccoli's wealth of the trace minerals, chromium, may be effective in preventing adult-onset diabetes in some people. Diabete's expert, have found that chromium boosts the ability of insulin to perform better in people with slight glucose intolerance. This food is low in Saturated Fat. It is also a good source of Pantothenic Acid, and a very good source of Dietary Fibre, Protein, Vitamin A, Vitamin C, Vitamin E (Alpha Tocopherol), Vitamin K, Thiamin, Riboflavin, Niacin, Vitamin B6, Folate, Calcium, Iron, Magnesium, Phosphorus, Potassium, Zinc and Manganese.

Broccoli Rabe also contains a phytochemical called lutein. Lutein is an antioxidant that protects the retina of the eye from oxidative damage caused by harmful free radical molecules. Studies have found that people who get the most lutein from their diet have a 20 to 50 per cent lower risk of cataracts compared with people who consume the least.

Lutein-rich vegetables such as Broccoli rabe, spinach, kale, romaine lettuce may even slow the onset of age-related macular degeneration, a disease that attacks the center of the retina and diminishes the ability to see fine details.

Broccoli Rabe is also good for your bones thanks to its sizable vitamin K content. A recent study from Tufts University found that women who consumed the most vitamin K had significantly better bone mineral density compared to those whose diets contained the least.

Health Benefits

Broccoli Rabe is highly nutritious and contains many of the same cancer-fighting elements as its cruciferous relatives. It is high in phytochemicals such as sulforaphane and indoles, which are believed to help the body naturally defend itself against certain cancers. Like all Brassicas, it's a rich source of glucosinolates, which our body converts to cancer-fighting sulforophanes and indoles. Studies show that these compounds are particularly effective against stomach, lung, and colon cancers, and promising research hints at protective effects against breast and prostate cancers as well.

A 3 1/2-ounce serving of broccoli rabe provides more than half your daily requirement of antioxidant-rich vitamins A and C, both of which fight off dangerous free radicals that can cause damage to your body's cells. The bitter green is also a good source of folate (a B vitamin that protects against birth defects and heart disease), not to mention potassium, fibre, and calcium.

1. **Rapini has cancer-fighting properties**:Rapini contains **powerful cancer fighting phytochemicals** called indole-3-carbonol (I3C). I3C helps prevent cellular damage caused by free radicals, maintains a healthy hormonal balance for both men and women and even reduce yeast infections in the body.
2. **Rapini helps keep bones strong**: Rapini's bone-strengthening properties are the outcome of its vitamin K content. One half cup serving contains 169 micrograms of vitamin K1, a daily dose enough to keep your bones from thinning.
3. **Rapini lowers the risk of heart disease**: Rapini contains strong anti-inflammatory nutrients, such as folate and vitamin C. Both nutrients reduce homocysteine, a type of amino acid that can damage the arteries causing coronary heart disease.
4. **Detox and heal with rapini**: Rapini is one of the many cruciferous vegetables that contain sulphur. Sulphur contains a specific compound called methyl solfonyl methane (MSM) that assists detoxification of the liver. Dr. Usha reported the effectiveness of MSM to reduce inflammation in arthritis patients.
5. **Improve insulin sensitivity with rapini**: Rapini contains two grams of fibre for every cup. Specifically, the fibre in rapini is soluble fibre, which slows transit and digestion time in the GI tract. Soluble fibre dissolves in water to form a gel-like material that lowers blood cholesterol and glucose levels. Combining rapini with a high carbohydrate meal (such as a pasta dish) will reduce the insulin response, which in turn will prevent both hyper and hypoglycemia.

BRINJAL (*Solanum melongena*)

Eggplant or Brinjal is a very low calorie vegetable and has healthy nutrition profile; good news for weight watchers! The veggie is popularly known as **aubergine** in the western world. Deep purple ones are considered to be most nutritious. The brinjal owes its colour to the presence of iodine and is recommended for people who suffer from goiter and thyroid disorders. Brinjal is high in bioflavanoids; plantains have high folic acid, vitamin C and potassium. Brinjal is reported to stimulate the intrapeptic metabolism of blood cholesterol. Leaf and fruit, fresh or dry produce a marked drop in blood cholesterol level. The de-cholesterolizing action is attributed to the presence of poly-unsaturated fatty acids(**Linoleic** and **Lenolenic**) which are present in flesh and seeds of the fruit in higher amounts(65.1 per cent). The presence of magnesium and potassium salts also help in de-cholesterolizing action. Aqueous extracts of fruit inhibit choline esterase activity of human plasma. It may not be a very glamorous food, but eggplant is packed with fibe and contains the whole gamut of B vitamins, which give you all the energy you need. Its deep purple colour is evidence that it has powerful antioxidants to protect brain cells and control lipid levels.

It is high in dietary fibre, low in fat contains an array of nutrients such as vitamins A, B1, B3, B6, folic acid and the minerals iron, calcium, magnesium, copper, manganese, phosphorus and especially potassium. It is also rich in anthocyanin flavonoids, particularly one called **nasunin,** a potent antioxidant and free- radical scavanger known to protect the cell membranes, whose integrity is crucial for proper metabolic activity and general health. **This** unique antioxidant in them like **nasunin** is a complex compound which defends your brain cells against damage. Many studies have proved that they are good for lessening your risk of stroke and dementia as they are packed with fibre and potassium. The best quality about eggplants is that they are pretty low in calories, thus they have a positive effect on one's heart health.

In addition, the mucilaginous fibre of brinjal can help to lower **cholesterol** levels, while its potassium content is useful in the prevention of water retension. It is reputed to reduce bleeding and has traditionally been used for the treatment of haemorrhoids and bleeding in general. It is an astringent food and is useful in the treatment of diarrhoea. It is also very suitable for inclusion in weight loss programes as one cup of cooked cubes contain only 27 calories, 3mg of sodium and a trace of fat, but due to its high fibre content, it provides a feeling of fullness.

Dry fruit is reported to contain goitrogenic principles. In native medicines, role of brinjal in treatment of liver diseases, cough due to allergy, rheumatism, colilithiasis, leucorrhea and internal worms has been mentioned(Gopalkrishnan, 2007).

Nandkarni (1927) has given the following uses of the brinjal. Pierced all over with a needle and fried in oil, the fruit is employed as cure for toothache. It has also been recommended as an excellent remedy for those suffering from liver complaints. The green leaves of brinjal plant are a source of the appetizer, aphordisiac, cardiotonic and beneficial in 'Vata and Kaph' etc. In Unani system of medicine, roots of brinjal are used to alleviate pain. Brinjal fruit is used as cardiotonic,

Laxative muturant and reliever of pain. The vegetable is high in water content and is very good resource of potassium. It is supposed to possess many medicinal properties, although it is not proven. It is still being used as a remedy for cancer, hypertension and diabetes.

Nutritional Profile and Medicinal Value of Brinjal

☆ **Reduced risk of cancer:** When used properly, eggplant can provide many health benefits, including being important in the treatment and prevention of cancer. Eggplant has been found to be especially useful in the treatment of colon cancer due to the high amount of fibre found within eggplant. Fibre is important in the treatment of colon cancer because it is a relatively porous nutrient, and because of this, as it moves through the digestive tract,it has the tendency to absorb toxins and chemicals that can lead to the development of colon cancer. For best results, individuals who are interested in reducing their risk for the development of color cancer should be sure to include the skin of the eggplant during consumption. Research has found that the skin of eggplant may contain more fibre that the actual eggplant itself (Moghul, 2013).The purple pigments in brinjals contain phenolic flavonoid phyto-chemicals (called anthocyanin) that have potential health effects against cancer and other problems like aging, inflammation and neurological disorders.

- ✰ **Eggplants help weight loss**: As stated above, eggplant contains high amounts of fibre, making it a great food in the fight against cancer. But, the fibre found in eggplant has other uses-namely, its ability to be a useful tool for people who are trying to lose weight.

  Fiber is a relatively "bulky" food, meaning that it takes up a lot of room in the stomach. Therefore, by eating eggplant in a salad or **appetizer** before a meal, dieters are likely to have a greater feeling of satiety, and generally eat fewer calories (thereby achieving a substantial weight loss with time). In addition, fibre is slow digesting, and takes a long time to move from the stomach to the digestive tract. Because of this, the eggplant keeps dieters feeling full for a longer period of time-and therefore, they won't be as tempted to snack between meals, which will again aid in weight loss.

- ✰ **Eggplants for better skin tone:** Finally, eggplant contains a high amount of water-and similarly, aside from fat and bone, our bodies are composed of almost primarily water. Water has important roles in a variety of pathways through the body, but has been found to be especially important in the maintenance of healthy skin and hair.

  Individuals who are dehydrated are more likely to exhibit hair that is thin, dry and has split ends, along with skin that appears to be flaky, dry, with a greater number of lines and wrinkles.

  Consuming adequate amount of water through either water itself or through food items such as eggplant can not only improve the quality of your hair and skin, but also the general performance of your body. For best results, eat the eggplant raw-research has found that cooking it removes some of the water that is so beneficial (Moghul, 2013).

- ✰ **Cholesterol**: Brinjals are effective in curing high blood cholesterol, hyper-cholesterol-lemia. It is rich in antioxidant monoterpenes and helps block the formation of free radicals to maintain cholesterol.

- ✰ **Cancer:** The purple pigments in brinjals contain phenolic flavonoid phyto-chemicals (called anthocyanin) that have potential health effects against cancer and other problems like aging, inflammation and neurological disorders.

- ✰ **Flatulence:** Roasted brinjal soup taken with asafeotida and garlic is very good for getting rid off gas and congestion within the body.

- ✰ **Hypertension:** The presence of potassium, an important intra-cellular electrolyte, helps to counteract the hypertension effects of sodium.

- ☆ **Metabolism:** Brinjals contain good quantity of vitamins B-complex (B1, B3, B5 and B6). These vitamins are essential for fat, protein and carbohydrate metabolism.
- ☆ **Dietary fibre:** This vegetable is extremely low in fat and calories and is rich in soluble fibre content. 100 gram bringals contain just 24 calories and 9 per cent of RDA of fibre.
- ☆ **Digestion**: Meshed brinjal and tomato soup taken with salt and pepper seasoning helps promote digestion and enhance appetite.
- ☆ **Insomnia:** If you have problem getting sound sleep, eat baked brinjal with honey daily before bed time. It will cure your insomnia after sometime.
- ☆ **Enlarged spleen**; It is a valuable remedy for treating enlarged spleen caused by malaria. Consume soft baked brinjal with raw sugar on empty stomach for curing it.
- ☆ **Stones:** Eating brinjals helps in destroying stones in their early stages.

Wind humor and phlegm can be cured by taking roasted, peeled brinjals with a dash of salt.

Health Benefits

- ☆ Eggplant is very low in calories and fats but rich in soluble fibre content. 100 g provides just 24 calories but contributes about 9 per cent of RDA of fibre.
- ☆ Research studies at the Institute of Biology of *Sao Paulo State University*, Brazil showed that eggplant is effective in the treatment of high blood cholesterol.
- ☆ It contains good amounts of many essential B-complex groups of vitamins such as pantothenic acid (vitamin B5), pyridoxine (vitamin B6) and thiamin (vitamin B1), niacin (B3). These vitamins are essential in the sense that body requires them from external sources to replenish and required for fat, protein and carbohydrates metabolism.
- ☆ It is also good source of minerals like manganese, copper, iron and potassium. Manganese is used as a co-factor for the antioxidant enzyme, *superoxide dismutase.* Potassium is an important intracellular electrolyte helps counter the hypertension effects of sodium.
- ☆ The peel or skin (deep blue/purple varieties) of aubergine has significant amounts of phenolic flavonoid phyto-chemicals called **anthocyanins.** Scientific studies have shown that these anti-oxidants have potential health effects against cancer, aging, inflammation, and neurological diseases.

Caution

Brinjal contains very small amount of nicotine than any other edible plant with a concentration of 0.01mg/100g. However, the amount of nicotine from eggplant or any other food is negligible and therefore shall not warrant against its usage. So, enjoy!

TOMATO (*Lycopersicon esculentum*)

Technically called a fruit, tomatoes are mostly eaten as a well-liked vegetable. It is one of the natures' attractive deobstruents. It removes disease particles from the body and even cleans the kidneys of toxins. Tomato is rich in calcium, phosphorus, vitamin C and carbohydrates. It actually triggers off alkaline reactions in the body though it tastes acidic. Its acidic taste is due to the presence of a little citric acid, and a little malic acid. Because of its alkaline nature it acts as an antacid. Tomato is helpful to those who are diabetic. It is said to be effective in controlling sugar in the urine in addition to being a good food because of its low carbohydrate concentration. The most important feature is the presence of glucoalkaloides, **Tomatine**[C_{50} H_{83}[81] O_{21} N and traces of **Solanine** [$C_{45}H_{71}$ O_{15} N]. The former is of a significant therapeutic value. Tomatine has an antibiotic property against fungal affliction of skin, particularly in cancer of mouth, stomatitis and chelilosis etc. **Tomatine** acts as a blood purifier and precipitating agent for cholesterol. Being rich source of vitamin C, it is a good anti scorbutic. Tomatoes help in the formation of gamma aminobutyric acid (GABA), a compound that can help bring down the blood pressure. Tomatoes have lycopene and red orbs which are well known for their cancer-defending properties. They are full of Vitamin A and Vitamin K which help in keeping your blood pressure levels under control.

Tomatoes have a good amount of vitamin A, and so are good for eyes and the optic system. The vitamin A found in tomatoes is ten times the amount found in a similar quantity of oranges. Though tomato juice contains a great deal of citric acid, it is not acid forming but gets alkaline within the body. It acts as an eliminator, and helps to remove uric acid from the body. They are also good for those who want to trim off that excess weight. Tomato juice helps digestive system to perk up and add pep to the liver functions. Sluggish liver, excessive formation of gas etc. are taken care of by the tomato. Asthmatic patients should ensure an all year supply of tomatoes for they reduce the congestion and keep a check on the increased secretion of mucous. Naturopaths believe that they also help in reducing spasms. Harvard Medical School found that consumption of tomato lower the risk of suffering from cancer and also provide general resistance. In tomato, the red colour is due to **lycopene**(Table 2.1) which is again a precursor for vitamin A, is a powerful antioxidant that can protect against harmful free radicals, and anthocyanin pigment in purple aubergines is also water insoluble. Eating over 10 tomatoes/week has shown to reduce cancer by 35 per cent and is also effective in reducing cholesterol. The red colour of tomato protects the human body from the ravages not only of cancer but cardiovascular diseases. Tomatoes have been ranked high as a protector against acute appendicitis. Research findings have revealed that eating tomatoes or tomato sauce four days a week reduces the risk of prostrate cancer and tomatoes have been ranked high as a protector against appendicitis(Anonymous(1993). Ronald G. Butterry, an Agricultural Research Service Scientist in Albany, California found that tomatoes kept in the refrigerator lose their freshness fast and further stated that such tomatoes had less of aroma- imparting chemical **(Z)-3-hexanol** than did tomatoes kept at room temperature(Khan, 1989).

Tomatoes contain modest levels of many nutrients including provitamin A, vitamin C and potassium as well as other components such as flavonoids and lycopene that may contribute to cardiovascular health (Willcox, *et al.*, 2003). Recent study demonstrated that a clarified tomato juice inhibits platelet aggregation in vitro and exvivo following consumption of tomato by the people with type 2 diabetes. These results support epidemiological data showing that consumption of tomato products are inversely associated with risk of cardiovascular disease. A number of good characteristics are also attributed to lycopene including the potential ability to reduce the risk of developing cancer. Lycopene is an effective antioxidant which is also potentially linked to preventing cardio-vascular diseases. Lycopene (2.1) content is normally correlated to colour, cartenoids and dry matter. These aspects are important quality factors of any cultivar released in to the market. However, the antioxidant levels in tomatoes can be influenced by environmental factors such as light and temperature, genetics and cultural practices. Tomato can be an effective alternative to drugs in lowering cholesterol and blood pressure and in preventing heart disease.

Scientists recently found that people in northern Italy who ate seven or more servings of raw tomato every week had 60 per cent less chance of developing colon, rectal or stomach cancer than those who ate two servings or less. Almost a day's requirement of calcium is present in a medium sized tomato and the protein content is almost the same as mother's milk. The vitamin C content of a tomato helps to form and maintain collagen which holds together all cells in the body. Collagen is concentrated in the ligaments, walls of the blood vessels, base of bones, developing teeth and gives all these structures strength and elasticity. Vegetables such as tomato, onion and cabbage have compounds that help to keep cholesterol down, arteries clear, bolster resistance to cancer in numerous ways. Tomatoes are high in vitamin A and C and are naturally low in calories. Being an excellent source of lycopene, which is the pigment that makes tomatoes red and has been linked to the prevention of many types of cancer. Lycopene is an antioxidant which fights free radicals that can interfere with normal cell growth and activity. These free radicals are what can potentially lead to cancer, heart disease and premature ageing. The best source of lycopene are found in processed tomato products, such as ketchups and tomato products. A medium sized tomato has about as much fibre as a slice of whole – wheat bread and only about 35 calories. It is recommended to have at least 16 milli grams of lycopene per day as they reduce the number of free radicals in the body and also help to retain the moisture in people who are used to sitting in air- condioned enviorns.

Tomatoes contain **lycopene**, one of the most powerful natural anti-oxidants, which, especially when tomatoes are cooked, has been found to help prevent prostate cancer in men, also reduces the chance of stomach and colourectal cancer. Lycopene is considered somewhat of a natural miracle antioxidant that may help to stop the growth of cancer cells. Tomato extract branded as **lycomato** is now also being promoted for treatment of high blood pressures. Not only tomatoes taste great but they also help the skin healthy. Lycopene has also been shown to improve the skins ability to protect against harmful U V rays, as it works as suncreen from within.

Natural genetic variation in tomatoes and their wild relatives has given a genetic treasure trove of genes that produce lycopene, carotene, anthocyanin and other anti-oxidants. Tomato varieties are available with double the normal vitamin C, 40 times normal vitamin A, high levels of anthocyanin and two to four times the normal amount of **lycopene** (numerous available cultivars with high crimson gene). Researchers at University of Toronto in Canada, who carried out the study, claimed that drinking two glasses of tomato juice containing 15mg of lycopene could be enough to strengthen brittle bones. Research coducted by Karin Ried (University of Adelaide- South Australia) revealed that if more than 25 milligrams of Lycopene is taken daily, it can reduce LDL (bad) cholesterol by up to 10 per cent. Half a litre of tomato juice taken daily or 50 grams of tomato paste, provide protection against heart disease. According to them, the key ingredient –lycopene, the anti- oxidant, already credited with cutting risk of prostrate cancer in men and protecting against heart diseases and led to a significant drop in levels of N- telopedtide in womem drinking either type of juice or taking the capsule. Due to potassium and vitamin B, tomatoes help to lower blood pressure and to lower high cholesterol levels, which in turn, could help prevent strokes, heart attacks and other potentially life threatening heart problems. Because of **vitamin A**, tomatoes are also an excellent food to help improve your vision, and help prevent night blindness. Vitamin A is good for eyes, skin, bones and keeps your hair strong and shining.

There are many reasons to toss a tomato in your diet, it adds to the content of collagen in the skin, thus preventing it from sagging. Also, tomatoes have higher antioxidants, this helps reduce fine lines on the skin. Also, eating tomato sauce and tomato juice can help clear up acne.

Plant Material Sources of Lycopene (mg/kg)

Vegetables (Fresh wt.)	*Scientific Name*	*Lycopene Content*
Asparagus (Cooked)	*Asparagus officinalis*	0.30
Carrots Orange	*Daucus carota*	0.02
Tomato Ripe	*Lycopersicon esculentum*	25.73
Red Cabbage	*Brassica oleracea*	0.20
Red Carrots	*Daucus carota*	61.00
Red peppers	*C. annum*	03.08
Water mellon	*Citrulla lanatus*	45.32
Swede	*Brassica rapa*	0.50

Tomatoes are a staple in every kitchen but hardly will you hear any one extolling its cosmetic benefits. Whether you want to cure large pores or reduce acne and rashes or sooth a nasty sun burn, or simply to revive the glow on dull skin, tomatoes are beneficial in many home made beauty treatments. Not only do tomatoes taste great but they keep the skin healthy. It is therefore, necessary to consume tomato as they have lycopene, which is an antioxidant and hence works as a sun screen from within. These antioxidants make tomato an anti- ageing product as they help in fighting cellular damage and reddening of skin. It is recommended to have at least 16 milligrams of lycopene per day as they reduce the number of free radicals in the body and also help to retain the moisture in the people who are used to sitting in air conditioned enviorns.

Health Benefits of this Rich Easily available Vegetable

- **Anti-oxidant**: Tomatoes contain a lot of vitamins A and C, mostly because of beta-carotene, and these vitamins act as an anti-oxidant, working to neutralize dangerous free radicals in the blood stream. Free radicals in the body can be flushed out with high levels of Lycopene, and the tomato is so amply loaded with this vital anti-oxidant that it actually derives its rich redness from the nutrient. These dangerous free radicals can cause cell damage. Also, remember that cooking destroys much of vitamin C, so stick to raw tomatoes for these benefits. So, do remember to carry a tomato-cucumber sandwich in your tiffin box.
- **Vision**: Because of all that vitamin A, tomatoes are also an excellent food to help improve your vision. This also means tomatoes can help your eyes be better about night blindness.
- **Cancer**: Various studies have shown that because of all that lycopene in tomatoes, the red fruit helps to lessen the chances of prostate cancer in men, and also reduces the chance of stomach cancer and colourectal cancer. Lycopene is considered somewhat of a natural miracle anti-oxidant that may help to stop the growth of cancer cells. And, interestingly enough, cooked tomatoes produce more lycopene than do raw tomatoes, so enjoy that tomato soup!
- **Heart troubles**: Due to potassium and vitamin B, tomatoes help to lower blood pressure and to lower high cholesterol levels. This, in turn, could help prevent strokes, heart attack and other potentially life-threatening heart problems.
- **Skin care**: Because of high amounts of lycopene, tomatoes are excellent for skin care. Make sure the inside of the tomato skins are against your skin, and let this be there for at least 10 minutes. Then wash off. You face will be cleaner and give a shiny look.
- **Hair**: Remember all that vitamin A in tomatoes? Well, it's good for keeping your hair strong and shiny, and its also good for your eyes, teeth, skin and bones.
- **Bones**: Tomatoes have a fair amount of vitamin K and calcium, both of which help to strengthen bones.
- Tomatoes are one of the low calorie vegetables containing just 18 calories per 100 g. They are also very low in any fat contents and have zero cholesterol levels.
- The antioxidants present in tomatoes are scientifically found to be protective against cancers including colon, prostate, breast, endometrial, lung, and pancreatic tumors.

- ☆ **Lycopene**, a flavonoid antioxidant, is the unique phytochemical present in the tomatoes. Red varieties are especially concentrated in this antioxidant. Together with carotenoids, it has the ability to protect cells and other structures in the body from harmful oxygen free radicals. Studies have shown that *lycopene* prevents skin damage from ultra-violet (UV) rays and offers protection from skin cancer.
- ☆ **Zeaxanthin** is another flavonoid compuond present abundantly in this vegetable. Zeaxanthin helps protect eyes from "age related macular disease" (ARMD) in the elderly persons by filtering harmful ultra-violet rays.
- ☆ The vegetable contains very good levels of vitamin A, and flavonoid anti-oxidants such as α and ß-**carotenes**, xanthins and lutein. Altogether, these pigment compounds are found to have antioxidant properties and are take part in vision, maintain healthy mucus membranes and skin, and bone health. Consumption of natural vegetables and fruits rich in flavonoids is known to help protect from lung and oral cavity cancers.
- ☆ In addition, they are also good source of antioxidant **vitamin-C** (provide 21 per cent of recommended daily levels per 100 g); consumption of foods rich in vitamin C helps body develop resistance against infectious agents and scavenge harmful free radicals.
- ☆ Fresh tomato is very rich in potassium. 100 g contain 237 mg of potassium and just 5 mg of sodium. Potassium is an important component of cell and body fluids that helps controlling heart rate and blood pressure caused by sodium.
- ☆ They contain moderate amounts of many vital B-complex vitamins such as folates, thiamin, niacin, riboflavin as well some essential minerals like iron, calcium, manganese and other trace elements.
- ☆ Tomatoes are rich in lycopene, flavonoids and other phytochemicals with anticarcinogenic properties
- ☆ Tomatoes are an excellent source of vitamin C (the vitamin C is most concentrated in the jelly-like substance that surrounds the seeds)
- ☆ They also contain vitamin A and B-complex vitamins, potassium and phosphorus
- ☆ A tomato grown in a hothouse has half the vitamin C content as a vine-ripened tomato
- ☆ Drink tomato juice with salt and pepper every morning on an empty stomach for controlling diabetes.

OKRA (*Abelmoschus esculentus*)

An annual, the vegetable is produced for its green pods which contain large amounts of vitamins A and C, calcium, phosphorus, potassium and magnesium. Okra is the number one vegetable source of manganese. The plant is also rich in mucilaginous substances, which lubricate the intestines and are recommended for soothing duodenal ulcers and reducing high cholesterol levels. Internally, the pods feature small, round, mucilaginous white coloured seeds arranged in vertical rows. The pods are harvested while immature and eaten as vegetable.

Health Benefits

- ☆ Very low in calories, provides just 30 cal per 100 g and contains no saturated fats or cholesterol; but is a rich source of dietary fibre, minerals, vitamins; recommended in cholesterol controlling and weight reduction programs.
- ☆ The rich fibre and mucilaginous content in Okra pods helps smooth peristalsis of digested food particles and relieve constipation condition.
- ☆ The pods contain healthy amounts of vitamin A, and flavonoid anti-oxidants such as beta-carotenes, xanthin and lutein. It is one of the *green* vegetable with highest levels of these anti-oxidants. These compounds are known to have antioxidant properties and are essential for vision. Vitamin A is also required for maintaining healthy mucus membranes and skin. Consumption of natural vegetables and fruits rich in flavonoids helps to protect from lung and oral cavity cancers.
- ☆ Fresh pods are good source of folates; provide about 22 per cent of RDA per 100 g. Consumption of foods rich in folates, especially during pre-conception period helps decrease the incidence of neural tube defects in the offspring.
- ☆ The pods are also an excellent source of anti-oxidant vitamin, vitamin-C; provides about 36 per cent of daily recommended levels. Consumption of foods rich in vitamin-C helps body develop immunity against infectious agents, reduce episodes of cold and cough and protects body from harmful free radicals.
- ☆ The veggies are rich in B-complex group of vitamins like niacin, vitamin B-6 (pyridoxine), thiamin and pantothenic acid. The pods also contain good amounts of vitamin K. Vitamin K is a co-factor for blood clotting enzymes and is required for strengthening of bones.
- ☆ The pods are also good source of many important minerals such as iron, calcium, manganese and magnesium.

CAPSICUM (*Capsicum annum* Peppers)

Whichever colour of bell peppers you may go for - red, orange or yellow- they are full of heart-healthy properties because of folic acid and lycopene present in them. Again, they have been substantially proved to lower your risk again cancer too. Unlike their fellow members, sweet peppers have characteristic bell shape with crunchy, thick fleshy skin. On comparison to other capsicum members, bell (sweet) peppers have very mild or zero hotness. The other important difference is that they are used worldwide as vegetables instead of spices. The Scientists at Central Institute for Medicinal and Aromatic Plants (CIMAP) have isolated 'Capsaicin,'the pigment responsible for pungency in the fruits of family Capsicum annum, which they reported has the capacity to cure common ailments like pain, arthritis, sprain and nerve reception. Capsaicin induces the process of apoptosis that destroys potential cancer cells and reduces the size of leukemia tumour cells considerably. It can be concluded that apart from setting our tongues on fire, chilli peppers can scare cancer pathogens off too.

Capsicum annum plant with fruits

Its colourful fruit – red, green and yellow peppers or capsicum – provide an excellent source of beta-carotene and vitamin C; they also contain high amounts of calcium, phosphorus, sodium and potassium. They also contain two other important carotenoids, **lutein** and **zeaxanthin,** both of which can help protect against age-related mascular degeneration (AMD), which can lead to a loss of central vision in the elderly. Peppers can improve appetite and digestion, promote circulation and reduce swellings.

Capsaicin (Methyl vanillyl nonenamid) the phytochemical present in chilli, impedes carcinogens such as nitrates and cigarette smoke from attaching to cellular material, thus preventing formation of cancer cells. It may also kill the bacteria that can cause ulcers. Capsaicin is shown to be a potent stressor agent and produces a rise in **corticosterone** level. When applied to human skin capsaicin produces erythema and burning without blistering. Intragastrically capsaicin promotes ulceration. Capsaicin at a dilution of 1: 1000 showed bacteriostatic activity against several organisms. Oleorsin red pepper- obtained from the longer, moderately pungent chillies chiefly used for flavouring.

Capsaicin wool, prepared by dissolving the **Oleoresin** in ether and pouring it on absorbent cotton wool, is useful in rheumatoid affections. **Capsicum** tincture has been employed externally in the treatment of chilblains. Chilli, in small doses, is a powerful stimulant, carminative and helps in increasing the metabolism. Capsaicin is a thermogenic food, so it causes the body to burn calories for 20 minuts after you eat the chillies. It stimulates the secretion of saliva and gastric juice, and increases peristaltic movements and motility of stomach. It is an excellent remedy in atonic and flatulence dyspepsia and dipsomania. Most of the capsaicin in a pungent (hot) pepper is concentrated in blisters on the epidermis of the interior ribs (septa) that divide the chambers of the fruit to which the seeds are attached. The amount of capsaicin in hot peppers varies significantly between varieties and is measured in Scoville heat units (SHO). Capsaicin in tropical form is promoted mainly for pain caused by conditions such as arthritis and general muscle soreness. There is some evidence that capsaicin may be useful in managing post-surgical pain from mastectomy, thoracotomy (chest surgery), amputation and other surgery related to main stream cancer treatment. Researchers have found that capsaicin may provide temporary relief for pain from mouth sores caused by chemotherapy and radiation. Some proponents claim that capsaicin has anti-oxidant properties which help to fight the carcinogen nitrosamine (a cancer causing agent). **An anti-oxidant is a compound that blocks the action of activated oxygen molecules, known as free radicals that damage cells.** Still others claim that it may prevent DNA damage and cancer of the lungs from cigarette smoke. Capsaicin is also active ingredient in several Tropical medications for temporary pain relief. A new study has found that chili peppers which usually make people 'tear up' may help people 'clear up' certain types of sinus inflammation. The study showed that participants who used a nasal spray with capsicum, reported a faster onset of action or relief, on an average within minute of using the spray.

Over the years the Capsicum annum or Capsicum frutescens has been used by alternative medicine practitioners as a remedy for a variety of conditions, such as upset stomach, menstrual cramps, head aches, shingles (herpes zoster), diarrhea, loss of appetite, stomach ulcers, poor digestion, sore throat, itching, alcoholism, motion sickness, tooth ache, malaria, yellow fever. While as some petitioners also claim it can prevent colds, heart diseases and stroke, increase sexual potency, foster weight loss and strengthen the heart, although available scientific evidence does not support these clams. Oleoresin of capsaicin (OC) is used in spray form as a way to incapacitate people in threatening situations. Usually this pepper spray causes burning of the eyes, nose, mouth and skin.

Health Benefits

- ☆ Capsicum contains impressive list of plant nutrients that are known to have disease preventing and health promoting properties. Unlike other chili peppers, it is very low in calories and fats. 100 g provide just 31 calories.
- ☆ Sweet (bell) pepper contains small levels of health benefiting an alkaloid compound capsaicin. Early laboratory studies on experimental mammals suggest that capsaicin has anti-bacterial, anti-carcinogenic, analgesic and anti-diabetic properties. When used judiciously it also found to reduce triglycerides and LDL cholesterol levels in obese individuals.
- ☆ Fresh bell peppers, red or green, are rich source of vitamin-C. This vitamin is especially concentrated in red peppers in highest levels. 100 g fresh red pepper provide about 127.7 mcg or about 213 per cent of RDA. Vitamin-C is a potent water soluble antioxidant. It is required for the collagen synthesis in the body. Collagen is the main structural protein in the body required for maintaining the integrity of blood vessels, skin, organs, and bones. Regular consumption of foods rich in vitamin C helps body protect from scurvy; develop resistance against infectious agents (boosts immunity) and scavenge harmful, pro-inflammatory free radicals from the body.
- ☆ It also contain good levels of vitamin-A. 100 g of sweet pepper has 3131 IU or 101 per cent of vitamin A. In addition, it contains anti-oxidant flavonoids

such as α and ß carotenes, lutein, zeaxanthin, and cryptoxanthin. Together, these antioxidant substances in capsicum helps to protect body from injurious effects of free radicals generated during stress and diseases conditions.

- Bell pepper has adequate levels of essential minerals. Some of main minerals in it are iron, copper, zinc, potassium, manganese, magnesium, and selenium. Manganese is used by the body as a co-factor for the antioxidant enzyme superoxide dismutase. Selenium is anti-oxidant micro-mineral that acts as co-factor for enzyme superoxide dismutase.
- Capsicum is also good in B-complex group of vitamins such as niacin, pyridoxine (vitamin B-6), riboflavin, and thiamin (vitamin B-1). These vitamins are essential in the sense that body requires them from external sources to replenish. B-complex vitamins facilitate cellular metabolism through various enzymatic functions.
- Capsicum is recommended for those suffering from constipation, because of its laxative effect.

Caution

The pungent level in bell peppers is almost zero "Scoville heat units" (SHU). Sometimes the seeds and central core may contain some capsaicin, which when eaten causes severe irritation and hot sensation to mouth, tongue and throat.

Note some of these points while handling capsicum annum members in general:

- Capsaicin in chilies, especially cayenne peppers, initially elicit inflammation when it come in contact with the mucus membranes of oral cavity, throat and stomach, and soon produces severe burning sensation that is perceived as 'hot' through free nerve endings in the mucosa. Eating cold yogurt helps reduce the burning pain by diluting capsaicin concentration and preventing its contact with stomach walls.
- Avoid touching eyes with pepper contaminated fingers. Rinse eyes thoroughly in cold water to reduce irritation.
- They may aggravate existing **gastro-esophageal reflux** (GER) condition.
- Capsaicin, the active ingrident in chilli, does cause tissue inflammation so that mucosa of the stomach or intestine might be damaged by a sufficiently large dose (John Prescott, 2011).

CHILLI (*Capsicum annum*)

Chilli plant is a perennial small shrub with woody stem growing up to a meter height and bears white coloured flowers. The pods are very variable in size, shape, colour and pungency. Depending on the cultivar type, they range from the mild, fleshy, Mexican bell peppers to the tiny, fiery, finger-like chili peppers, commonly grown in Indian subcontinent. The hotness of chili is measured in "Scoville heat units" (SHU). On the Scoville scale, a sweet bell pepper scores 0, a jalapeno pepper around 2,500-4,000 and a Mexican habaneros 200,000 to 500,000. Chilies can be

picked up while they are green or when they reach complete maturity and dried in the plant. Usually, the fruits are picked up by hand when they are matured and turned red. They are then left to dry which causes them to shrivel. Chilies have strong spicy taste that comes to them from the active alkaloid compounds *capsaicin, capsanthin* and *capsorubin*.

Its taste is extremely used as a local stimulant for the tonsils, internally it is irritant and large doses produce gastroenteritis. Its fruit is digestive, stimulant, prevents heart diseases and curative for rheumatic troubles. According to boffins, capsaicin, the stuff that gives chili, peppers their kick, may cause weight loss and fight build up by trigging certain beneficial protein changes in the body. As laboratory studies have hinted that capsaicin may help fight obesity, by decreasing calorie intake, shrinking fat tissue and lowering fat levels in the blood (India times. com)#. Foods containing chillies are said to be as foods that burn fat. Chilli peppers activate sympathetic nervous system, which helps control most of the body's internal organs, to expend more energy, so that the body burns more calories when the the same food is eaten with chilli peppers. Eating chillies is associated increases in metabolic rate and thermogensis. Chillies contain capsaicin, a chemical which also increases blood flow and triggers the release of mood-enhancing endorphins that naturally pump up your libido.

According to Zhiming Zhu of third Military Medical University (China) chilli might set your mouth on fire but it also leads blood vessels to relex. Long term dietary consumption of capsaicin could reduce blood pressure in genetically hypertensive rats. Chilli also contain a compound called **dihydrocapsiate** that is a fat buster. Chillies are rich in vitamin A, B, C and E and also manganese, potassium and copper-excellent for negating the effects of harmful free radicals in the body. The capsaicin in chillies also offer great pain relief and thanks to the burning sensation associated with them are a great help in relieving nasal congestion. It is recommended that you consume chillies three times a day to enjoy their complete health benefits. Needless to say this should not be a problem given that green and red luscious chillies are present in almost everything we eat.

According to AYURVED- it suppress Kapha, Vata, promotes bile and is lekhan (scraping). It purifies blood, cures gastric problems and sleeplessness, stimulates the heart and cleanses the urinary system. It is anti-pyretic and especially effective in controlling dangerous fever. Due to its hot, sharp and bitter nature, it induces saliva secretion thus helps digestion, but in large amounts may cause burning sensation. Food scientists say that capsaicin actually causes your body to heat up. However, researchers have found growing evidence that the body- heat generating power of peppers might even lend a hand in our quest to lose those extra inches accumulating around the waistline.

#-Appears in AC'S monthly Journal of Proteome Research.

Health Benefits of Chili Peppers

- ☆ Chilli pepper contains chemical compounds that are known to have disease preventing and health promoting properties.
- ☆ Chillies contain health benefiting an alkaloid compound *e. g.* **capsaicin** which gives strong spicy pungent character. Capsaicin has anti-bacterial, anti-carcinogenic, analgesic and anti-diabetic properties. It also found to reduce LDL cholesterol levels in obese individuals.
- ☆ Fresh chilli peppers, red or green, are rich source of **vitamin-C**.
- ☆ They are also good in other antioxidants like vitamin A, and flavonoids like beta-carotene, alpha carotene, lutein, zeaxanthin, and cryptoxanthins. These antioxidant substances in capsicum helps to protect body from injurious effects of free radicals generated during stress, diseases conditions.
- ☆ Chillies contain good amount of minerals like potassium, manganese, iron, and magnesium. Potassium in an important component of cell and body fluids that helps controlling heart rate and blood pressure. Manganese is used by the body as a co-factor for the antioxidant enzyme superoxide dismutase.
- ☆ Chillies are also good in B-complex group of vitamins such as niacin, pyridoxine (vitamin B-6), riboflavin and thiamin (vitamin B-1). These vitamins are essential in the sense that body requires them from external sources to replenish.

Medicinal Uses

- ☆ Chilli peppers contain chemical compound capsaicin. Capasicin and its co-compounds used in the preparation of ointments, rubs and tinctures for their astringent, counter-irritant and analgesic properties.
- ☆ These formulations have been in use in the treatment of arthritic pain, post herpetic neuropathic pain, sore muscles etc.
- ☆ Capsaicin has anti-bacterial, anti-carcinogenic, analgesic and anti-diabetic properties. It also found to reduce LDL cholesterol levels in obese persons.

Caution

- ☆ Chilli peppers contain active component in them, capsaicin, which gives strong spicy pungent character which when eaten causes severe irritation and hot sensation to mouth, tongue and throat.
- ☆ Capsaicin in chillies initially elicit inflammation when it come in contact with the delicate mucus membranes of oral cavity, throat and stomach, and soon produces severe burning sensation that is perceived as 'hot' through free nerve endings in the mucosa. Eating cold yogurt helps reduce the burning pain by diluting capsaicin concentration and also preventing its contact with stomach walls.

- ☆ Avoid touching eyes with chili contaminated fingers. Rinse eyes thoroughly in cold water to reduce irrittion.
- ☆ Chilies may aggravate existing gastro-esophageal reflux (GER) condition.
- ☆ Certain chemical compounds like aflatoxin (fungal mold), found in spoiled chilies have been known to cause stomach, liver and colon cancers.

ONION (*Allium cepa*)

Including onions in your daily food intake, gives your body's immunity levels a royal boost. They are full of peptide known as GPCS which helps in keeping your body's calcium level under control. Packed with Vitamin C and folate, onions promote good cardiovascular and gastrointestinal health.

The healing properties of onions are now scientifically established and are attributed to their many important components, such as **quercetin**, **mustard oils** and sulphur-containing compounds. Onions can reduce blood stickiness, preventing blood clots and heart attacks. Crude extracts of onions have been shown to lower high blood pressure and cholesterol levels. Onions are also used to expel phlegm and alleviate coughs, colds and inflammations of the nose and throat. Onion soup can help cure hangovers and is good for coughs, colds and bronchitis.

It contains an acid volatile oil(0.05 per cent) with a pungent smell. The oil is rich in sulphur and contains a variety of aliphatic disulphide, dipropyl disulphide, methyl propyl disulphide and their trisulphide. Onions contain an enzyme called "**alliinase**" which converts anlliins to disulphide oxides. These oxides are allicin type compounds. The bactericidal action of onion and garlic juice is due to allicin type compounds. The allicins and disulphides interact with -SH(Thiol) compounds like cysteine and prevent their incorporation into proteins, which inhibits the growth of bacteria. The hypoglycemic [blood sugar lowering) effects of onion and garlic extracts has long been recognized. Recently, another anti hypoglycemic agent present in onion has been identified *viz.* **diphenylamine**, a non sulphur compound to have hypoglycemic action. Diphenyl amine found in onion is also effective against diabetes. Compound, dially disulphide limits the amount of cartilage damaging enzymes when introduced to human. The doctors at Kings College London have further found that those whose consume a healthy diet with a high intake of fruits and vegetables particularly alliums such as garlic there was less evideence of early osteoarthritis in the hip joints.

The dry **onion skin,** *are 'rich in compounds that are beneficial for human health, a*ccording to the researchers, the brown skin and external layers are rich in fibre and flavonoids, while the discarded bulbs contain sulphurous compounds and fructans.

According to the study, the brown skin could be used as a functional ingredient high in **dietary fibre** (principally the non-soluble type) and phenolic compounds, such as quercetin and other flavonoids (plant metabolites with medicinal properties). The two outer fleshy layers of the onion also contain fibre and flavonoids.

Eating fibre reduces the risk of suffering from **cardiovascular disease**, gastrointestinal complaints, colon **cancer, type-2 diabetes** and obesity", added the researcher. Phenolic compounds, meanwhile, help to prevent coronary disease and have anti-carcinogenic properties. The high levels of these compounds in the dry skin and the outer layers of the bulbs also give them high antioxidant capacity.

Charak, Sushruta and other leading lights of Ayurveda have described onion to be carminative, digestive, liver stimulant, anti- inflammatory, expectorant, diureti, emmanagogue (that enhances the menstural flow) and aphordisiac in effect. Scientific studies have also found that onion is an anti-oxidant, anti- rheumatic, anti-tumour, lipid lowering and bronchodilatory agent. Many of its therapeutic effects match to those of garlic. The use of onion has been found to be effective in conditions like anorexia, indigestion, flatulence, bronchitis, asthma, otalgia, osteoporosis and skin blemishes. Ancient reference indicate that regular use of onion enhances glow on one's face and keeps premature changes of age at bay. Researchers have found that under controlled studies onion inhibit platelet aggregation and increased the bleeding time. It was observed that an addition of onion to a fatty diet can protect against changes in serum cholesterol. It tends to keep fibrinolytic activity at normal levels in ischemic heart patients, thereby preventing further risk of arterial thrombosis. Res\asonable consumption of the allium group of herbs like onion and garlic has been found to be protective against arthritic conditions and cancer of the stomach and the colon (Vatsyayan, 2010). Onion, garlic, shallot, leek chives, bunching onion etc. regulate the blood sugar in the body to moderate levels thereby, promote the death of the cancer cells in colon, breast, lung and prostate cancers.

According to traditional medicinal literature they are a source of many vitamins and are useful in fever, dropsy, catarrh, and chronic bronchitis, mixed with common salt, onions are a domestic remedy in colic and scurvy. Rosted or otherwise they are applied as a poultice to indolent boils, bruises, wounds etc. to relieve hot sensation and applied to the navel for dysentery and fever. Raw onion has a antiseptic value though out the alimentary canal that is more effective than when roasted or cooked. Eaten raw it is also a diuretic and emmenagogue. Warm juice is dropped into the ear to relieve each ache, applied hot to the soles of feet in convulsive disorders. It is sniffed in epistaxis. It is applied to eyes in dimness of vision and locally to ally the irritation of insect bites and scorpion stings and of skin diseases. It is given to as an antidote in tobacco poisoning. Mixed with mustard oil in equal proportions it is good application for rheumatic pains and other inflammatory swellings. Onion eaten with unrefined sugar stimulate the growth of children. Oil contained in the bulbs is a stimulant, diuretic and expectorant, Mixed with vinegar, onions are useful in case of soar throat. Cooked with vinegar they are given in jaundice, spleenic enlargement, and dyspepria. In malaria fevers they are eaten twice a day with remarkable relief. Rosted onions mixed with cumin, vinagar candy, butter oil are a demulcent of great benefit in piles((Nandkarni, 1954). Fresh onion juice is moderately bactericidal. The essential oil(0.05 per cent of bulbs) contain a heart stimulant, increases pulse volume, and frequency of systolic pressure and coronary flow and stimulates the intestinal smooth musculature and the uterus.

As per Bhava Prakash, onion is tasty, eradicates gastric problems, promotes physical strength, improves pacifies and is heavy. Red onion is cool, alleviates pitta, improves digestive functions and works like sedative. Onion seeds are aphrodisiac, cure tooth infections and urinary disorders. As per Ayurveda, onion pacifies vata, enhances pitta and kapha, relieves pain, cures oedema and swelling, subsides inflammation and cures skin disorders. It increases the rajas and tama guna and so is amedhya (reduces intellect). It enhances digestive functions, improves digestion, anuloman and purifies urinary system and increases potency. It is spermatogenic, coagulates blood, emenogogue, aphrodisiac and improves physical strength and cures itching (Acharya Balkrishna, 2008). Scientists in Hong Kong have discovered that red onions help remove bad cholesterol- which can cause heart attacks and strokes-from the body. At the same time, red onions retain the body's good cholesterol, which help and protect against heart disease.

Health Benefits of Onions

- ☆ Onions are rich in **quercetin**, a powerful antioxidant that may reduce the risk of cancer. Like garlic, onions also contain the amazing compound **allicin**. Red and purple onions contain **anthocyanins**, the same **antioxidants** that make berries so robust in healing powers. In addition to being extraordinary at preventing and healing cancer, the quercetin contained in onion makes them a safe therapy for allergies; it also helps prevent heart disease and reduce high blood pressure.
- ☆ Onions are very low in calories and fats; but rich in soluble dietary fibre.
- ☆ Onion phyto-chemical compounds ***allium*** and ***Allyl disulphide*** convert to **allicin** by enzymatic reaction when the bulb disturbed (crushing, cutting etc. these compounds have anti-mutagenic (protects from cancers) and anti-diabetic properties (helps lower blood sugar levels in diabetics).
- ☆ *Allicin* reduces cholesterol production by inhibiting *HMG-CoA reductase* enzyme in the liver cells. Further, it also found to have anti-bacterial, anti-viral, and anti-fungal activities.
- ☆ *Allicin* also decreases blood vessel stiffness by release of nitric oxide (NO); thereby bring reduction in the total blood pressure. It also blocks platelet clot formation and has fibrinolytic action in the blood vessels which, helps decrease overall risk of coronary artery disease (CAD), peripheral vascular diseases (PVD), and stroke.
- ☆ They are rich source of chromium, the trace mineral that helps tissue cells respond appropriately to insulin levels in the blood; thus helps facilitate insulin action and control sugar levels in diabetes.
- ☆ Onions are also good source of antioxidant flavonoid *quercetin*, which is found to have anti-carcinogenic, anti-inflammatory, and anti-diabetic functions.
- ☆ They are also good in anti-oxidant vitamin, vitamin-C and mineral manganese which is required as co-factor for anti-oxidant enzyme *superoxide dismutase*.

- Onions are also good in B-complex group of vitamins like pantothenic acid, **pyridoxine**, folates and thiamin.
- Onions contain sulphur compounds as thiosulfinates which exhibit antimicrobial properties.
- Onion is an aphrodisiac food, it helps in strengthening the reproductory organs.
- Onions are an excellent antioxidant, and they contain anti-allergy, antiviral and antihistamine properties.
- Sulphur compounds in onions help to detoxify the body.
- Onions aid in cellular repair.
- Onions are a rich source of quercetin, a potent antioxidant.
- To obtain the maximum nutritional benefits, onions should be eaten raw or lightly steamed

SPRING ONIONS

Nutritionally, green onions have a combination of the benefits of onions and greens. They are an excellent source of vitamin K and vitamin C, and a very good source of vitamin A too. Spring onions can be added to dal and make it a much tastier dish. Similarly, you can add it to vegetables like cauliflower and potatoes and it will make an excellent dish. Mushrooms go very well with spring onions and interestingly spring onions can be added to soy nuggets while making a Chinese dish. Spring onions are used in salads as the flavour tends to be milder than other onions. It is used widely in oriental food both as an ingredient and as garnish. Spring onion is a nutritious plant, and therefore it provides a host of health benefits to us. It is a rich source of vitamins and minerals which aids in curing the various ailments. It is seen to have helped in reducing the harmful impacts of various diseases.

Some of the health benefits of spring onions are as follows:

- Spring onion lowers the blood sugar level.
- It is a support against gastrointestinal problems.
- It is often used as a medicine for common cold.
- It is used as an appetizer as it helps digestion.
- It speeds up the level of blood circulation in the body.

Health Benefits of Onions

- Onions are very low in calories (just 40 cal per 100 g) and fats; but rich in soluble dietary fibre.
- Onion phyto-chemical compounds *allium* and *Allyl disulphide* convert to allicin by enzymatic reaction when the bulb disturbed (crushing, cutting etc). Studies have shown that these compounds have anti-mutagenic (protects from cancers) and anti-diabetic properties (helps lower blood sugar levels in diabetics).

- ☆ Laboratory studies show that *allicin* reduces cholesterol production by inhibiting *HMG-CoA reductase* enzyme in the liver cells. Further, it also found to have anti-bacterial, anti-viral, and anti-fungal activities.
- ☆ *Allicin* also decreases blood vessel stiffness by release of nitric oxide (NO); thereby bring reduction in the total blood pressure. It also blocks platelet clot formation and has fibrinolytic action in the blood vessels which, helps decrease overall risk of coronary artery disease (CAD), peripheral vascular diseases (PVD), and stroke.
- ☆ They are rich source of chromium, the trace mineral that helps tissue cells respond appropriately to insulin levels in the blood; thus helps facilitate insulin action and control sugar levels in diabetes.
- ☆ They are also good source of antioxidant flavonoid *quercetin*, which is found to have anti-carcinogenic, anti-inflammatory, and anti-diabetic functions.
- ☆ They are also good in anti-oxidant vitamin, vitamin-C and mineral manganese which is required as co-factor for anti-oxidant enzyme *superoxide dismutase*. In addition, *isothiocyanate* anti-oxidants in them help provide relief from cold and flu by exerting anti-inflammatory actions.
- ☆ Onions are also good in B-complex group of vitamins like pantothenic acid, pyridoxine, folates and thiamin. Pyridoxine or vitamin B-6 helps keep up GABA levels in the brain, which works against neurotic conditions.

Caution

Raw onions can cause irritation to skin, mucus membranes and eyes. This is due to release of *allyl sulphide* gas while chopping or slicing them. The gas when mixed with moisture (water), convert to sulphuric acid. Allyl sulphide is concentrated more at the ends, especially at the root end. Its effect can be minimized by immersing the trimmed bulb in cold water for few minutes before you chop or slice it.

GARLIC (*Allium satium*)

Vegetables belonging to Allium family like garlic bulb, Chinese leek, Chinese chive, scallion and shallot bulb have been shown to possess strong antioxidant activity. Where as garlic is likely the most important herb widely known for its medicinal properties. The purported health benefits include chemoprotective, antibiotic and cholesterol lowering properties. It is known both for its culinary flavour and medicinal value, can protect against radiation damage to tissues. The findings indicate that in future garlic may protect against natural back ground and man made source of radiation *i. e.* patients under going radiotheraphy. Scientists speculate that eating garlic regularly could help traffic policemen and industrial workers, who are particularly exposed to lead poisoning. Garlic is a common spice food, reduces blood sugar, unhealthy fats and blood cholesterol in man. It has been reported to reduce hypertension as it helps break the cholesterol in the blood thus preventing the hardening of arteries. It have anti - bacterial and anti tumour properties. Those who are figure conscious, garlic reduces the production of body fat. Most of the effects of garlic have been linked to its **Sulphydryul** compounds - the non volatile non protein sulphur amino acid that are responsible for its pungent flavour. These include S-Allyl cysteine Sulphoxide (SACS). This chemical either increase the secretion of insulin by the pancreas or enhance the transport of glucose to the peripheral tissues. Continuous SACS intake, the concentration of serum, lipids, blood glucose, activities of various enzymes and liver glucose decreases considerably. The results established strong anti- diabetic and the already proved hypoglycemic effects of SACS. SACS-reduces the body lipids by decreasing their synthesis and increases their excretion through the intestinal tract. SACS, has potent- anti hyper chloles - trotemic effect which considerably lowers the blood cholesterol level in body and extra cholesterol synthesis in the liver. Garlic helps lower cholesterol, acts as a natural antibiotic and blood thinner. It has the simple effect of Asprine. It is effective fat burning food. Garlic is an excellent source of manganese, a very good source of vitamin B6 and vitamin C and a good source of selenium. Garlic oil is rich in **Sulphydryl** compounds that are known to protect against tissue damaging effects of irradiation, chiefly by destroying a harmful group of chemicals called free radicals. It works as vasodialator and reduces hypertension. Anti septic properties are attributed to garlic which can ward off germs and diseases thus improving one's general health. It stimulates stomach and gall blader secretions and kills intestinal worms. Chinese doctor used intravenous garlic treatment for cryptococcal meningitis- in which the fungus can invade the nervous system. It is known that some bacteria convert food into nitrosamines which is a carcinogenic. Potent organosulphur compounds in garlic inhibit bacterial growth in the stomach. Garlic compounds were found to slow the growth of breast cancer cells and prostrate cells. Garlic also inhibit growth of tumors of colon, rectum, esophagus and skin in rodents. It also reduces the stickiness of platelets that clamps together to from clots, which can lead to heart attacks. Garlic may also cut down on thickening and hardening of arteries known as atherosclerosis by preventing them from adhering to the lining of blood vessels. An animal study, involving rats, conducted by scientists from the Indian Institute of Chemical Biology in Kolkata, has found evidence to prove that garlic helps fight arsenic poisoning, caused by drinking contaminated ground water. Garlic proved effective because it is rich in sulphur, which scavenged arsenic from tissues and blood(Kounteya Sinha, 2008).

Modern scientific research confirms these ancient uses of garlic, including its ability to lower cholesterol and blood pressure. It is the sulphur-containing compounds of garlic that lend the herb its spicy aroma and are responsible for many of its healing properties. These compounds lower cholesterol by stimulating the release of bile by the gall bladder and by decreasing the production of cholesterol in the liver. In addition, garlic compounds gently lower blood pressure by slowing the production of hormones related to blood pressure. This wonder drug aids digestion, relieves earache, acts as an expectorant and as an antiseptic. Garlic also increases the flow of urine thereby killing thread worms and roundworms in your body. Besides, its ability to expel mucous by liquefying it, can help you control your cold, cough and other respiratory tract infections. This amazing herb has also demonstrated the ability to protect against a variety of environmental toxins. Garlic's sulphur compounds are potent antioxidants which protect cell membranes and DNA (deoxyribonucleic acid) from damage. "Lipid peroxide plays a detrimental role in all cancers, including skin carcinogenesis," a paper submitted at an International Symposium on New Frontiers in Hermatology and Oncology, said the report, titled "Garlic Suppresses oxidative Stress during DMBA- induced skin carcinogensis. As the tests done on Swiss Albino mice showed that garlic suspension inhibited the oxidative modification of lipids, thus protecting cells from injury by oxidized molecule. It is said that the chemo-preventive action was observed in mice in which garlic treatment was performed before and after the induction of skin carcinogenesis (Anonymous, 2009). While you do not want to be stinking of garlic during a passionate lip lock, but allicin, an ingredient present increases blood flow to the sexual organs. This 'wonder-drug' contains more than 100 biologically useful chemicals. It also has various anti-bacterial, anti-viral, anti-fungal and anti-oxidant properties.

"Garlic injection delays formation of papillomas (kind of rash) in animals and simultaneously decrease the size and number of papillomas which was also reflected in skin histology of mice treated," the report, by the researchers at Department of Cancer Chemo prevention, Chittrangan National Cancer Institute (Anonymous, 2009). Garlic oil possesses more than 20 substances that have significant potential for propecting heart from diabetes- induced cardiomyopathy.

The Good

This food is very low in Saturated Fat, Cholesterol and Sodium. It is also a good source of Calcium, Phosphorus and Selenium, and a very good source of Vitamin C, Vitamin B6 and Manganese.

Garlic is one of the world's most potent medicines, and its potent smell is what makes it so powerful. The active ingredient allicin turns into **organosulfurs**, which are the compounds that keep the cells safe from all the destructive cellular processes that can cause major chronic diseases. Garlic is a natural antiseptic; it prevents cancer, fights infection, and prevents colds. Research also states that garlic may prevent or decrease chronic diseases associated with age, such as **atherosclerosis**, stroke, cancer, immune disorders, brain aging, cataracts, and **arthritis**.

Therapeutic and Medicinal Values

Garlic is a wonder 'drug' contains more than 100 biologically useful chemicals. It is a hot stimulant, carminative, antirheumatic and has antoxidant properties. Garlic oil is a powerful antiseptic. It is used as vermifuge for expelling round worms and has long been recommended for cure of a number of ailments, *viz.* wounds, foul ulcers, pneumonia, bronchitis, atopic dyspepsia and gastro-intestinal disorders. Insecticidal, anti-bacterial, anti-fungal, anti-tumor, hypoglycemic, hypolipidemic, anti-atherosclerotic, fibrinolytic and anti-platelet.- aggregation effects of garlic have have been reported by modern scientists. It is best used in conditions like chest congestion, raised lipids, sciatica, degenerative disease of nervous system and various types of joint pains such as rheumatoid arthritis and osteoarthritis. It is also a major immuno- strengthening and 'rasayan' drug used in geriatrics. A controlled and standardised dose of garlic has been found to be effective in hypertensive and atherosclerotic paticients with a significant decrease in the blood cholesterol level. But uncontrolled use of onion and garlic may also lead to anaemia as all these vegetables contain compounds which remove cysteine from the system and inhibit protein synthesis. Therefore, only 50-80g of onion and 7-10 g per day of garlic is advised. However, these vegetables may enhance the effect of insulin and reduce hyperlipidemia, obesity and resistance to insulin treatment. Also, these vegetables may prevent the incidence of altheroselerosies and related diseases of the heart. Researchers have reported that men who ate the most vegetables containing allium - the pungent, sulphur- based compound blamed for the antisocial effects of garlic and onions had a 50 per cent lower risk of having prostate cancer than those who ate the least. Garlic is most important than onion in these aspects. Garlic is very good for cleansing the body of toxic wastes and should be added to most of the meals for it`s alkalising effect. It helps in lowering cholesterol, bronchial congestion, aids in digestion, controls blood pressure, arthritis, cancerous growths and whooping cough. Garlic kills viruses responsible for colds and the flu, according to tests by James North, a microbiologist at Brigham Young University, Eat garlic when you feel a sore throat coming on, he says, and you may not even get sick.(Eat garlic when you`re stuffed up, too: It acts as a decongestant). Other studies suggest that garlic revive up immune functioning by stimulating infection- fighting **T- cells**. Eating one or two cloves of raw garlic a day is recommended to people with chronic or recurrent infection, says Andrew Weil of the University of Arizona College of Medicine, author of Natural Health, Natural Medicine. Garlic oil because of its selective bactericidal action, is a part of the current revival of folk type medicines. Garlic oil acts selectively in food systems such as salami to prevent the growth of unwanted bacteria while permitting desirable microbes to flourish and produce the desired flavour and colour.

Ayurveda describes garlic as a digestive, carminative, cardiac stimulant, expectorant, analgesic, anti-oxidant and an aphrodisiac agent. It contains numerous secondary metabolites and sulphur containing compounds which are responsible for its peculiar colour and numerous healing properties. Eating green chilies with garlic is an old (tried and tested) way of enjoying sex for a longer period. Peel off its top layers, crush cloves and then fry in butter and your partner is ready to be a nutter.

Health Benefits

Here's a list of some of it's proven properties.

- ☆ Studies have shown that taking 600 mgs of garlic daily, lowers cholesterol levels considerably.
- ☆ Garlic's ability to fight infections and bacteria makes it an effective cure for warts and skin problems.
- ☆ It's a great aphrodisiac. It helps improve blood circulation and also helps prolong an erection.

- ☆ Regulates blood sugar levels by increasing the release of insulin in diabetics.
- ☆ Garlic's antibacterial and anesthetising properties can help cure toothaches.
- ☆ Strong flavoured, garlic cloves contain many noteworthy minerals, vitamins, anti-oxidants, and phyto-nutrients that have proven health benefits.
- ☆ Its bulbs contain organic thio-sulfinites such as *diallyl disulfide, diallyl trisulfide* and *allyl propyl disulfide* that can form **allicin** by enzymatic reaction, which is activated by disruption of bulb (like crushing, cutting etc). Allicin reduces cholesterol production by inhibiting *HMG-CoA reductase* enzyme in the liver cells.
- ☆ Allicin also decreases blood vessel stiffness by release of nitric oxide (NO); thereby bring reduction in the total blood pressure. It also blocks platelet clot formation and has fibrinolytic action in the blood vessels, which helps decrease the overall risk of coronary artery disease (CAD), peripheral vascular diseases (PVD) and stroke.
- ☆ Consumption of garlic is associated with possible decrease in the incidence of stomach cancer.
- ☆ *Allicin* and other essential volatile compounds in the garlic also found to have anti-bacterial, anti-viral, and anti-fungal activities.
- ☆ Garlic is an excellent source of minerals and vitamins that are essential for optimum health. The bulbs are one of the richest sources of potassium, iron, calcium, magnesium, manganese, zinc, and selenium. **Selenium** is a heart-healthy mineral, and is an important cofactor for anti-oxidant enzymes in the body. Manganese is used by the body as a co-factor for the antioxidant enzyme, *superoxide dismutase*. Iron is required for red blood cell formation.
- ☆ It contains many flavonoid anti-oxidants like carotene beta, zea-xanthin, and vitamins like vitamin-C. Vitamin C helps body develop resistance against infectious agents and scavenge harmful, pro-inflammatory free radicals.
- ☆ Studies have shown that taking 600 mgs of garlic daily, lowers cholesterol levels considerably.
- ☆ Garlic juice when applied on face for an hour at least once a week, kills the acne causing bacteria. But do not keep it for too long as it can aggarvate sensitive skin.
- ☆ **Garlic** helps remove toxins, stimulates blood circulation and normalises intestinal flora.
- ☆ The allicin in garlic has been shown to lower blood pressure. Garlic has been proven to lower LDL cholesterol and raise HDL cholesterol.
- ☆ Garlic is not only known to help in digestive disorders but it is also known to naturally detox your body. Eating garlic stimulates the liver into creating detoxifying enzymes that help clean out toxins which remain in your digestive system.

- ☆ If sore throat is accompanied by fever have **raw garlic** throughout the day. It's a natural **antibiotic**, effective against bacteria and viruses.

Undesirable Effects

The sulfide compounds in the garlic metabolized to allyl methyl sulfide, which is excreted through sweat and breathe producing unpleasant odour and breath (halitosis).

Side-Effects

Raw garlic is very strong, so eating too much could produce problems, for example irritation of or even damage to the digestive tract.

There are a few people who are **allergic to garlic**. Symptoms of garlic allergy include skin rash, temperature and headaches. Also, garlic could potentially disrupt anti-coagulants, so it's best avoided before surgery

Caution

- ☆ Its cloves contain allicin that acts as blood thinner. It is therefore, advised to avoid in patients on anticoagulants like *warfarin* as the resultant combination might cause excessive bleeding.
- ☆ Garlic-in-oil, as in the preparations of pickles favours growth of *clostridium botulism*, which may result in a condition known as **botulism** (paralysis of nervous system). It is therefore, advised that garlic preparations should be preserved inside the refrigerator and should be used as quickly as possible.

SHALLOT (*Allium cepa* var. *aggregatum*)

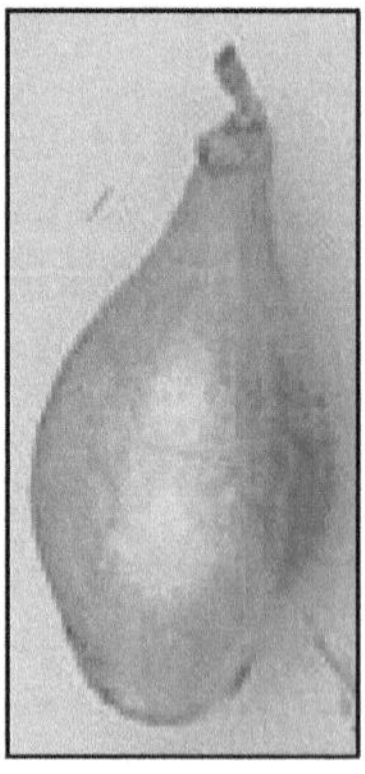

Shallots or *eschalots* are long slender bulbs in the Allium family of root vegetables. In general, they differ from onions in being smaller, and grow in clusters of bulbs from each plant root system. Like in other allium species, shallots top greens and flower heads are also eaten in many parts of world. The bulbs have less sharp in pungency than that of onions and garlic which makes them one of the favorite vegetable all around the world

Health Benefits of Shallots

- ☆ Shallots have better nutrition profile than onions. Have more anti-oxidants, minerals, and vitamins on weight per weight basis than onions.
- ☆ They are rich source of flavonoid anti-oxidants such as quercetin, kemferfol etc. Further, they contain sulphur anti-oxidant compounds such as *diallyl disulfide, diallyl trisulfide* and *allyl propyl disulfide*. These compounds convert to allicin through enzymatic action following disruption of their cell surfaces by actions like crushing, cutting etc.

- ✰ Research studies show that *allicin* reduces cholesterol production by inhibiting *HMG-CoA reductase* enzyme in the liver cells. Further, it also found to have anti-bacterial, anti-viral, and anti-fungal activities.
- ✰ Allicin also decreases blood vessel stiffness by releasing vasodialator chemical nitric oxide (NO) and thereby help bring reduction in the total blood pressure. Further research studies suggest that allicicn blocks platelet clot formation in the blood vessels that helps decrease overall risk of coronary artery disease (CAD), peripheral vascular diseases (PVD), and stroke.
- ✰ The phyto-chemical compounds *allium* and *Allyl disulphide* in onion have anti-mutagenic (protects from cancers) and anti-diabetic properties (helps lower blood sugar levels in diabetics).
- ✰ Shallots have several fold more concentration of vitamins and minerals than in onions, especially vitamin A, pyridoxine, folates, thiamin, vitamin C etc. Pyridoxine (B-6) helps raises GABA chemical levels in the brain that help sooth nervous irritability. In addition, 100 g fresh shallots have 1190 IU (35 per cent RDA) of vitamin A. Vitamin A is a powerful antioxidant that helps protect from lung and oral cavity cancers.
- ✰ They are also good in minerals and electrolytes than onions particularly **iron**, calcium, copper, potassium, and phosphorus

Caution

Although less in severity than onions, raw shallots can cause irritation to skin, mucus membranes, and eyes. This is due to release of *allyl sulphide* gas while chopping or slicing them which when mix with wet surface becomes sulphuric acid. Allyl sulphide is concentrated more at the ends, especially at the root end. Its effect can be minimized by immersing the trimmed bulbs in cold water for a few minutes before you chop or slice them.

LEEK (*Allium ampeloparsum* var. *porrum*)

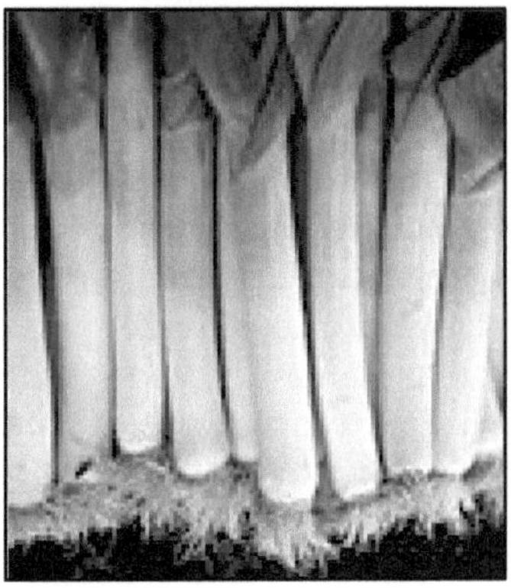

Delicate flavoured leeks are cylindrical, leafy stalks obtained from the onion-like plant in the *Allium* family. They are biennial, tall, slender plants with long cylindrical stem composed of overlapping leaves. They are commonly used as vegetables in many parts of Europe, America and Asia. It has mild astringent qualities which makes it helpful in the treatment of diarrhoea and internal bleeding.

Health Benefits

- ✰ Leeks contain many noteworthy flavonoid anti-oxidants, minerals, and vitamins that have proven health benefits.
- ✰ Leeks are low in calories. 100 g fresh stalks contain 61 calories. Further, their elongated stalks provide good amounts of soluble and insoluble fibre.

- Though leeks contain proportionately less thio-sulfinites than that in garlic, they still possess significant amounts of these anti-oxidants such as *diallyl disulfide, diallyl trisulfide* and *allyl propyl disulfide*. These compounds convert to allicin by enzymatic reaction when the stalk disturbed (crushing, cutting etc).
- Laboratory studies show that *allicin* reduces cholesterol production by inhibiting *HMG-CoA reductase* enzyme in the liver cells. Further, it also found to have anti-bacterial, anti-viral and anti-fungal activities.
- Allicin also decreases blood vessel stiffness by release of nitric oxide (NO); thereby bring reduction in the total blood pressure. It also blocks platelet clot formation and has fibrinolytic action in the blood vessels which, helps decrease overall risk of coronary artery disease (CAD), peripheral vascular diseases (PVD), and stroke.
- Leeks are great source of minerals and vitamins that are essential for optimum health. Their leafy stems indeed contain several vital vitamins such as pyridoxine, folic acid, niacin, riboflavin, and thiamin in healthy proportions. 100 g fresh stalks provide 64 µg of folates. Folic acid is essential for DNA synthesis and cell division. Their adequate levels in the diet during pregnancy can help prevent neural tube defects in the newborn babies.
- In addition, leeks are one of the good source of vitamin A (1667 IU or 55 per cent of RDA per 100 g) and other flavonoid phenolic anti-oxidants such as carotenes, xanthin, and lutein. They also have some other essential vitamins such as vitamin C, K, and vitamin E. Vitamin C helps body develop resistance against infectious agents and scavenge harmful, pro-inflammatory free radicals.
- Further, its stalks have small amounts of minerals such as potassium, iron, calcium, magnesium, manganese, zinc, and selenium.

CHIVES (*Allium schoenoprasum*)

The sweet, mild-onion flavoured chives are fresh top greens of lily family vegetables. The leaves are round and hollow, similar to onions, but smaller in diameter, somewhat appear like grass from the distance. They should not be confused to green onions, which are top greens of young, immature onion plants or scallions, which are top greens of *Allium fistulosum* (welsh onion). Unlike in onions and garlic, their small underground bulbs have unpleasant taste and not used in cooking. Flower stems, which rise directly from the base, grow slightly taller than leaves and bears small clusters of mauve or purple flower heads. Chives are one of the most sought after ingredients used for flavouring and garnishing recipes in many parts of the world. They have subtle onion flavour, besides imparting bright green colour to the recipes. Generally, fresh leaves tied in bunches are sold along with other leafy-greens.

Health Benefits

- ☆ Chives are very low in calories; 100 g of fresh leaves provide just 30 calories, but they contain many flavonoid anti-oxidants, fibre, minerals, and vitamins.
- ☆ Like in scallions, they contain more plant derived dietary fibre than fellow allium members like onions, shallots, leeks etc. 100 g fresh leaves provide 2.5 g or 7 per cent of daily-recommended levels of dietary fibre.
- ☆ Chives too possess thio-sulfinites anti-oxidants. Thio-sufinites such as *diallyl disulfide, diallyl trisulfide* and *allyl propyl disulfide* convert to **allicin** by enzymatic reaction when its leaves disrupted (crushing, cutting etc). It also have anti-bacterial, anti-viral, and anti-fungal activities.
- ☆ Allicin also decreases blood vessel stiffness by release of nitric oxide (NO); thereby bring reduction in the total blood pressure. It also blocks platelet clot formation and has fibrinolytic action in the blood vessels which, helps decrease overall risk of coronary artery disease (CAD), peripheral vascular diseases (PVD), and stroke.
- ☆ Chives have exceptionally more vitamin A than any other allium family member vegetables. In addition, their green leaves have other flavonoid phenolic anti-oxidants such as carotenes, zeaxanthin, and lutein. Together, they help body protect from lung and oral cavity cancers.
- ☆ They also have some other essential vitamins such as vitamin C and K. Chives are one of the richest sources of **vitamin K**, slightly more than that of scallions. Vitamin K has potential role in bone health by promoting osteotrophic (bone formation and strengthening) activity. Adequate vitamin-K levels in the diet helps limiting neuronal damage in the brain; thus, has established role in the treatment of **Alzheimer's disease.**
- ☆ Fresh chives are rich source of folates. Folic acid is essential for DNA synthesis and cell division. Adequate folate levels in the diet during pregnancy may help prevent neural tube defects in the newborn babies.
- ☆ Their leaves are packed with other B-complex vitamins as well as some essential minerals such as copper, iron, manganese, zinc, and calcium. The leafy greens contain several vital vitamins such as pyridoxine, pantothenic acid, niacin, riboflavin, and thiamin in healthy proportions.

BUNCHING ONION (*Allium fistulosum* L)

Bunching onion does not form a real bulb. Oriental bunching onion has a green leaf portion and a long blanched white stalk portion. The blanched portion can be from a few centimetre to 50 centimetres, depending the varieties. The long-stalk onions are blanched by earthing up during growth. The long-stalk onions are very tender

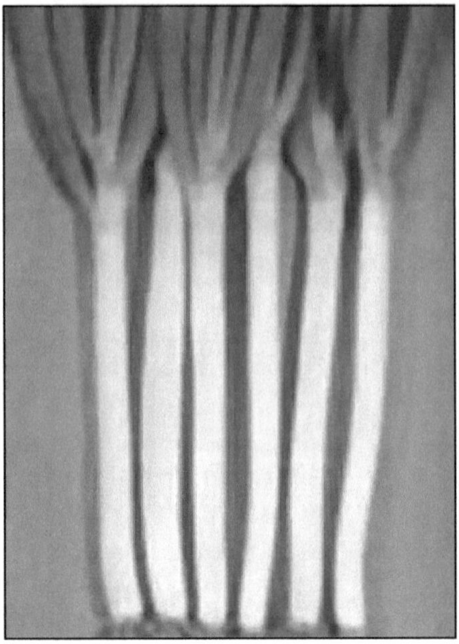

and well flavoured, excellent for stir-fry, sukiyaki, tempura and many Japanese dishes. Plants can be harvested for vegetable use at any growing stage. It is thought that it improves eye sight and internal organ function, enhances metabolism, and prolongs life. It is further said that it is effective in aiding digestion, perspiration, recovery from the common cold, as treatment for head aches, wounds and sores. The odour of Japanese bunching onions is attributed to volatile allyl sulfides.

Health Benefits

- ✰ They are very low in calories; 100 g of fresh leaves provide just 31 calories and flavonoid anti-oxidants, plant fibre, minerals, and vitamins that have proven health benefits.
- ✰ Being greens they naturally contain more plant derived dietary fibre than fellow allium members like onions, shallots, leeks etc. 100 g fresh spring onions provide 2.6 g or 7 per cent of daily-recommended levels of dietary fibre.
- ✰ Like in leeks, they possess proportionately less thio-sulfinites anti-oxidants than that in the garlics. Thio-sufinites such as *diallyl disulfide, diallyl trisulfide* and *allyl propyl disulfide* convert to **allicin** by enzymatic reaction when its leaves disrupted (crushing, cutting etc
- ✰ Allicin also decreases blood vessel stiffness by release of nitric oxide (NO); thereby bring reduction in the total blood pressure. It also blocks platelet clot formation and has fibrinolytic action in the blood vessels, which helps decrease overall risk of coronary artery disease (CAD), peripheral vascular diseases (PVD), and stroke.
- ✰ Spring onions contain healthy composition of vitamin-A and other flavonoid phenolic anti-oxidants such as carotenes, zeaxanthin, and lutein. Together, they help body protect from lung and oral cavity cancers.
- ✰ They also have some other essential vitamins such as vitamin C and K. In fact, scallions are one of the richest sources of **vitamin K**. 100 g of fresh greens provide **207 µg or about 172 per cent** of daily-recommended intake. Vitamin K has potential role in bone health by promoting osteotrophic (bone formation and strengthening) activity. Adequate vitamin-K levels in the diet helps limiting neuronal damage in the brain; thus, has established role in the treatment of **Alzheimer's disease.**

☆ Spring onions are plentiful in B-complex vitamins as well as some essential minerals such as copper, iron, manganese, and calcium. The leafy greens contain several vital vitamins such as pyridoxine, **folic acid**, niacin, riboflavin, and thiamin in healthy proportions. 100 g fresh leaves provide 64 µg of folates.

LETTUCE (*Lactuca sativa*)

It is a rich source of cellulose and its intake increases the bulk of the intestinal contents, so it is highly effective in warding off constipation. As the leaves are rich in iron, chomping on salad leaves can increase the haemoglobin content of the blood. They act as a tonic for patients suffering from anemia. Eating lettuce together with other green vegetables like spinach, asparagus and cauliflower significantly increase the folic acid content of the body. The fresh and crisp leaves have a high juice content full of magnesium which is vital for the brain and the nurves. It is valuable source of vitamin A and folic acid. Folic acid and vitamin B12 are involved in the formation of healthy red blood cells that are so essential for carrying life giving oxygen to different parts of the body. In addition, folic acid is vital for pregnant women as it protects the unborn child from damage to spinal cord, and also lowers incidence problems as cleft lip and palate and vitamin B12 helps to maintain nervous tissues. Lettuce is a fat free and low calorie food. **Lactucarium** (or "Lettuce opium") is a mild opiate like substance that is contained in all types of lettuce. Both Romans and Egyptians took advantage of this property eating lettuce at the end of a meal to induce sleep. Depending on variety, the nutrients in lettuce vary. Romaine lettuce is said to be the most healthy one boasting of vitamin A, B and C, folic acid, manganese and chromium. It makes for excellent salad as it is low in calories. Lettuce is also a good source of iron which helps with blood formation.

Health Benefits

☆ Lettuce leaves are the store house of many phyto-nutrients that have health promotional and disease prevention properties.

☆ Vitamins in lettuce are plentiful. Fresh leaves are an excellent source of several Vitamin A and beta-carotenes. Just 100 g of fresh, raw-lettuce provides 247 per cent of daily vitamin A, and 4443 mcg of beta-carotene (Carotenes convert to vitamin A in the body; 2 mcg of carotene is considered equivalent to 1 IU of vitamin A). These compounds have antioxidant properties. Vitamin A is required for maintaining healthy mucus membranes and skin, and is also essential for vision. Consumption of natural fruits and vegetables rich in flavonoids helps to protect body from lung and oral cavity cancers.

☆ Zeaxanthin (1730 mcg per100) an important dietary carotenoid in lettuce is selectively absorbed into the retinal macula lutea where it is thought to provide antioxidant and protective light-filtering functions, thus it offers some protection against age related macular disese (ARMD) in the elderly.

- It is a rich source of vitamin K, Vitamin K has potential role in the increase of bone mass by promoting osteotrophic activity in the bone. It also has established role in Alzheimer's disease patients by limiting neuronal damage in the brain.
- Fresh leaves contain good amounts folates and vitamin C. Folates require for DNA synthesis and therefore, vital in prevention of neural tube defects in-utero fetus during pregnancy. Vitamin C is a powerful natural antioxidant; regular consumption of foods rich in vitamin C helps body develop resistance against infectious agents and scavenge harmful, pro-inflammatory free radicals.
- It also contain good amounts of minerals like iron, calcium, magnesium, and potassium which are very essential for body metabolism. Potassium in an important component of cell and body fluids that helps controlling heart rate and blood pressure. Manganese is used by the body as a co-factor for the antioxidant enzyme *superoxide dismutase.* Copper is required in the production of red blood cells. Iron is essential for red blood cell formation.
- It is rich in B-complex group of vitamins like thiamin, vitamin B-6 (pyridoxine), riboflavins.
- Regular inclusion of lettuce in salads is known to prevent osteoporosis, iron deficiency anemia and believed to protect from cardiovascular diseases, ARMD, Alzheimer's disease and cancers.

Caution

Pesticides are commonly used in lettuce crops. The most common pesticides found in the leaves are organo-phosphates, Permethrin and DDT. Wash them thoroughly in cold water before consumption. However, the organic forms are believed to be free from these toxins and safe for consumption.

METHI (*Trigonella foenum-graecum*)

It is one of the oldest-known herbal remedies. The seeds are used as a spice and can be used to expel mucous from nasal passages. Its seeds are very rich in iron, calcium sulphur, chlorine and vitamin C. The most important nutrient contained in methi is beta-carotene, the 21st century's wonder food. The high Sulphur and chlorine content gives it a cleansing property. Fresh tender leaves and shoots of fenugreek contain large quantities of fatty acids, proteins, fats, vitamins(A and C), carbohydrates and minerals such as calcium, phosphorus and iron and so are considered very nutritive. In fact it is equated to quinine in its efficacy in reducing fevers. Its alkaline nature makes it a coolant to the system. Therefore, it is that in instances of dyspepsia, indigestion, acidity, flatulence and even colic, fenugreek leaves come in very handy. The high iron content helps in the formation of blood. It is thus recommended for adolescent girls and young women particularly after childbirth, and its beta-carotene content gives the micro nutrients so essential, for balanced growth. Its seeds have

proven to be very useful to diabetic patients. A tea made from fenugreek seeds was traditionally known to increase milk secretion in nursing mothers. As fenugreek contains an alkaloid, the **trigonelline** and its seeds contain **saponins**, and these saponins are **diosegenin, gitogenin, trigogenin**, which are thought to stimulate production of male sex horomones in including testosterone which in turn causes the rise in male libido. Fenugreek has a variety of medicinal properties. Since ages it has been used to cure fatigue, small pox, gastric trouble, gout, flatulence, loss of appetite and irregular menses in women. It is also used in colic flatulence, dysentery, diarrhea, chronic cough, enlargement of spleen and liver, rickets and diabetes. National Institute of Nutrition Hyderabad has reported that fenugreek reduces blood sugar levels in diabetic patients. The alkaloids of seeds are said to stimulate the nervous system. The **diosgenin** present in its seeds is a starting material in the synthesis of sex hormones and oral contraceptives. Its seeds are also used by Indian women for promoting lactation. It is often ground and mixed with cotton seeds and is fed to the cattle to increase the milk production. In Java, it is used in hair tonic preparations and as a cosmetic. The mucilagenous watery extract of the seed when mixed with the dye imparts glossiness and durability. It is thus recommended for adolescent girls and young women, particularly after child birth and its beta-carotene content gives the micro nutrient so essential, for the balanced growth. Its leaves contain several flatus and anti nutritional factors. Dry methi leaves contain so much of protein that nutritionists equate it to pulses in food value. Dried methi leaves also supplement lysine deficiencies. Lysine is an amino acid which is a fundamental constituent of all proteins.

Fenugreek seeds are said to resemble cod liver oil and can be taken instead of cod-liver oil for complaints of anemia, arthritis and diabetes. The alkaloids of the seeds are said to stimulate the nervous system and useful for diabetic patients. Besides, the seed also contains a lactating factor (Duggal, 1995). Studies have shown that adding fenugreek to our diet reduces fasting blood sugar and improves after meal glucose tolerance significantly. Fenugreek extract promotes insulin secretion and inhibits the rise of blood glucose, thus helping to reduce body fat production. Active ingredients in fenugreek seeds exhibits a specific effect on the Islets of Langerhans in the pancreas. These cells are directly responsible for insulin production. Most significantly, the effect of active ingredients is glucose dependent. The higher the level of blood glucose, the greater the insulin promoting response elicited by this active ingredient. Thus the active ingredient in fenugreek seeds exhibits a significant regulating effect, which corresponds with the insulin needs of the body at any given time, making this compound "adaptogenic." The saponins and flavonoids present in fenugreek extract has following benefits:

- ☆ Powerful pancreatic stimulator.
- ☆ A natural insulin secretion promoter.
- ☆ Scientifically proven blood cholesterol and blood glucose lowering properties
- ☆ Significant reductions in total cholesterol, LDL cholesterol and triglyceride levels, but not HDL cholesterol levels.
- ☆ Proven non-toxic.
- ☆ Its seeds was traditionally known to increase milk secretion in nursing mothers.
- ☆ Used to lower blood sugar levels.
- ☆ Fenugreek seeds contain compounds called saponins which are thought to stimulate production of male sex hormones including testosterone.
- ☆ Fenugreek seeds contain compounds called saponins which are thought to stimulate production of male sex hormones including testosterone.

Health Benefits of Fenugreek Seeds

- ☆ They are known to increase milk production and stimulate breast tissue growth.
- ☆ Methi seeds help lower blood sugar and reduce the need for insulin in the body.
- ☆ Methi subdues the assimilation of cholesterol and lowers the amount that is manufactured by the liver.
- ☆ Fenugreek helps induce labour, treat hormonal disorders and reduce menstrual pain to a certain extent.
- ☆ For skin inflammations, like abscesses, boils, burns, cuts, eczema and gout, fenugreek is a big help.
- ☆ It works as an effective tropical treatment for skin problems.
- ☆ It is recognized as a treatment for diabetes- A glass of water in which a table spoon of fenugreek seeds has been soaked overnight is drunk each morning.
- ☆ Fenugreek seeds are rich source of minerals, vitamins, and phytonutrients.
- ☆ The seeds are very good source of soluble dietary fibre. Soaking the seeds in water makes their outer coat soft and mucilaginous.
- ☆ Non-starch polysaccharides (NSP) which constitute major fibre content in the fenugreeks include *saponins, hemicelluloses, mucilage, tannin,* and *pectin.* These compounds help lower blood LDL-cholesterol levels by inhibiting

bile salts re-absorption in the colon. They also bind to toxins in the food and helps to protect the colon mucus membrane from cancers.

- NSPs (non-starch polysaccharides) increase the bulk of the food and augments bowel movements. Altogether, NSPs assist in smooth digestion and help relieve constipation ailments.
- Amino-acid **4-hydroxy isoleucine** present in the fenugreek seeds has facilitator action on insulin secretion. In addition, fibre in the seeds help lower rate of glucose absorption in the intestines thus controls blood sugar levels. The seeds are therefore recommended in diabetic diet.
- The seeds contain many phytochemical compounds such as ***choline, trigonelline diosgenin, yamogenin, gitogenin, tigogenin*** and ***neotigogens***. Together, these compounds account for the medicinal properties of fenugreeks.
- Excellent source of minerals like copper, potassium, calcium, iron, selenium, zinc, manganese, and magnesium. Potassium is an important component of cell and body fluids that helps control heart rate and blood pressure by countering action on sodium. Iron is essential for red blood cell production and as a co-factor for cytochrome-oxidases enzymes.
- Rich in many vital vitamins including thiamin, pyridoxine (vit. B-6), folic acid, riboflavin, niacin, vitamin-A and vitamin-C that are essential nutrients for optimum health.

Medicinal Uses of Fenugreek Seed

- Its seeds have been used in many traditional medicines as laxative, digestive, and as a remedy for cough and bronchitis.
- If used regularly, fenugreeks may help control cholesterol, triglyceride as well as high blood sugar (glycemic) levels in diabetics.
- Fenugreek seeds added to cereals and wheat flour (bread) or made in gruel, given to the nursing mothers to increase milk synthesis.
- Amino-acid 4-hydroxy isoleucine present in the fenugreek seeds has facilitator action on insulin secretion. In addition, fibre in the seeds help lower rate of glucose absorption in the intestines thus controls blood sugar levels. The seeds are therefore recommended in diabetic diet.
- The seeds contain many phytochemical compounds such as ***choline, trigonelline diosgenin, yamogenin, gitogenin, tigogenin*** **and** ***neotigogens***. Together, these compounds account for the medicinal properties of fenugreeks.
- Excellent source of minerals like copper, potassium, calcium, iron, selenium, zinc, manganese, and magnesium. Potassium is an important component of cell and body fluids that helps control heart rate and blood pressure by countering action on sodium. Iron is essential for red blood cell production and as a co-factor for cytochrome-oxidases enzymes.

☆ Rich in many vital vitamins including thiamin, pyridoxine (vit. B-6), folic acid, riboflavin, niacin, vitamin-A and vitamin-C that are essential nutrients for optimum health.

SPINACH (*Spinacia oleracea*)

Spinach is an excellent source of iron, calcium, chlorophyll, beta-carotene (provitamin A), vitamin C, riboflavin, sodium and potassium. It is a diuretic, natural laxative, intestine-toner and cleaner and can also be used to stop minor haemorrhaging such as nose bleeds. As it is rich in iron and chlorophyll, spinach builds the blood, while its sulphur content helps to clean the liver and relieve herpes irritations. In addition, its vitamin A content can help to prevent night blindness. Although spinach provides an exceptional source of calcium, it is also high in oxalic acid, which can partly interfere with the absorption of the calcium. It contains more nutrients than lettuce. It is a good source of calcium and so intrinsically good for growing children as well as older people whose bones need strengthening. It is a valuable gum toner and can actually aid bleeding gums and prevent cavities. However, it contains small amounts of oxalic acid so those who suffer from gout or liver dysfunctions should control their intake. Both carrots and spinach are rich in beta-carotene or vitamin-A, which appears to help prevent the built up of cholesterol on blood vessel walls. Cholesterol deposits eventually can block blood flow through an artery causing heart attack or stroke. Spinach contains iron, carotenoids and chlorophyll and is an especially good source of lutin. It is also loaded with vitamins like C and A that help with good eye sight and glowing skin. The high level of potassium makes it very beneficial for high blood pressure patients. Further it strengthens muscles and helps those suffering from anaemia. Since it is rich in iron, spinach is nature's answer to anaemia. It is good for eyes because of its vitamin A content. A regular cupful, with its nutrients intact, is a food proof remedy. Because of this, it is a must for pregnant mothers because it helps their own haemoglobin to remain steady. Even after the birth of the baby, continued intake improves the quality of the mother's milk. Two nutrients found in spinach and other leafy green vegetables offer some protection against the most common cause of blindness amongst the elderly. The two nutrients, Lutein and zeaxanthin, are both carotenoids- compounds that give many fruits and vegetables a yellow colour. They help ward off the condition apparently by allowing the eyes to filter harmful short-wavelength light and by curtailing other damaging effects to the macula, or the center of the eye's retina. Dark leafy greens viz spinach are a great source of iron- a nutrient that may help protect against the sleep robber known as restless legs syndrome. Well, it won't be wrong if we tag spinach as a natural Viagra. This green leafy vegetable is loaded with sexual benefits. It is rich in Vitamin E, which is a major catalyst in the production of sex hormones in the body. It is also rich in manganese, which facilitates the production of the female hormone estrogen. A deficiency of magnesium also affects a woman's fertility levels. Green leafy veggies are also loaded with zinc, which is known for its libido and sperm production qualities in men.

It is believed that spinach contains around 13 flavonoid compounds which keep us away from cancer, heart diseases and osteoporosis. 1/2 a cup of this lutein-rich food, daily, guards us against heart attacks.

Health Benefits

- ☆ **Best Muscle Building Foods:** Researchers at Rutgers University (2008) showed that the **phytoecdysteroids** contained in spinach may increase muscle growth up to 20 per cent. This makes it a great muscle building food!
- ☆ Spinach is rich in beta carotene, which the body transforms into vitamin A, triggering your immune response to keep you well. Spinach prevents cancer and heart disease and is rich in the disease-fighting mineral zinc. The vitamin C helps you resist colds and infection and keeps your skin healthy; the B vitamins keep you calm and more energetic.
- ☆ Spinach is store house for many phyto-nutrients that have health promotional and disease prevention properties.
- ☆ Very low in calories and fats (100 g of raw leaves provide just 23 cal). It contains good amount of soluble dietary fibre; no wonder greeny spinach is one of the vegetable source recommended in cholesterol controlling and weight reduction programs!
- ☆ Fresh 100 g of spinach contains about 25 per cent of daily intake of iron; one of the richest among green leafy vegetables. Iron is an important trace element required by the body for red blood cell production and as a co-factor for oxidation-reduction enzymes *cytochrome-oxidases* during the cellular metabolism.
- ☆ Fresh leaves are rich source of several vital anti-oxidant vitamins like vitamin A, vitamin C; and flavonoid poly phenolic antioxidants such as lutein, zea-xanthin and beta-carotene. Together these compounds help act as protective scavengers against oxygen-derived free radicals and reactive oxygen species (ROS) that play a healing role in aging and various disease processes.
- ☆ Zea-xanthin, an important dietary carotenoid, is selectively absorbed into the retinal macula lutea in the eyes where it is thought to provide antioxidant and protective light-filtering functions; thus helps protect from "age related macular disease" (ARMD), especially in the elderly.
- ☆ Vitamin A is also required for maintaining healthy mucus membranes and skin and is essential for vision. Consumption of natural vegetables and fruits rich in vitamin A and flavonoids helps body protect from lung and oral cavity cancers.

- 100 g of Spinach provides 402 per cent of daily vitamin-K requirements. Vitamin K plays vital role in strengthening bone mass by promoting osteotrophic (bone building) activity in the bone. It also has established role in patients with Alzheimer's disease by limiting neuronal damage in the brain.
- This greeny leafy vegetable also contain good amounts of many B-complex vitamins like vitamin- B6 (pyridoxine), thiamin (vitamin B-1), riboflavin, folates and niacin. Folates help prevent neural tube defects in the offspring.
- 100 g of farm fresh spinach has 47 per cent of daily recommended levels of vitamin C. Vitamin C is a powerful antioxidant which helps body develop resistance against infectious agents and scavenge harmful oxygen free radicals.
- The leaves also contain good amount of minerals like potassium, manganese, magnesium, copper and zinc. Potassium in an important component of cell and body fluids that helps controlling heart rate and blood pressure. Manganese and copper are used by the body as a co-factor for the antioxidant enzyme *superoxide dismutase.* Copper is required in the production of red blood cells. Zinc is a co-factor in many enzymes that regulate growth and development, sperm generation, digestion and nucleic acid synthesis.
- It is also rich source of **omega-3 fatty acids**.
- Regular consumption of spinach in the diet helps prevent osteoporosis (weakness of bones), iron deficiency anemia and is believed to protect from cardiovascular diseases and colon and prostate cancers.

Caution

- People with a tendency to kidney stones should eat spinach sparingly since its high oxalic acid can tend to form calcium -oxalate kidney stones in susceptible individuals.
- Reheating of spinach left-over may cause conversion of nitrates in to nitrites and nitrosamines by certain bacteria that thrive on pre-prepared nitrate-rich foods, such as spinach and many other green vegetables. These poisonous compounds may be harmful to health, especially in children.
- Phytates and dietary fibre present in the leaves may interfere with the bio-availability of iron, calcium and magnesium.
- Because of its high vitamin K content, patients taking anti-coagulants such as "warfarin" are encouraged to avoid spinach in their food since it interferes with drug metabolism.
- Spinach contains oxalic acid, a naturally occurring substance found in some vegetables which may crystallize as oxalate stones in the urinary tract in some people. It is therefore, people with known oxalate urinary tract stones are advised to avoid eating certain vegetables belonging to amaranthaceae and brassica family. Adequate intake of water is therefore advised to maintain normal urine output.

☆ It may also contain goitrogens which may interfere with thyroid hormone production and can cause thyroxin hormone deficiency in individuals with thyroid dysfunction.

TURMERIC (*Curcuma longa* L.)

Turmeric is a staple ingredient in most Indian cuisines. It is the king of spices when it comes to dealing with cancer diseases, besides it adding a zesty colour to our food on the platter. It contains the powerful polyphenol Curcumin that has been clinically proven to retard the greowth of cancer cells causing prostrate cancer, melanoma, breast tumour, brain tumour, pancreatic cancer and leukemia amongst a host of others. Curcumin promotes Apoptosis-(programmed cell death\ cell suicide) that safely eliminates cancer breeding cells without posing a threat to the development of other healthy cells.

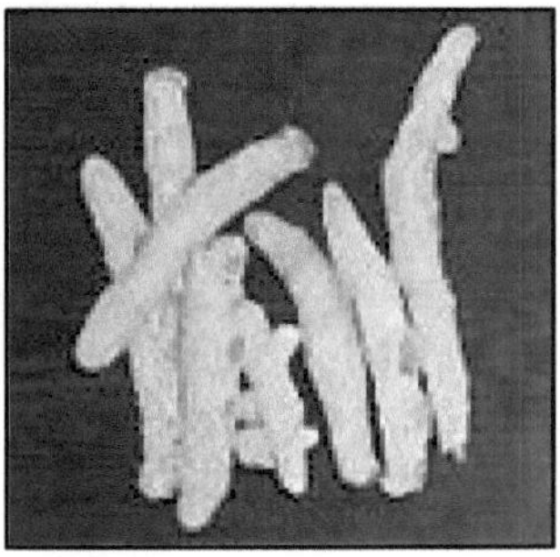
Dried Turmeric -Root

It has many therapeutic effects. Curcumin, the active component of turmeric is an object of research owing to its properties that suggest they may help to turn off certain genes that cause scarring and enlargement of heart. Regular intake may help reduce low-density lipoprotein or bad cholesterol and high blood pressure, increase blood circulation and prevent blood clotting, helping to prevent heart attacks.

Turmeric is often referred to as the spice of life in ancient Indian lore. Being bitter and sharp it cures cough. It refines blood and normalizes blood cell levels. It is tasty and is an astringent. It accumulates urine and makes it clear. It is very effective in curing urinary diseases. Its paste relieves swelling and oedema. It is anti-dermatoses, cleanses and cures wounds and ulcers. Its fumes are useful for curing hiccoughs, bronchial asthma, and is an antidote to poison. It cures cold and cough, hepatitis, diarrhea, urinary diseases, jaundice, skin disorders and eye ailments (Acharya Balkrishna, 2008). Turmeric contains 5 - 8 per cent of a volatile oil and the main active components are **curcumin** – a fragrant compound oil and **turpenide**. Curcumin derived from turmeric root can help reduce cancer risk among post- menopausal women exposed to hormone replacement therapy. Studies suggest that a combined oestrogen and progestin hormone replacement therapy increases post menopausal women's risk of developing progestin acceleration breast tumours. Curcumin prevents the appearance of gross morphological abnormalities in the mammary glands. Progestin accelerates the development of certain tumours by increasing production of molecule called **VEGF** that helps to supply blood to the tumour. And blocking the production of **VEGF** could potentially reduce the proliferation of breast cancer cells (Menopause-A journal of North America Menopause Society).

The new and ongoing research according to nutritionist have found that besides flavouring the palate, the most commonly used spices possibly reduces the risk of cancer and cardio-vascular diseases. Research now has revealed that its main component, "**curcumin** is a phyto-chemical and inhibits chronic diseases. Phytochemicals, otherwise known as bioactive chemicals have important physiological and bio-chemical impact on the body's processes. A new Saint Louis University study has revealed that a chemical that gives curry its zing holds promise in preventing or treating liver damage from an advanced form of a condition known as fatty liver disease called non-alcoholic steatohepatitis (NASH).

Besides turmeric, curry leaves, pepper, cumin(zeera), coriander, fenugreek (methi) have now been found to be able to prevent tissue damage and decrease glucose and cholesterol in the blood. Studies have shown that curcumin, the chemical in turmeric that gives it the yellow colour has effective anti-malarial qualities. The turmeric therapy in combination with other therapies is expected to redefine the whole anti-malaria fight. Scientists at the University of California, Los Angeles, have found that turmeric holds the potential to fight against Alzheimer's disease, one of the most common form of illness among the elderly across the world. Their studies indicated that turmeric could be more effective than not only many of the existing medications, but also the new therapies under test. Turmeric not only inhibited the accumulation of beta amyloid, a protein, in the brain of the Alzheimer's patients, but also broke up the existing plaques. It could be used for both treatment and prevention(Sunderajan, 2005). Turmeric curry once or twice a week could keep Alzheimers disease and dementia at bay, thanks to a magic ingredient curcumin found in the spice according to an Indian American Phychiatrist.

Turmeric curry once or twice a week could keep Alzheimer's disease and dementia at bay, thanks to its magic ingredient curcumin, according to Murali Doraiswamy Psychiatry Professor at the Mental Fitness Laboratory of the Duke University Medical Centre (DUMC) North Carolina. He told delegates at Royal College of Psychiatrists annual meeting Liverpool that curcumin prevented the spread of **Amyloid plaques**. These plaques are thought to contribute to the degradation of the wiring in brain cells and lead to the subsequent symptoms of Alzheimer's disease. There is thus solid evidence that curcumin binds to plaques (IANS, 6th June2009). Researchers at Cork Cancer Research Centre in Ireland treated esophageal cancer cells with **curcumin-** a chemical in the spice, which gives curries a distinctive yellow colour and found that it started to kill cancer cells within 24 hours. The cells also began to digest themselves reports the BBC. Dr Sharon Mc Kenna said " Scientists have known for a long time that natural compounds have the potential to treat faulty cells that have become cancerous and we suspected that curcumin might have therapeutic value". They are of the opinion that it opens up the possibility that natural chemical found in turmeric could be developed into new treatment for oesophageal cancer (Anonymous, 2009). According to a study, curcumin that we take is quickly destroyed by the digestive juice in the gastro intestinal tract and very little actually gets into the blood. Therefore, to overcome this problem, a team of scientists is developing nano-sized capsule that would boost the body's uptake of curcumin and help fight several diseases. Trials are under way to test its safety and effectiveness in fighting colon cancer, psoriasis and Alzheimer's disease. Scientists are working on developing nano-sized capsules containing the curry ingredient in an effort to improve its absorption and effectiveness in the body (Anonymous, 2009). Researchers at UCLA'S Jonsson Comprehensive Cancer Centre have found that curcumin suppresses a cell- signalling pathway in human saliva that drives the growth of head and neck cancer. The study shows that curcumin can work in the mouths of patients with head and neck malignancies and reduce activity that promote cancer growth. "And it not only affected the cancer by inhibiting a critical cell signalling path way, it also affected the saliva itself by reducing pro- inflammatory cytokins within the saliva."(Anonymous, 2011). If you have an uneven skin tone and would do anything to get that perfect peaches and cream complexion, then turmeric's what you need. Turmeric evens out the skin tone by reducing the traces of pigmentation. Mix a little turmeric powder alongwith cucumber juice or lemon juice and apply on skin. Wash it off after 15 minutes. But remember to do this everyday to notice a significant change in your skin colour. Turmeric makes for an awesome body scrub. This is why most Indian brides used to smear a paste of turmeric and besan on their bodies before their wedding to get skin that is super smooth, baby soft and glowing. Application of turmeric and a little milk cream also makes the skin supple and improves its elasticity. Indians have been using turmeric to beautify themselves since ages now.

If you have an uneven skin tone and would do anything to get that perfect peaches and cream complexion, then turmeric's what you need. Turmeric evens out the skin tone by reducing the traces of pigmentation. Mix a little turmeric powder alongwith cucumber juice or lemon juice and apply on skin. Wash it off after 15 minutes. But remember to do this everyday to notice a significant change in your skin colour.

Health Benefits

- ☆ Turmeric is also used in Indian traditional medicine systems. It contains the flavonoid curcumin, known for its anti-inflammatory properties. It detoxifies the liver, fights allergies, stimulates digestion and boosts immunity. Traditionally, women used to have a bath using a paste of turmeric and oil as it is excellent for skin.
- ☆ The root has been in use since antiquity for its anti-inflammatory (painkiller), carminative, anti-flatulent and anti-microbial properties in India for ages.
- ☆ The herb contains many health benefiting essential oils such as *termerone, curlone, curumene, cineole,* and *p-cymene.*
- ☆ ***Curcumin,*** a poly-phenolic compound, is the principal pigment that imparts deep orange colour to the turmeric. In vitro and animal studies have suggested the curcumin may have anti-tumor, antioxidant, anti-arthritic, anti-amyloid, anti-ischemic, and anti-inflammatory properties.
- ☆ This popular herb contains no cholesterol; but is rich in anti-oxidants and dietary fibre, which helps to control blood LDL or "bad cholesterol" levels.
- ☆ It is very rich source of many essential vitamins such as **pyridoxine** (vitamin-6), choline, niacin, and riboflavin etc, which are essential for optimum health. 100 g herb provides 1.80 mg or 138 per cent of daily-recommended levels of pyridoxine. Pyridoxine is used in the treatment of homocystinuria, sideroblstic anemia and radiation sickness. Niacin helps prevent "pellagra" or dermatitis.

- Fresh root contains very good levels of **vitamin-C.** It is a water-soluble vitamin and a powerful natural anti-oxidant; helps body develop immunity against infectious agents, helps fight type-1 diabetes and remove harmful free oxygen radicals from the body.
- Turmeric contains very good amounts of minerals like calcium, **iron**, potassium, manganese, copper, zinc, and magnesium. Potassium is an important component of cell and body fluids that helps controlling heart rate and blood pressure. Manganese is used by the body as a co-factor for the antioxidant enzyme, *superoxide dismutase.* Iron is an important co-factor for *cytochrome oxidase* enzymes at cellular level metabolisms and required for red blood cells productions.
- It's powder is used in all vegetables and curries. It contains curcumin that helps nip pain
- It enhances the effectiveness of chemotheraphy and reduces tumour growth.

Medicinal Uses

- Research studies have suggested that *Curcumin,* a poly-phenolic compound, found in this herb may inhibit the multiplication of tumor cells including multiple myeloma, pancreatic and breast cancer, leukemia and colon cancer (Ipshita Mitra, 2012).
- It contains many health benefiting essential oils such as *termerone, curlone, curumene, cineole, and p-cymene.* These compounds have applications in cosmetic industry.
- *Curcumin,* along with other antioxidants, has been found to have anti-amyloid and anti-inflammatory properties. Thus; is effective in preventing or at least delaying the onset of **Alzheimer's disease.**
- This root herb contains no cholesterol; but is rich in anti-oxidants and dietary fibre, which helps to control blood cholesterol levels, preventing coronary artery disease and stroke risk.
- Early laboratory studies have been suggestive that turmeric is liver protective, anti-depressant, anti-retroviral effects.
- It has been in use since a very long ago as an important ingredient in traditional Chinese and ayurvedic medicines for its anti-microbial, anti-inflammatory, carminative, anti-flatulent and anti-microbial

GINGER (*Zingiber officinale*)

This humble spice boasts of medicinal qualities that help lowering cholesterol, boosts metabolism and kills cancer cells. It *has long been used in food for its wonderful medicinal properties. If you are having cold, having a cup of 'kadak adraki chai' can give you instant relief.* Similarly, on a cold winter morning, you can boil water with ginger, basil leaves, mint leaves and then add a spoon of honey. Even in traditional Indian dishes, ginger is added for giving the dish a unique flavour. Ginger's health benefits

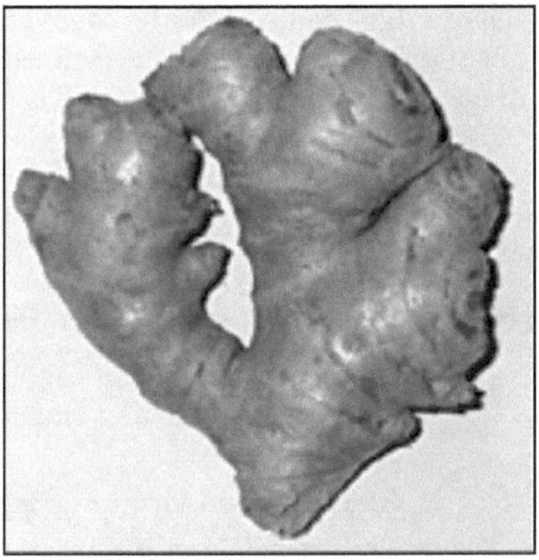

has made it an essential ingredient in Indian kitchens. It is the best source of gingerol. It is known to increase the production of substances that protect the stomach lining, preventing the formation of ulcers; stimulates gastric activity, causing the stomach to empty more quickly, stimulates gall bladder, promoting healthy digestion and helps alleviate nausea caused by pregnancy and motion sickness. Ginger has anti-inflammatory properties and will also protect the stomach from irritation. Migraines and motion sickness can also be relieved by ginger juice. Ginger is stimulant that relieves flatulence, discharges mucus, strengthens digestion, stimulates glandular secretions and relieves vomiting.

According to Ayurveda, ginger (saunth) eradicates cold, cures swelling, and oedema and relieves pain. It strengthens nerves and cures vata. It quenches thirst, is purgative, enhances digestion and cures all digestive disorders; it makes wind go down and cures piles. It improves functioning of the circulatory system and purifies blood. Ginger is bitter and unctuous and so cures cough and bronchial asthma. It also stimulates sexual desire and is anti-pyretic.

British Journal of Nutrition has reported that an international team has carried out the study and found that curcumin, the major yellow constituent of turmeric, helps in reducing inflammation in many people suffering from bowel disease (Anonymous, 2009).

Our ancient Ayurvedic texts have described raw ginger as acrid, carminative, digestive and thermogenic agent whereas its dried form has been additionally cited to be possessing stomachic, expectorant, anthelmintic, rubefacient (counter-irritant if applied on the outer skin), anti-rheumatic, cardiac stimulant, aphrodisiac and anti-oxddant properties. Ginger has also been described as one of the best ama-pachak, meaning a substance helping to nullify toxins produced in the body due to improper digestion. Ginger increases sexdrive and stimulates sexual performance (misha. paul @ times group. com).

Du to its anti-vata effect the use of ginger has been indicated in rheumatoid arthritis, chronic backache, inflammations, minor respiratory infections and cough and could. As a digestive aid no other herb seems to be a match to ginger, as it helps the digestion from the begning till the end. Wide ranging, acute and chronic problems of the digestive tract such as anorexia, nausea, vomiting, flatulence, diarrhoea, dysentery and colitis are known to be relieved by the judicious use of ginger (Vatsyayan, 2011).

Its root features knotty finger like projections that grow downward from the ground surface. Fresh raw root has silver gray outer surface. Cut sections feature creamy white, yellow, or red coloured crunchy flesh depending upon the variety. The root often contains fibrils running through its center, especially in over-matured. Its pungent, spicy and aromatic smell is due to essential oils and phenolic compounds such as **gingerols** and **shogaols.** Ginger is the best pain killer having analgesic properties like the popular ibuprofen. It contains a quartet, gingerols, paradols, shogaols, and zingerone which are active ingredients to reduce pain. Drink ginger tea in monsoons and winter to get relief from that recurring pain.

Health Benefits of Ginger Root

- ✰ Ginger has long been used as a natural treatment for colds and the flu. Many people also find ginger to be helpful in the case of stomach problems or food poisoning.
- ✰ Ginger is good for curing digestive problems.
- ✰ Ginger has anti-inflammatory properties and is a powerful natural painkiller.
- ✰ Ginger has long been used as a natural heartburn remedy. It is most often taken in the form of tea for this purpose.
- ✰ Research has shown that ginger may provide migraine relief due to its ability to stop prostaglandins from causing pain and inflammation in blood vessels.
- ✰ Ginger is known to be beneficial in reducing the symptoms of morning sickness.
- ✰ It is beneficial for treatment of rheumatoid arthrities.
- ✰ To seek relief from menstrual cramps and heart burns – to sip on ginger tea.
- ✰ It provides a great remeady for cold and cough.
- ✰ To curb motion sickness chew ginger candy.
- ✰ Ginger aids in digestion. Either buy powdered ginger in capsule forms or add the herb to your recipes.
- ✰ Ginger has been in use since ancient times for its anti-inflammatory, carminative, anti-flatulent, and anti-microbial properties.
- ✰ The root contains many health benefiting essential oils such as *gingerol, zingerone, shogaol, farnesene and small amounts of* β*-phelladrene, cineol,* and *citral.*
- ✰ **Gingerols** help improve the intestinal motility and has anti-inflammatory, painkiller (analgesic), nerve soothing, anti-pyretic as well as anti-bacterial properties. Studies have shown that it may reduce nausea induced by motion sickness or pregnancy and may help relieve migraine headaches.
- ✰ **Zingerone,** which gives pungent character to the ginger root, has been found to be effective against E. coli induced diarrhea, especially in children.
- ✰ This herb root is low in calories and contains no cholesterol, but is very rich source of many essential nutrients and vitamins such as pyridoxine (vitamin B-6), pantothenic acid (vit. B-5) that are essential for optimum health.

☆ It also contains good amount of minerals like potassium, manganese, copper, and magnesium. Potassium is an important component of cell and body fluids that helps controlling heart rate and blood pressure.

Medicinal Uses

☆ Ginger root slices boiled in hot water with added lemon or orange juice and honey is a popular herbal drink in ayurvedic medicine to relieve common cold, cough, and sore throat.

☆ It is also used as vehicle in many ayurvedic decoctions to mask bitterness and alter taste.

☆ Gingerols increase the motility of the gastrointestinal tract and have analgesic, sedative, anti-inflammatory, and antibacterial properties. Studies have shown that it may reduce nausea caused by motion sickness or pregnancy and may relieve migraine.

BEANS (*Phaseolus vulgaris*)

Bean Plants with Pods

Broadly, tender beans are classified depending on their growth habits as "bush beans" which stand erect without support and "Pole Beans" that need climb supports. The other varieties of unripe fruits of beans family include Shell beans, Pinto or mottled beans, White beans, Red or kidney beans, Black beans, Pink beans and Yellow wax beans. All most all the varieties are available year around across the world.

Beans are not just vegetables- beans are a nutrient rich power house. Soybean is the best source of **genistein** (an iso-flavonoid), phytosterols and saponins. It suppresses the growth of cancer cells in the large intestine and enhances immunity against many cancers. Other compounds in soybean also reduce blood cholesterol level. Beans and other legumes provide a nice sampling of B-vitamins, including **B6,** BI2 and folic acid, all of which help the body regulate sleep cycles and produce relaxing **Serotonin**. In fact, studies have shown that boosting B vitamins may help people with insomnia. Beans have more fibre and less sugar. Hence they play an important role in bringing down the risk of diabetes while simultaneously controlling the body's cholesterol level.

Many edible beans, including broad beans and soybeans, contain **oligosaccharides** (particularly **Raffinose** and **Stachyose**) a type of sugar molecule also found in cabbage. An anti- oligosaccharide enzyme is necessary to properly digest these sugar molecules. As a normal human digestive tract does not contain any anti-oligosaccharide enzymes, consumed oligosaccharides are typically digested by bacteria in the large intestine. This digestion process produces flatulence- causing gases as a by product. Because of their nutrient content, beans and pea are the only foods that appear in two food groups. Meat and Beans and Vegetable's.

Beans are in the Meat and Beans group, because they are a good source of protein and iron. Even better, beans provide a low fat, saturated fat- free, and cholesterol free source of protein.

Beans are listed in the Vegetable's group because they are a plant based food that provides fibre, folate, potassium and anti-oxidants. Researchers in Canada have found that proteins from a common garden pea could be used as a food additive or new dietary supplement to fight high blood pressure and chronic kidney disease (CKD). Researchers where of the view that the study is the first to report that a natural food product can relieve symptoms of CKD. "In people with high blood pressure, our protein could potentially delay or prevent the onset of kidney damage. In people who already have kidney disease, our protein may help them maintain normal blood pressure levels so they can live longer (Anonymous, 2009)."

Bean sprouts are sprinkled on salads or eaten as between meals and snacks. Sprouts are a great way of getting some nutrients. They are rich in vitamin C and a handful of fresh sprouts can provide you with three quarters of your daily vitaminC requirement. In addition they also contain vitamin A, B and E as well as calcium, iron and potassium. Being high in protein, beans should be a regular feature in your diet. They are also very lean. Beans also provide the fatty acids your nails need to stay strong and prevent splitting.

Excellent for the heart, ½ a cup of beans added in our diet (in the form of any dish or soup), helps in lowering cholesterol levels by 8 per cent. You should try black, kidney, or pinto beans; each one of them supplies about one-third of your day's fibre needs.

Health Benefits of Green Beans

- ☆ **Protects your heart**: Studies have shown that the soluble fibre present in the beans is helpful in reducing the bad cholesterol levels. In addition, it also reduces inflammation and blood pressure.
- ☆ **Speeds up blood clotting**: The vitamin K present in green beans helps speed up the healing process in your body. For example, if you are injured, vitamin K can help stimulate the clotting of blood which in turn stops excessive bleeding. Making beans a part of your daily diet means more intake of vitamin K. Vitamin K helps absorb calcium which in turn can help in prevention of bone density loss and osteoporosis?
- ☆ **Premenstrual Syndrome**: Green beans are packed with manganese which helps alleviate PMS symptoms such as irritability and mood swings. In

addition, eating beans also helps in relieving symptoms of osteoarthritis and osteoporosis.

- ☆ **Delay aging**: Want to delay aging, avoid wrinkles and fine lines? Add green beans to your daily diet. Green beans are a great source of vitamin A and equally great for your skin. So, eat more beans and get rid of that dull skin and age spots now!
- ☆ **Rich dietary fibre**: Green beans are rich in fibre and if you are suffering from problems such as constipation or irritable bowel syndrome, this vegetable is a must in your diet. Green beans also help in regulating your blood sugar levels and so if you are diabetic have some every day.
- ☆ **Low fat**: If you are trying to lose weight or maintain it, green beans are ideal as they are low in fat. Add green beans to your diet as it gives you all the essential nutrients without any extra calories. You can make a side dish to go with your rotis or add them to your curries and soups.
- ☆ **For pregnant women**: Green beans is said to be good for pregnant women as it helps in the healthy development of the baby's heart. Studies show that green beans also help protect babies from conditions such as asthma.
- ☆ Beans have less sugar and more fibre and protein than any other vegetable. Hence they play an important role in bringing down the risk of diabetes while simultaneously controlling the body's cholesterol level.
- ☆ Beans are good to excellent sources of eight important nutrients that many people do not get enough of their diet.
- ☆ Beans are naturally cholesterol and saturated fat free.
- ☆ Research suggests that beans can play a role in promoting health and preventing diseases, including reducing cancer risk, improving cardiovascular health, controlling blood sugar and managing weight.
- ☆ Beans contain a lot of fibre that help reduce the absorption of fat in the body.
- ☆ Beans also contain high amounts of protein that is beneficial in building body muscle. Body muscle cells are known to burn calories at much higher rate.
- ☆ While beans are often associated with the gastrointestinal disturbances therefore, need to be cooked thoroughly because our digestive tracks are not adapted to breaking down some proteins that are contained in certain beans.
- ☆ Some of the best kinds of beans to eat are:
  1. Navy beans
  2. Kidney beans
  3. White beans
  4. Lima beans

Green Beans Nutrition Facts

- ☆ Tender, flexible green beans are delight of vegetarian lovers for their wholesome nutritional qualities. They are unripe or immature pods obtained from the bean plant belonging to common *fabaceae* family and known scientifically as *Phaseolus vulgaris*.
- ☆ Fresh green beans are very low in calories (31 kcal per 100 g of raw beans) and contain no saturated fat; but are very good source of vitamins, minerals, and plant derived micronutrients.
- ☆ They are very rich source of dietary fibre (9 per cent per100g RDA) which acts as bulk laxative that helps to protect the mucous membrane of the colon by decreasing its exposure time to toxic substances as well as by binding to cancer causing chemicals in the colon. Dietary fibre has also been shown to reduce blood cholesterol levels by decreasing re-absorption of cholesterol binding bile acids in the colon.
- ☆ Green beans contain excellent levels of vitamin A, and many health promoting flavonoid poly phenolic antioxidants such as lutein, zeaxanthin and ß-carotene in good amounts. These compounds help act as protective scavengers against oxygen-derived free radicals and reactive oxygen species (ROS) that play a role in aging and various disease process.
- ☆ Zea-xanthin, an important dietary carotenoid in the beans, selectively absorbed into the retinal macula lutea in the eyes where it thought to provide antioxidant and protective UV light filtering functions. It is, therefore, green beans offer some protection in preventing *age related macular disease* (ARMD) in the elderly.
- ☆ Fresh snap beans are good source of folates. 100 g fresh beans provide 37 µg or 9 per cent of folates. Folate along with vitamin B-12 is one of the essential components of DNA synthesis and cell division. Good folate diet when given during preconception periods and during pregnancy helps prevent from neural-tube defects in the offspring.
- ☆ They also contain good amounts of vitamin-B6 (pyridoxine), thiamin (vitamin B-1), and vitamin-C. Consumption of foods rich in vitamin C helps body develop resistance against infectious agents and scavenge harmful oxygen free radicals.
- ☆ In addition, beans contain healthy amounts of minerals like **iron**, calcium, magnesium, manganese, and potassium, which are very essential for body metabolism. Manganese is a co-factor for the anti oxidant enzyme *superoxide dismutase,* which is a very powerful free radical scavenger. Potassium is an important component of cell and body fluids that helps controlling heart rate and blood pressure.
- ☆ Fresh beans contain vitamin A, B-complex vitamins, calcium and potassium.
- ☆ Green beans are diuretic and may be used to treat diabetes.
- ☆ A fresh bean should snap crisply and feels velvety to the touch.

Caution

Green beans contain oxalic acid, a naturally occurring substance found in some vegetables, which, may crystallize as oxalate stones in the urinary tract in some people. It is, therefore, people with known oxalate urinary tract stones are advised against eating vegetables belong to brassica and fabaceae family. Adequate intake of water is therefore advised to maintain normal urine output to minimize the stone risk.

PEA (*Pisum sativum*)

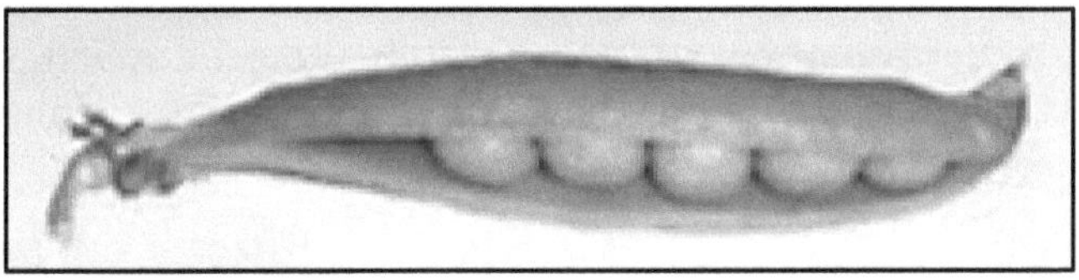

Fresh peas are sweet and juicy and one of the most digestible and non-gassy legumes. They contain 78 per cent water, with only traces of fat, are low in sodium and provide a good source of iron and vitamin A, C, B1, B2 and niacin. Peas have a diuretic and mildy laxative effect and are recommended for strengthening digestion, reducing water retention and helping to promote elimination. They are popular as a summer vegetable and also make a nutritious addition to salads, soups and casseroles. Pea tendrils are also edible. They are tender top shoots of young pea plants, have similar pea taste and favoured in cooking as well in salads.

Health Benefits of Green Peas

- ☆ Peas are one of the most nutritious leguminous vegetable, rich in health benefiting phyto-nutrients, minerals, vitamins and anti-oxidants.
- ☆ Peas are relatively low in calories when compared with beans, and cowpeas. 100 g of green peas provide only 81 calories, contain good amount of soluble and insoluble fibre but contains no cholesterol.
- ☆ Fresh pea pods are excellent source of folic acid. 100 g provides 65 mcg or 16 per cent of recommended daily levels of folates. Folates are B-complex vitamins required for DNA synthesis inside the cell. Well established research studies suggest that adequate folate rich foods in expectant mothers would help prevent neural tube defects in the newborn babies.
- ☆ Fresh green peas are very good in ascorbic acid (vitamin C). Contain 40 mcg/100 g or 67 per cent of daily requirement of vitamin C. Vitamin C is a powerful natural water-soluble anti-oxidant. Vegetables rich in this vitamin helps body develop resistance against infectious agents and scavenge harmful, pro-inflammatory free radicals from the body.
- ☆ Peas contain phytosterols especially ß-sitosterol. Studies suggest that vegetables like legumes, fruits and cereals rich in plant sterols help lower cholesterol levels in the body.
- ☆ Garden peas are also good in vitamin K. 100 g of fresh leaves contain about 24.8 mcg or about 21 per cent of daily requirement of vitamin K-1 (phylloquinone). Vitamin K has found to have potential role in bone mass

building function by promoting osteo-trophic activity in the bone. It also has established role in Alzheimer's disease patients by limiting neuronal damage in the brain.

- ✰ Fresh green peas also contain adequate amounts of anti-oxidants flavonoids such as carotenes, lutein and zeaxanthin as well as vitamin-A (provide 765 IU or 25.5 per cent of RDA per 100 g). Vitamin A is essential nutrient which is required for maintaining healthy mucus membranes and skin and is also essential for vision. Consumption of natural fruits rich in flavonoids helps to protect from lung and oral cavity cancers.
- ✰ In addition to folates, peas are also good in many other essential B-complex vitamins such as pantothenic acid, niacin, thiamin, and pyridoxine. Furthermore, they are rich source of many minerals such as calcium, iron, copper, zinc and manganese.

CARROT (*Daucus carota*)

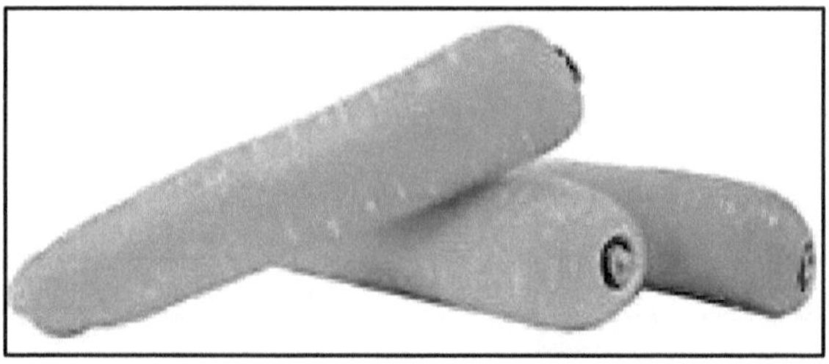

Naturally sweet, delicious and crunchy, carrots are healthy additions you can make to the vegetable list in your diet. Indeed these root vegetables comes with wholesome health benefiting compounds such as beta-carotenes, vitamin A, minerals and anti-oxidants in ample amount. Carrots are full of nourishing properties for eyes, hair and skin. They are the best source of pro-vitamin and A carotenes. Known for improving night vision, carrots are rich in Vitamin C, Vitamin A which helps in regulating healthy blood sugar levels - good for the heart. Carrots are considered a goldmine of natural vitamins and nutrients. Among other raw vegetable juices, carrot juice is a sure shot winner. Carrot juice is known for its excellent cleansing effects and, in addition, it was considered a good remedy for constipation and fatigue. As we all know, no other vegetable contains as much beta-carotene as carrots. In our body, beta-carotene is converted into Vitamin A, which assists in improving our eyesight and the functioning of our immune system, strengthening our bones and teeth, preventing possible problems with the functioning of the thyroid glad. Carrot juice has very good anti -inflammatory, anti-cancerous and anti-aging properties. Carotene is a known anti-oxidant, which can prevent degenerative processes in the cells and has great anti-aging effects. Also, drinking raw carrot juice can be helpful for lowering the symptoms of stresses and normalizing the function of your digestive system.

Carrot is a good of source beta-carotene which our body converts into vitamin A and minerals such as potassium and manganese. It contains some amount of sodium, iron, zinc, copper and calcium. It contains about 87 per cent of water, rich in mineral salts and Vitamin (B, C, D, E). Raw carrots are an excellent source of vitamin A and potassium. Even though the colour of original carrots is orange it grows in other colours including white, yellow, red or purple. Carrot juice reduces the risk of many different types of cancer including skin and breast cancer.

Carrot juice is immensely beneficial. It is diuretic in action and helps in case of scanty urination and menstruation. Its juice eliminates unwanted uric acid from the blood, is thus helpful to gout patients. Carrots contain a high level of carotenoids and antioxidants. Psoralins and coumarins found in umbelliferous vegetables such as celery, parsley and parsnip bind to hepatic enzymes and inhibit mutagenicity in the experimental animals. Carotenoids - α- carotene, β- carotene, lycopene, lutein and zeaxanthin, present in carrots, tomatoes and corn have been shown to be anti carcinogenic in nature. Micro nutrients such selenium, and zinc repair the damage done by mutations likewise thiamine and vitamin- B_6 inhibit human platelet aggregation. Tannins and phenolic compounds found in vegetables are antimutagenic in humans [Hunang and Toshihiko, 1994). In **carrots** and other yellow or brown- fleshed vegetables such as pumpkin, amaranthus, capsicum, sweet potato, watermelon and the like, the colour is imparted by **carotenoids**, of which only vitamins B and C are water soluble and vitamins A, D, E and K are fat soluble. While as the orange-coloured carrots contains a high concentration of **betacarotene**, which is a substance that is converted to vitamin- A in the human body. A half cup serving of cooked carrots contain four times the recommended daily intake of vitamin A in the form of protective **betacarotene.** Betacarotene is also a powerful antioxidant effective in fighting against some forms of cancer, especially lung cancer and protect against stroke and heart disease. Carrots are also very good source of dietary fibre, vitamin C, vitamin K, foliate, vitamin B6, pantothenic acid, iron and manganese. But to obtain maximum benefit it is best to eat them raw. Carrot juice is a excellent source of beta-carotene, a vitamin A percursor. They are also excellent sources of vitamin B1 and vitamin B2 and of the minerals potassium, sodium, silicon and anti-cancer nutrients, including **phthalide** and **glutathione** (antioxidants). A new Super-A carrots, intensely orange with sweet taste, 40 per cent richer in beta-carotene, an oxidant the body converts into vitamin and boosts health were developed by growers in Shropshire-London (Anonymous, 2009).

Carrots are excellent source of anti-oxidant compounds and the richest vegetable source of the provitamin A carotenes. Carrot's oxidant compounds help protect against cardiovascular disease and cancer and also promote good vision, especially night vision, inhibit cataracts and treat indigestion. Research demonstrated that high carotenoid diets are associated with a reduced risk of heart disease. After beta-carotene is converted to vitamin A in the liver, it travels to the retina where it is transformed into **rhodopsin** – a purple pigment that is necessary for night vision. Plus beta-carotene's powerful anti-oxidant actions help provide protection against muscular degeneration and the development of senile cataracts, the leading cause of blindness in the elderly. Carrots are by for one of the richest source of carotenenoids- just one cup provides **16,679** IUs of beta-carotene and **3,432** REs (retinol equivalent), or roughly **686.3 per cent** the RDA for vitamin A. A high carotene intake has been linked with a 20 per cent decrease in post menopausal breast cancer and upto 50 per cent decrease in the incidence of cancers of bladder, cervix, prostate, colon, larynx and oesophagus. Research has suggested that physiological level's as well as dietary intake of carotenoids may be inversely associated with insulin resistance and high blood sugar levels. According to Acharya Balkrishna (2008), carrot enhances body heat and promotes digestion. It cures bleeding diathesis, piles, sprue and irritable bowel syndrome and cough. It also cures acidity and gas formation. While according to Raj Ninghantu, it cures flatulence. It is anthelmintic and cures burning sensation and pain. It quenches thirst and cures pitta.

Carrots also contain a phyto-nutrient called **falcarinol** that may be responsible for the recognized epidemiological association between frequently eating carrots and a reduced risk of cancers. **Falcarinol** provides protection against colon cancer a study published in the Journal of Agricultural and Food Chemistry. Howover, massive over consumption of carrot's, can cause hyper- carotenemia, a condition in which the skin turns orange (although hyper-carotenemia is not itself dangerous, unlike over dose of vitamin A, which can cause liver damage). Researchers at the University of Newcastle found that "boiled- before-cut" carrots contain 25 per cent more of anti-cancer compound **falcarinol** than those that were chopped up first. And the sugars which give the carrots its distinctively sweet flavour were also found in higher concentrations in the carrot that had been cooked whole, so the vegetable tasted better as well being healthier (Jeremy Laurance, 200p).

Carrots are nutritional heroes; they store a gold mine of nutrients. No other vegetable contains as many carotene as carrots, which our body coverts into vitamin A. For maintaining optimum health and weight loss carrots are ideal. ½ cup carrot peeled and boiled have carbs 4, fibre 2gms, energy 80kj. Cooked carrots are an excellent source of vitamin A, it contains Vitamin B6. Copper, Folic acid. The high level of beta-carotene is very important and gives carrots their distinctive orange colour. In order to assimilate the greatest quantity of the nutrients present in carrots, it is important to chew them well.

Benefit's of Carrot in Our Body

- Boosting immunity (especially among older people), strengthening bones and teath, preventing possible problems with the functioning of the thyroid gland.
- Reducing photosensitivity (beta-carotene protects the skin from sun damage).
- Helping to heal minor wounds and injuries.
- Reducing the risk of heart disease and high blood pressure.
- Cleansing the liver when consumed regularly, can help the liver excreate fats and bile
- Fighting bronchitis.
- Fighting infection (vitamin A helps cell membranes healthy, making them stronger against disease Causing micro-organisms).
- Carrots boast of alpha-carotene, beta-carotene, and betacryptoxanthin. Eat them raw in the form of salad.
- Improves muscle, flesh and skin health.

- ☆ Its juice has very good anti-inflammatory, anti- cancer and anti- ageing properties.
- ☆ Carrot juice can be helpful for lowering the symptoms of stress.
- ☆ Helping fight enaemia.
- ☆ Reducing acne
- ☆ Improving eye health *i. e.* help night vision, inhibit cataracts.
- ☆ Useful in treating infections of lung, digestive system and urinary tract.
- ☆ Diet of raw or cooked carrots can improve skin appearance and calcium absorption.
- ☆ Sweet and succulent carrot are notably rich in anti-oxidants, vitamins and dietary fibre; however, they provide only 41 calories per 100 g, negligible amount of fat and no cholesterol.
- ☆ They are exceptionally rich source of **carotenes and vitamin-A**. 100 g fresh carrot contain 8285 mcg of beta-carotene and 16706 IU of vitamin A. Studies have found that flavonoid compounds in carrots help protect from skin, lung and oral cavity cancers.
- ☆ Carotenes are converted in to vitamin A in the liver. Beta-carotene is the major carotene that is present in these roots. Beta-carotene is one of the powerful natural anti-oxidant helps protect body from harmful fee radical injury. In addition, it also has all the functions of vitamin A such as vision, reproduction (sperm production), maintenance of epithelial integrity, growth and development.
- ☆ Carrots are rich in **poly-acetylene anti-oxidant falcarinol**. Research study conducted by scientists at University of Newcastle on laboratory animals has found that *falcarinol* in carrots may help fight against cancers by destroying pre-cancerous cells in the tumors.
- ☆ Fresh roots are also good in vitamin C; provide about 9 per cent of RDA. Vitamin C is water soluble anti-oxidant. It helps body maintain healthy connective tissue, teeth and gum. Its anti-oxidant property helps body protect from diseases and cancers by scavenging harmful free radicals.
- ☆ This root vegetable is especially contain good amounts of many B-complex group of vitamins such as folic acid, vitamin B-6 (pyridoxine), thiamin, pantothenic acid, etc that acts as co-factors to enzymes during substrate metabolism in the body.
- ☆ It also has healthy levels of minerals like copper, calcium, potassium, manganese and phosphorus. Potassium is an important component of cell and body fluids that helps controlling heart rate and blood pressure by countering effects of sodium. Manganese is used by the body as a co-factor for the antioxidant enzyme, *superoxide dismutase*.

Caution

It is recommended that the intake of carrot juice should be limited to no more than four cups a day since over- consumption can cause Yellowing of the skin, a condition known as Xanthosis.

RADISH (*Raphanus sativus*)

Radishes come in different forms varying in size, colour and duration of required cultivation time. Radishes can be broadly categorized into four main types- summers, fall, winter, and spring while growers classify them by shapes, colours, and sizes, such as black or white coloured radishes, with round or elongated roots. Top greens can also be used as food. Radish are suggested as an alternative treatment for a variety of ailments including whooping cough, cancer, coughs, gastric discomfort, liver problems, constipation, dyspepsia, gallbladder problems, arthritis, gall stone, kidney stones and intestinal parasites. Radish are rich in ascorbic acid, folic acid and potassium. They are a good source of vitamin B6, riboflavin, magnesium, copper and calcium. One cup of sliced radish bulbs provide approximately 20 calories or less coming largely from charbohydrates, making radishs relative to their size, a very filling food for their caloric value.

Radishe and their juice are an old home remedy for cough, rheumatism and gall bladder problems. The leaves of the plant, which are usually discarded, are very nutritious, containing almost 10 times as much vitamin C as the roots. In Yemenite folk medicine, radishes are used to eliminate kidney stones. Half a cup of fresh radish juice each morning on an empty stomach usually dissolves even stubborn stones, passing then out of the body in the urine.

A close look reveals that it is largely made up of water (more than 90 per cent) yet contains as much as potassium as banana's and about half the ascorbic acid of oranges. It is an excellent source of vitamin C and K. It has a excellent levels of copper, manganese, potassium. It is a good source of calcium, magnesium, iron, phosphorus, zinc and sodium. Radish is low in saturated fat and very low in cholesterol but very good source of dietary fibre. It has anti-bacterial and anti-viral properties. It is beneficial for cough, respiratory problems, digestive disorders, asthama, bronchitis and liver and gall bladder troubles. According to Ayurved (Traditional System of Medicine developed over millennium by sages and seers of ancient India) radish promotes digestion, is anthelmintic, cures digestive disorders, piles, and all kinds of swelling. Good for heart, hiccoughs, dermatoses, cholera and in secondary amenorrhoea. It also cures tuberculosis, phthisis, bronchial asthama, eye problems, abdominal pain and kapha and pitta disorders. It also cures blood disorders and enhances appetite. Radishes have antibacterial and anti-fungal properties. They are a member of the cabbage family. Radishes contain vitamin C, potassium and other trace minerals

Health Benefits

- ☆ It is an astringent and diuretic and is used to promote **bile** flow.
- ☆ Radish is low in saturated fat and cholesterol.

- High in dietary fibre, vitamin C, folate, vitamin K, riboflavin, vitamin B6, calcium, magnesium, copper and manganese.
- The nutritional value of radish makes it ideal for maintaining optimum health and weight loss.
- Radish helps in preventing Urinary Disorders. It is a diurectic *i. e.* increases production of urine. It is helpful in the treatment of dysuria or painful urination.
- It is beneficial as a carminative for relieving gastric discomforts, and also serves as a laxative and stimulant.
- Avoid including to many radishes in your diet if you are interested in weight gain.
- Radishes are very low calorie root vegetables; contains only 16 calories per 100 g. However they are very good source of anti-oxidants, minerals, vitamins and dietary fibre.
- Fresh Radishes are rich in vitamin C; provide about 15 mg or 25 per cent of DRI of vitamin C per 100 g. Vitamin C is a powerful water soluble anti-oxidant required by the body for synthesis of collagen. Vitamin C helps body scavenge harmful free radicals, prevention from cancers, inflammation and helps boost immunity.
- In addition, they contain adequate levels of folates, vitamin B-6, riboflavin, thiamin and minerals such as iron, magnesium, copper and calcium.
- They contain many phytochemicals like indoles which are detoxifying agents and zeaxanthin, lutein and beta-carotene which are flavonoid antioxidants.

- Radishes are very low calorie root vegetables; contains only 16 calories per 100 g. However they are very good source of anti-oxidants, minerals, vitamins and dietary fibre.
- Fresh Radishes are rich in vitamin C; provide about 15 mg or 25 per cent of DRI of vitamin C per 100 g. Vitamin C is a powerful water soluble anti-oxidant required by the body for synthesis of collagen. Vitamin C helps body scavenge harmful free radicals, prevention from cancers, inflammation and helps boost immunity.
- In addition, they contain adequate levels of folates, vitamin B-6, riboflavin, thiamin and minerals such as iron, magnesium, copper and calcium.
- They contain many phytochemicals like indoles which are detoxifying agents and zeaxanthin, lutein and beta-carotene which are flavonoid antioxidants

Accoding to Ayurved (Traditional System of Medicine developed over millennium by sages and seers of ancient India) radish promotes digestion, is anthelmintic, cures digestive disorders, piles, and all kinds of swelling. Good for heart, hiccoughs, dermatoses, cholera and in secondary amenorrhoea. It also cures tuberculosis, phthisis, bronchial asthama, eye problems, abdominal pain and kapha and pitta disorders. It also cures blood disorders and enhances appetite.

Caution

Radishes are not recommended for people with gastro- intestinal inflammations or ulcers. Radishes may contain goitrogens, a plant based compounds found in cruciferous and brassica family vegetables like cauliflower, broccoli etc. Goitrogens may cause swelling of thyroid gland and should be avoided in individuals with thyroid dysfunction. However, they may be used liberally in healthy persons.

TURNIP (*Brassica campestris*)

Although this bulbous root that is widely eaten; it is its top fresh greens that are more nutritious; several times richer in vitamins, minerals, and anti-oxidants. The root is rich in vitamins A, B and C and minerals such as sulphur, calcium, potassium, sodium and phosphorus. It is also rich in fibre and contains far fewer calories than potato. A serving of 100g has only 22 calories. Turnip greens supply many times the nutritional content of the root. They are excellent sources of vitam, ins and minerals. Turnips detroxify the body and alkalize the blood, promoting sweating, releasing mucous and improving the appetite. They are also generally neneficial in conditions such as indigestion, diabetes and jaundice. Turnip greens have traditionally been used in Asia in the treatment of lung congestion, bronchitis, asthma and sinus problems. Although raw turnips have a somewhat pungent smell, this is neutralized in cooking.

Health Benefits

- ☆ Turnips are very low calorie root vegetables; contains only 28 calories per 100 g. However, they are very good source of anti-oxidants, minerals, vitamins and dietary fibre.
- ☆ Fresh roots are indeed one of the vegetables rich in vitamin C; provide about 21mg or 35 per cent of DRA of vitamin C per 100 g. Vitamin-C is a powerful water-soluble anti-oxidant required by the body for synthesis of collagen. It also helps body scavenge harmful free radicals, prevents from cancers, inflammation, and helps boost immunity.
- ☆ Turnip greens are the storehouse of many vital nutrients, in fact several times than the roots. The greens are very rich in antioxidants like vitamin A, vitamin C, carotenoids xanthins and lutein. In addition, the greens are excellent source of vitamin K.
- ☆ Top greens are also very good source of B-complex group of vitamins such as folates, riboflavin, pyridoxine, pantothenic acid and thiamin.
- ☆ Fresh greens are also excellent sources of important minerals like calcium, copper, iron and manganese.

Caution

Turnips and top greens are generally very safe including in pregnant women. However, the roots and top greens contain small amounts oxalic acid (0.21 g per 100 g), a naturally occurring substance found in some vegetables belonging to brassica family which may crystallize as oxalate stones in the urinary tract in some people. It is therefore, those with known oxalate urinary tract stones may have to avoid eating them. Adequate intake of water is therefore advised to maintain normal urine output in these individuals to minimize the stone risk.

HORSE RADISH (*Armoracia rusticana*)

Horseradish is a long, tapering root used as a condiment in the kitchens. The root has strong, hot, and sharp flavour, which can be only be described after experiencing its unique taste! The plant is a small perennial herb grown as annual field crop for its thick, rough, fleshy roots in many parts of Europe, America, and Asia.

Horse radish roots - Long coarse tapering roots

Valued for the pungent effect of its long, white root, horse radish is put to good use as a flavouring and as a medicinal agent, especially in Japan. It contains mustard oil **(allylisothiocyanate)**, which in fact an irritant. It alleviates colds, snotty noses and sinusitis by promoting nasal discharge and releasing phlegm from lungs.

Health Benefits of Horseradish

- ☆ Horseradish is low in calories and fat; but contains good amount of dietary fibre, vitamins, minerals, and anti-oxidants. The active principles in the root found to have anti-inflammatory, *diuretic* (increase urine output), and nerve soothing effects.
- ☆ The root contains many volatile phyto-chemical compounds, which give its much-famed pungent character. Some of the major constituents in the root are *allyl isothiocyanate, 3-butenyl isothiocyanate, 2-propenylglucosinlate (sinigrin), 2-pentyl isothiocyanate,* and *phenylethyl isothiocyanate.* It has been found that these compounds have anti-oxidant as well as de-toxification functions.
- ☆ It is a potent gastric stimulant; increases appetite, and aids in digestion. The volatile phyto-chemical compounds in the root stimulate salivary, gastric, and intestinal glands to secrete digestive enzymes, thereby facilitate digestion.

- ☆ Horseradish has good amounts of **vitamin-C** which is a powerful water soluble anti-oxidant.
- ☆ The root spice has some of vital minerals in moderation like sodium, potassium, manganese, iron, copper, zinc, and magnesium.
- ☆ In addition, the root has small amounts of essential vitamins such as **folate**, vitamin B-6 (pyridoxine), riboflavin, niacin, and pantothenic acid

Caution

Horseradish can cause irritation to skin, mucus membranes, and eyes. This is due to release of allyl sulphide gas (allyl-isothiocyanate) while chopping, crushing, or grating the root. Disruption of cell wall activates enzyme myrosinase which when reacts with glucosinolates to form allyl isothiocyanates. Lemon citrus or vinegar stops this reaction and stabilizes the flavour. Its effect can be minimized by using blender/mixer in well-ventilated place and wearing protective gloves and mask.

CELERY (*Apium graveolens* Dulce)

It provides iron, calcium, sodium, potassium, phosphorus and traces of other minerals, as well as vitamin A, B1, B2, C as well as silicon. It is specially rich in potassium and sodium, four fresh stalks of celery provide 341 mg potassium and 126mg sodium and their juice makes a great electrolyte replacement drink. It has been recommended for rheumatism, arthritis, and nervousness and sciatica to liver disorders. Iranian boil the aromatic seed oil and inhale the vapour as a cure for headache(Levey, 1966). In Egypt, wild celery seed is used as a diuretic, digestive aid, and emmemagogue, a substance that induces menses in non-pregnant women(Levey, 1966). Celery juice contains the anti-cancer nutrients phthalid and polyacetylene (antioxidants). Being rich in potassium and sodium helps in lowering blood pressure. Celery is valuable in weight loss diets, where it provides low calorie fibre bulk. Bergapten in the seeds can increase photosensitivity, so the use of seed of essential oil externally in bright sun shine should be avoided. The oil and large doses of seeds should be avoided during pregnancy. They can act as a uterine stimulant. May help prevent certain cancers, regular blood pressure and reduce cholesterol. It contains vitamin C and several other active compounds that promote health, including phalides, which may help lower cholesterol and coumarins that may be useful in cancer prevention.

Celery is an excellent source of vitamin C, a vitamin that helps to support the immune system. Vitamin C rich food like celery help reduce cold symptoms or severity of cold systems. Celery contains compounds called **pthalides**, which can help relax the muscles around arteries and allow those vessels to dilate. With more space inside the arteries, the blood can flow at a lower pressure. Pthalides also reduce stress harmones, one of whose effects is to cause blood vessels to constrict. Celery has a reputation among some persons being a high-sodium vegetable and blood pressure reduction is usually associated with low- sodium foods. Celery contains compounds called **coumarins** that help prevent free radicals from damaging cells, thus decreasing the mutations that increase the potential for cells to become cancerous. Coumarins also enhance the activity of certain white blood cells, immune defenders that target and eliminate potentially harmful cells, including cancer cells. Coumarin compounds in the celery appear to be useful in toning the heart and blood vessels. These compounds can also be useful in cases of migraine. One of them, 3-n-butyl phthalide, was found to significantly lower blood pressure. In addition, compounds in celery called **acetylenics** have shown to stop the growth of tumor cells. The stalks of the plant, which are rich in iron, magnesium and carotene, are usually eaten raw in salads. Leaves are thought to stimulate the appetite and also increase urination and bring on menstruation. Celery juice has been used to alleviate oedema, treate rheumatism and clear skin problems. In addition to rich source of vitamin C and fibre, it contains negative calories *i. e.* you burn caloreies to digest them and this property of celery is very beneficial for those who want to lose weight.

According to Acharya Balkrishna (2008), it suppresses Kapha and Vata, elevates pitta and is a pain reliever, but gives heartburn. It stimulates heart, increases urine flow, but it is slightly aphrodisiac and excites the womb. It is beneficial in curing hiccoughs, vomiting, annul pain and whooping cough. Since it treats digestive system it is one of the main medicines for stomach disorders. It is also effective in curing piles and stones in kidney or gallbladder.

Celery may not be the food of choice that you dream of when thinking of sex, but it is a great source food for sexual stimulation. It contains andro-sterone, an odourless hormone released through male perspiration and turns women on. Celery is best eaten raw, wash and cut some and much away in bed just before you get down and dirty. Celery is the best vegetable source of naturally occurring sodium. It is high in potassium. The high water content in celery makes it ideal for vegetable juicing. As an easy way to reduce grains in your diet, spread peanut butter on celery rather than bread.

Health Benefits

- ☆ One of the very low calorie herbal plant, celery leaves contain only 16 cal per 100 g weight and lots of non-soluble fibre which when combined with other weight loss regimens may help to reduce body weight and blood cholesterol levels.
- ☆ Celery is a functional food. Its leaves are rich source of flavonoid antioxidants such as *zeaxanthin, lutein* and *beta- carotene,* which have anti-oxidant, cancer protective, and immune-boosting functions.
- ☆ It is also good source of vitamin-A. Vitamin-A and beta-carotene are natural flavonoid antioxidants. Vitamin A is also required for maintaining healthy mucus membranes and skin, and for vision. Consumption of natural foods rich in flavonoids helps body to protect from lung and oral cavity cancers.
- ☆ The herb is also rich in many vital vitamins including folic acid (provides 9 per cent of RDA), riboflavin, niacin and vitamin-C, which are essential for optimum metabolism.

- ☆ Fresh celery is an excellent source of **vitamin-K**, provides about 25 per cent of DRI. Vitamin-K help increase bone mass by promoting osteotrophic activity in the bones. It also has established role in Alzheimer's disease patients by limiting neuronal damage in the brain.
- ☆ The herb is very good source of minerals like potassium, sodium, calcium, manganese, and magnesium. Potassium is an important component of cell and body fluids that helps control heart rate and blood pressure.
- ☆ Its leaves and seeds contain many essential volatile oils that include terpenes, mostly **limonene** (75 to 80 per cent), and the sesquiterpenes like β-selinene (10 per cent) and humulene; but its characteristic fragrance is due to chemical compounds known as *phthalides* (butylphthalid and its dihydro derivate sedanenolid) in them.
- ☆ Essential oil obtained from extraction of celery plant has been used in soothing remedies for nervousness, osteoarthritis, and gouty-arthritis conditions. In addition, its seeds, and root has diuretic (removes excess water from body through urine), galactogogue (help breast milk secretion), stimulant, and tonic properties.
- ☆ Root has been used in the preparation of soups and sauces.

Medicinal Uses

- ☆ Wild celery has been used in complementary medicines to reduce blood pressure, to relieve indigestion and as an anti-inflammatory agent. It also used as a diuretic to remove excess water from the body.
- ☆ The essential oils in seeds, herb, and root have use as carminative, emmenagogue, galactogogue (help breast milk secretion), nervous system ailments such as headache and nervous irritability.
- ☆ The herb has also been claimed useful in rheumatism and gouty conditions.

Caution

The herb especially wild celery can cause severe anaphylactic reactions in some sensitive individuals. It is also should not be eaten by pregnant women. People on diuretic medications and anticoagulant medications should use this herb sparingly. It is also very high in soluble as well as insoluble fibre contents. Eating recipes with too much fibres may cause stomach-pain, indigestion and bloating, and sometimes complicates existing constipation condition.

PARSLEY (*Petroselinum crispum*)

Wonderfully nutritious parsley is a popular culinary as well as medicinal herb, which is recognized as one of the functional food for its unique anti-oxidants and disease preventing properties. Is a very popular biennial herb with aromatic leaves that are used as a culinary flavouring. The leaves are either eaten fresh in salads or steeped for tea. Due to its significant therapeutic benefits, parsley was first used as herbal medicine and later as food. Parsley is a wonderful aid to digestion. It is known to have laxative effects, relieves flatulence, reduces stomach scrams and stimulates appetite. It has been known to have the ability to shrink small blood vessels and is helpful in treating piles. It is also known to have liver protection properties. A recent animal study reported that degenerative changes observed in the liver cells of diabetic rates were significantly reduced or disappeared on treatment with parsley. Its anti septic properties make it useful in cystitis (inflammation of urinary bladder). Parsley has a tonic effect on reproductive system and is sometimes used during labour. It is known to be effective emmenagogue: a substance that induces or hastens menstrual flow and helps to regulate the menstrual cycles. It appears to increase diuresis by inhibiting the Na+\K+ ATPase pump in the kidney, thereby enhancing sodium and water excretion while increasing potassium re-absorption. It is also valued as an aquaretic. When crushed and rubbed on the skin, parsley can reduce itching in mosquito bites. When chewed, parsley can freshen bad breath. Folk medicine has a tradition of using parsley to treat health conditions. Tea made from leaves or roots (of parsley) can be used to treat jaundice, coughs, menstrual problems, rheumatism, kidney stones and urinary infection. Juice from parsley sooths conjunctivitis and other eye inflammations.

Both seeds and dried roots of parsley are used as spices. The essential oil, distilled mainly from seeds has a warm, spicy, herbaceous scent. The parsley leaf oil is also used extensively for garnishing and seasoning. Parsley leaves are used as a flavouring and as garnish either dried or fresh, in soups, meat dishes, poultry, fish and as stuffing. Parsley is an excellent diuretic and useful for kidney disorders and the elimination of stones and gravel. Both the fresh juice and a tea made from the leaves or seeds can be used to relieve water retension (oedema) and strengthen digestion. Parsley also stimulates delayed menstruation and promotes menstrual flow.

Health Benefits

- ☆ The herb contains no cholesterol; but is rich in anti-oxidants, vitamins, minerals and dietary fibre which helps control blood cholesterol levels, prevents constipation, protects body from free radicals mediated injury and from cancers.
- ☆ Parsley contains many health benefiting essential volatile oils that include *myristicin, limonene, eugenol,* and *alpha-thujene.*
- ☆ The essential oil, **Eugenol,** present in this herb has been in therapeutic use in dentistry as a local anesthetic and anti-septic agent for teeth and gum diseases. *Eugenol* has also been found to reduce blood sugar levels in diabetics, however, further detailed studies required to establish its role.
- ☆ Parsley is rich in poly-phenolic flavonoid anti-oxidants including *apiin, apigenin, crisoeriol, and luteolin;* and has been rated as one of the plant source with **highest anti-oxidant** activities.
- ☆ The herb is a good source of minerals like potassium, calcium, manganese, iron, and magnesium. Potassium in an important component of cell and

body fluids that helps control heart rate and blood pressure by countering the effects of sodium. Iron is essential for heme production inside red blood cells. Manganese is used by the body as a co-factor for the antioxidant enzyme, *superoxide dismutase.*

- ☆ It is also rich in many antioxidant vitamins including **vitamin-A**, beta-carotene, **vitamin-C**, vitamin-E, zea-xanthin, lutein, and cryptoxanthins. The herb is also an excellent source of vitamin-K and folates. **Zea-xanthin** helps prevent age related macular degeneration (ARMD) in the retina of the eye in the old age population through its anti-oxidant and ultra-violet light filtering functions.
- ☆ Fresh herb leaves are also rich in many essential vitamins such as pantothenic acid (vitamin B-5), riboflavin (vitamin B-2), niacin (vitamin B-3), pyridoxine (vitamin B-6) and thiamin (vitamin B-1). These vitamins are essential during carbohydrates, fat and protein metabolism by acting as co-enzymes.
- ☆ It is probably the richest of the entire herb source for **vitamin K;** provides 1640 mcg or 1366 per cent of recommended daily intake. Vitamin K has been found to have potential role in bone health by promoting osteotrophic activity in the bones. It has also established role in the treatment of *Alzheimer's disease* patients by limiting neuronal damage in the brain.

Medicinal Uses

- ☆ The leaves, stems and roots of this herb plant have antiseptic and carminative properties.
- ☆ *Eugenol* also has been found to reduce blood sugar levels in diabetics, however, further detailed studies required to establish its role.
- ☆ The extraction from the herb has been found to have diuretic effects.
- ☆ Parsley is useful as a digestive aid.
- ☆ It helps to purify the blood and stimulate the bowels.
- ☆ Parsley is an anticarcinogen.
- ☆ It contains three times as much vitamin C as oranges, and twice as much iron as spinach.
- ☆ Parsley contains vitamin A and is a good source of copper and manganese.
- ☆ For a natural breath freshener, try a sprig of parsley!

PARSNIP (*Pastinaca sativa*)

Parsnips are closely related to carrots family of vegetables; grown for their sweet, succulent underground taproots. The root vegetable has similar appearance and growth characteristics as other apiaceae family members like **carrots, parsley, celery, cumin, dill etc. A sweet culinary root vegetable and mildly diuretic,**

it lubricates the intestines and is benefical for soothing stomach and intestinal irritations. It helps to clean the liver and gall bladder and is used in soups or teas. It promotes perspiration and is beneficial for coughs, cold, headaches and arthritis. Its fleshy, stout roots appear like that of carrots, but are white or cream in colour and sweeter than that of carrots. Good winter frost is essential for better crop, as frost converts much of the starch to sugars and helps develop long, firm parsnips. The roots are generally harvested when they reach about six to ten inches long, by pulling the entire plant with its root (uprooting) as in carrots.

Health Benefits

- ☆ Generally, parsnip contains more sugar than carrots, radish, turnips. In general, it has calories (100 g provide 75 calories) comparable to that of some fruits like banana, grapes etc. Nonetheless, its sweet, juicy root is rich in several health-benefiting phyto-nutrients, vitamins, minerals, and fibre.
- ☆ It is one of the excellent sources of soluble and insoluble **dietary fibre**. 100 g root provides 4.9 mg or 13 per cent of fibre. Adequate fibre in the diet helps reduce blood cholesterol levels, obesity and constipation conditions.
- ☆ As in carrots and other members of apiaceae family vegetables, parsnip too contains many **poly-acetylene** anti-oxidants such as *falcarinol, falcarindiol, panaxydiol,* and *methyl-falcarindiol*.
- ☆ Several research studies from scientists at University of Newcastle at Tyne found that these compounds have anti-inflammatory, anti-fungal, and anti-cancer function and offer protection from colon cancer and acute lymphoblastic leukemia (ALL).
- ☆ Fresh roots are also good in vitamin C; provide about 17 mg or 28 per cent of RDA. Vitamin C is a powerful water-soluble anti-oxidant, easily available to us from natural sources. It helps body maintain healthy connective tissue, teeth, and gum. Its anti-oxidant property helps protect from diseases and cancers by scavenging harmful free radicals from the body.
- ☆ Further, the root is rich in many B-complex groups of vitamins such as **folic acid,** vitamin B-6 (pyridoxine), thiamin, and pantothenic acid as well as vitamin K and vitamin E.
- ☆ In addition, it also has healthy levels of minerals like iron, calcium, copper, potassium, manganese and phosphorus. Potassium is an important component of cell and body fluids that helps controlling heart rate and blood pressure by countering effects of sodium.

Caution

Parsnip plant and its parts may cause hypersensitivity reactions like contact dermatitis and oral allergy syndrome (OAS) when handled in some sensitive individuals. The reaction symptoms may include rash and skin lesions. Some of common OAS symptoms may include itching or burning sensation in the lips, mouth, and throat. In severe cases swelling of the lips, tongue, and redness in eyes, and breathing difficulty may be observed. Individuals with known history of allergy to birch category pollen agents like walnuts, fig, carrots, parsley etc may develop cross-sensitivity to parsnip and should be avoided. Only the root part of the plant should be eaten. Parsnip leaves are poisonous.

PURSLANE (*Portulaca oleracea*)

Purslane is widely grown in many Asian and European regions as staple leafy vegetable. Its leaves appear thick, contain mucilaginous substance, and have a slightly sour and salty taste. Leaves and tender stems have slightly sour and salty taste. In addition to succulent stems and leaves, its yellow flower buds are also edible. It contain more **Omega-3 fatty acids (alpha- linolenic acid** in particular) than any other leafy vegetable plant. This soft, succulent **Purslane** has more omega-3 fatty acids than in some of fish oils. If you are a vegetarian and want to avoid all forms of animal products, then here is the answer! Go for this healthy dark green leafy vegetable and soon you will forget fish! Purslane has 0.1mg per gram of EPA and this is extra ordinary amount of EPA for land based vegetable source. EPA is an **Omega- 3 fatty acid** normally found mostly in fish and algae. It also contains vitamins, mainly vitamin A, vitamin C and some vitamin B and carotenoids as well as dietary minerals, such as magnesium, calcium, potassium and iron. Also present are two types of betalain alkaloid pigments, the reddish **betacyanins** (visible in the colouration of the stems) and the yellow **betaxanthins** (noticeable in the flowers and in the light yellowish cast of the leaves). Both of these pigment types are potent anti -oxidants and have been found to have anti-mutagenic properties in laboratory studies. 100 grams of fresh purslane leaves (about 1 cup) contains 300-400mg of **alpha-linolenic acid.** One cup of cooked leaves contain 90mg of calcium, 561mg of potassium and more than 2000 IU's of vitamin A.

Health Benefits

- ☆ This wonderful green leafy vegetable is very low in calories (just 16 kcal/ 100g) and fats; but is rich in dietary fibre, vitamins, and minerals.
- ☆ Fresh leaves contain surprisingly more Omega-3 fatty acids (α-linolenic acid) than any other leafy vegetable plant. 100 grams of fresh purslane leaves provides about 350 mg of α-linolenic acid. Research studies shows that consumption of foods rich in ω-3 fatty acids may reduce the risk of coronary heart disease, stroke, and help prevent development of ADHD, autism, and other developmental differences in children.
- ☆ It is an excellent source of Vitamin A, which is one of the highest among green leafy vegetables. Vitamin A is a known powerful natural antioxidant and is essential for vision. This vitamin is also required to maintain healthy mucus membranes and skin. Consumption of natural vegetables and fruits rich in vitamin A is known to help to protect from lung and oral cavity cancers.

- Purslane is also a rich source of vitamin C, and some B-complex vitamins like riboflavin, niacin, pyridoxine and carotenoids, as well as dietary minerals, such as iron, magnesium, calcium, potassium, and manganese.
- Also present in purslane are two types of betalain alkaloid pigments, the reddish *beta-cyanins* and the yellow *beta-xanthins*. Both of these pigment types are potent anti-oxidants and have been found to have anti-mutagenic properties in laboratory studies. [Proc. West. Pharmacol. Soc. 45: 101-103 (2002)]

Caution

Purslane contains oxalic acid, a naturally occurring substance found in some vegetables, which may crystallize as oxalate stones in the urinary tract in some people. 100 g fresh leaves contain 1.31 g of oxalic acid, more than in spinach (0.97 g/100 g) and cassava (1.26 g/100 g). It is therefore, people with known oxalate urinary tract stones are advised to avoid eating purslane and certain vegetables belonging to amaranthaceae and brassica family. Adequate intake of water is therefore advised to maintain normal urine output.

CHICORY (*Cichoium intybus*)

Chicory is used as a vegetable, most often found in a green salad, as a coffee alternative, by roasting and grinding its roots, or as a medicinal remedy. As a supplement, chicory contains cardioactive, probiotic and sedative properties that may be helpful in the treatment of cardiovascular disease, gastrointestinal issues such as diarrhea, constipation, dyspepsia and cancer. The leaves of the chicory plant can be eaten raw in salads or cooked like greens. The roots are usually roasted, dried and ground into coffee. Chicory contains many nutrients, including vitamins A, B6, C, E, and K, as well as zinc, folate, calcium, magnesium, potassium, iron and manganese. Chicory can be found in infusions or extracts; it is used as a supplement that may have laxative, diuretic, cardioprotective and anti-inflammatory properties.

Especially the flower- It is variously used as a tonic and appetite stimulant and as a treatment for gallstone, gastroenteritis, sinus problems and cuts and bruises (Howard, 1987). Root chicory contains oils similar to those found in plants in the related Tanacetum which includes Tansy and is likewise effective at eliminating intestinal worms. All parts of the plant contain these volatile oils, with the majority of the toxic compounds concentrated in the plant root. Chicory is well known for its toxicity to internal parasites. Studies indicate that ingestion of chicory by farm animals result in reduction of worm burdens which has prompted its wide spread use as a forage supplement. It has a bitter spicy taste, which mellows when it is grilled or roasted. It can also be used to add colour and zest to salads. The young leaves can be used in salads and the root can also be boiled and eaten like a vegetable. It is also grown for cattle feed in Europe.

Radicchio, like other chicory class vegetables, is low in calories, 100g fresh leaves provide just 23 calories. The bitter principle in radicchio is Lactucopicrin (intybin), a sesquiterpene. Lactucopicrin is a potent-anti-malarial agent and has a sedative and analgesic effect.

Infusion or decoction of root or flowers can stimulate the appetite, aid digestion, promote bile secretion and help to relieve the pain and discomfort caused by gallstones. Dried ground chicory is used as a coffee substitute, either on its own or combined with cereals i n cereal coffee. Chicory contains inulin, which helps diabetics regulate their blood sugar levels. It is closely related to lettuce and dandelion but is a member of the sunflower family. It may be cleansing to the liver and gallbladder. Chicory is beneficial for digestion, the circulatory system and the blood. Its leaves are a good source of calcium, vitamin A and potassium.

Chicory is also known to contain a number of nutritional and health benefits. It is also high in Vitamin A, calcium and potassium. It also contains iron, niacin, phosphorus, and other important vitamins and minerals. Chicory has been found to contain a substance called inulin, which is very beneficial to diabetics, as it helps to regulate blood sugar levels. As chicory is also utilized as a sweetener, with a taste comparable to sugar, it makes a great sugar alternative for diabetics as well. Other health benefits of chicory include its effectiveness at fighting intestinal worms and parasites, aiding in digestion and colon cleansing, and helping to cleanse the blood, liver and circulatory system of harmful toxins.

Many other health benefits have been associated with chicory, or, more specifically, chicory root, when used as a tea. Chicory tea can be made by simmering a tablespoon of chicory root in a cup of water for approximately 20-30 minutes. It is recommended to drink a cup of chicory tea every day in order to truly benefit from the health and nutritional properties inherent in the chicory root. Some herbal tea companies also sell a chicory root tea, or an herbal tea with chicory as an ingredient. Other ailments that have utilized the chicory root tea as a treatment include cramps, sore throats, and tuberculosis, among others.

GLOBE ARTICHOKE (*Cyanara scolymus*)

Artichoke globe measures about 6-10 cm in diameter and weigh about 150 g. Fuzzy, immature florets in the centre of the bud constitute "choke". These are inedible in older, larger flowers. Edible portion of the buds consists primarily of the fleshy lower portions of the involucres bracts (triangular scales) and the base, known as the "heart". The flower heads are commonly eaten as vegetables but extracts of leaves and roots were once used to treat liver disorders, jaundice and atherosclerosis. Artichokes provide a number of vitamins and minerals and one of their key active ingredients, **Cynarin** was shown by clinicalk studies to stimulate bile flow from gall bladder, thus mobilising fatty stores and reducing cholesterol. Other studies found that its extracts can help digestive complaints such as poor appetite, nausea and abdominal discomfort.

The flower heads are considered useful in the dietary of diabetics. The leaves are bitter and considered to be useful in dropsy and rheumatism. Dried and fresh leaves and \or stems of cynara are used to increase bile production. **Cynarin,** an active constituent in cynara causes an increase in bile flow. Globe artichoke has (cynara) has a powerful effect on the production of bile and fat indigestion enzymes, stimulating liver functions and lowering cholesterol levels. Seemingly this latter action is due to increased bile production and reduced absorption of cholesterol in the intestine, with less cholesterol being synthesized in the liver and more being eliminated. Artichoke can also be made in to herbal tea, artichoke tea is produced as a commercial product in the Dalat region of Vietnam. Other studies found that artichoke extracts can help digestive complaints such as poor appetite, nausea and abdominal discomfort.

Health Benefits

- ☆ Artichoke is low in calories and fat, but is a rich source of dietary fibre; provides 5.4 g per 100 g, about 14 per cent of RDA. Dietary fibre helps control constipation conditions, decrease bad or "LDL" cholesterol levels by binding to it in the intestines and helps prevent colon cancer risks by preventing toxic compounds in the food from absorption.
- ☆ Scientific studies have shown that bitter principles, **cynarin** and **sesquiterpene- lactones** in artichoke extraction have overall cholesterol reduction action in the body by inhibiting its synthesis and increasing its excretion in the bile.
- ☆ Fresh artichoke is an excellent source of vitamin **folic acid**; provides about 68 mcg per 100 g (17 per cent of recommended daily allowance). Folic acid acts as a co-factor for enzymes involved in the synthesis of DNA. Scientific studies have proven that adequate levels of folates in the diet during pre-conception period and during early pregnancy help prevent from neural tube defects in the newborn baby.
- ☆ It is also rich in B-complex group of vitamins such as niacin, vitamin B-6 (pyridoxine), thiamin, and pantothenic acid that are essential for optimum cellular metabolic functions.
- ☆ Fresh globes also contains good amounts of anti-oxidant vitamin, **vitamin-C**. Provides about 20 per cent of recommended levels per 100 g. Regular consumption of foods rich in vitamin C helps body develop resistance against infectious agents and scavenge harmful, pro-inflammatory free radicals from the body.
- ☆ It is one of the vegetable sources for vitamin K; provides about 12 per cent of DRI. Vitamin K has potential role bone health by promoting osteotrophic

(bone formation) activity. Adequate vitamin-K levels in the diet helps limiting neuronal damage in the brain; thus, has established role in the treatment of patients suffering from Alzheimer's disease.

- It is also good source of anti-oxidants such as **silymarin, caffeic acid and ferulic acid,** which help body protect from harmful free-radical agents.
- It is also rich source of minerals like copper, calcium, potassium, iron, manganese and phosphorus. **Potassium** is an important component of cell and body fluids that helps controlling heart rate and blood pressure by countering effects of sodium. Manganese is used by the body as a co-factor for the antioxidant enzyme, *superoxide dismutase*. **Copper** is required in the production of red blood cells. **Iron** is required for red blood cell formation.
- In addition, it also contains adequate levels of anti-oxidant flavonoid compounds like carotene-beta, lutein and zea-xanthin levels.

Caution

This veggie is very well tolerated in general population. However, its leaves are thought to exacerbate gall stone disease.

CUMCUMBER (*Cucumis sativus*)

It has a very cooling effect on body. It is rich in potassium, due to which it helps combat fatigue and muscle weakness. Its juice cleanses the kidneys, lowers high blood pressure and improves skin problems. Cucumber is a natural diuretic which is used in the fitness world by body builders and people trying to reduce their weight. It is also said to help digestion and constipation. Although less nutritious than most fruit and vegetables, the fresh cucumber is still a very good source of vitamin C and the mineral molybdenum. It is also a good source of vitamin A, potassium, manganese, folate, dietary fibre and magnesium. What you may not know is that this crisp, refreshing fruit also contains compounds called **sterols**, which have been shown to lower cholesterol in animals. The heaviest concentration of **sterols** is in the skin of the cucumber, so you should not remove the peel before eating. Many people use cucumbers on the skin as a beauty aid as well.

By taking a glass of cucunber juice every day, the skin tone gets improved and the skin starts glowing. It also contains innumerable foliate, besides Vitamin A and C. Cucumber has diuretic properties, which can help to eliminate water from the body and it also contains an enzyme that splits protein and cleanses the intestines. Its juice is beneficial to internal inflammations such as stomach and kidney inflammationns and sore throats.

Cucumbers are best eaten with skin, which is rich in chlorophyll and silicon. Externally, a blend of juiced cucumber with equal parts of glycerine and rose water makes a soothing lotion for chapped hands and lips. These are excellent as anti-ageing food and improve tired and dehydrated skin. The high water content and high amounts of silica present helps to get a smooth and glowing skin. This wonderful, low calorie vegetable indeed has more nutrients to offer than just water and electrolytes.

Nutritive Value

Cucumber	*Partion*	*Carbohydrate*	*Fibre*	*Fat*	*Energy*
Raw (un-peeled)	5 slices	(g)	(g)	(g)	(kj)
	45g	3	0.5	0	15

There is 1 calorie in 1 slice of cucumber (1 stick, 10cm long)

Cucumber is:

- ✰ Low in saturated fat, cholesterol and sodium.
- ✰ High in vitamin C, vitamin K, potassium and vitamin A, pantothenic acid, magnesium, phosphorus and manganese.
- ✰ Maintaining good general health and losing weight.
- ✰ But do not eat too much cucumbers if you are interested in gaining weight.

Health Benefits

- ✰ It is one of the very low calories vegetable; provides just 15 calories per 100 g. It contains no saturated fats or cholesterol. Cucumber peel is a good source of dietary fibre that helps reduce constipation, and offers some protection against colon cancers by eliminating toxic compounds from the gut.
- ✰ It is a very good source of potassium, an important intracellular electrolyte. Potassium is a heart friendly electrolyte; helps reduce blood pressure and heart rates by countering effects of sodium.
- ✰ It contains unique anti-oxidants in good ratios such as β-carotene and α-carotene, vitamin-C, vitamin-A, *zea-xanthin* and *lutein*. These compounds help act as protective scavengers against oxygen-derived free radicals and reactive oxygen species (ROS) that play a role in aging and various disease processes.
- ✰ Cucumbers have mild diuretic property probably due to their high water and potassium content, which helps in checking weight gain and high blood pressure.
- ✰ They are surprisingly have high amount of vitamin K, provides about 17 µg of this vitamin per 100 g. Vitamin-K has been found to have potential role in bone strength by promoting osteotrophic (bone mass building) activity. It also has established role in the treatment of *Alzheimer's disease* patients by limiting neuronal damage in their brain.

ASH GOURD (Petha) (*Benincasa hispida*)

The health benefits of ash-gourd have long been known. The ash-gourd is mentioned in ancient Ayurvedic texts like *Charaka Samhita* and *Ashtanga Hridaya Samhita* for its many nutritional and medicinal properties. Ash-gourd is loaded with nutrients. It's an excellent source of vitamin B1 (thiamine), a good source of vitamin B3 (niacin), and vitamin C. It is also rich in many minerals like calcium. Its high potassium content makes this a good vegetable for maintaining a healthy blood pressure. Containing almost 96 per cent water, this gourd is a dietitian's delight. Be sure to include this vegetable in your weght-loss diet.

In Ayurveda and other traditional eastern medicine, ash-gourd is used as a general tonic for its restorative properties. It is also used as brain food - to treat mental illnesses and nervous disorders such as epilepsy, paranoia, and insanity in Ayurveda. Ash-gourd is alkaline in nature and hence has a cooling and neutralizing effect on stomach acids and as such used effectively for treating digestive ailments like hyperacidity, dyspepsia, and ulcers. Ash-gourd juice is a popular home remedy for peptic ulcers. Ash-gourd juice is also used to treat diabetes. Ash-gourd is also useful in treating respiratory disorders like asthma, blood-related diseases, and urinary diseases like kidney stones.

Every part of this fruit is useful. Ash-gourd leaves are rubbed on bruises to heal them, while the seeds are used for expelling intestinal worms. The ash made from burning the rind and seeds are mixed with coconut oil and used to promote hair growth and to treat dandruff.

Inexpensive and versatile, ash-gourd is a healthful vegetable that should definitely be a part of any nutritious diet.

Traditional Vaidyasalas in Kerala use a specific ash gourd variety called Vaidya Kumbalam or Nei Kumbalam for preparing ayurvedic medicine called Kushmand rasayana. It is used as a tonic and health rejuvenator for treating diabetes, fever and also as a laxative.

Health Benefits

- ☆ The dilate juice of ash gourd is useful in the treatment of peptic ulcer. The juice squeezed out of grated out of ash gourd with equal amount of water added to it, should be taken daily in the morning on empty stomach. No food should be consumed for 2 to 3 hours after wards. This also relieves inflammation in the alimentary canal.

- A delicious sweet prepared from pulp of the fruit by boiling its pieces in water and adding sugar syrup to it is used as a medicine to increase weight, in tuberculosis, weakness of the heart, heat in the body, thinness of semen and anemia.
- Its peel and seeds, when boiled in coconut oil are useful in hair growth; prevent dandruff and dryness of the scalp.
- Ash gourd acts as a blood coagulant. From ancient times, its fresh juice mixed with a teaspoon of goose berry or lime juice is used as a specific medicine to stop profuse bleeding from lungs and presence of blood cells in the urine.
- Due to its diuretic action, it increases the output of urine and washes out waste products from the body.
- Shelled seeds of ash gourd are anabolic- that promotes tissue growth-especially when taken with coconut milk, they expel tape worm and other worms from the intestines.
- The fruit is used to treat summer fevers. In Ayurveda, the fruit is used to treat epilepsy, lung diseases, asthama, coughs, urine retention and internal hemorrhage. The rind is used to treat diabetes and the seeds to expel tapeworms. The juice of the fruit is effective in cases of mercury poisoning and snakebites. Being extremely low in calories, the ash gourd is used to treat obesity. A delicious sweet made by boiling the pulp in sugar syrup is used to treat general debility, to increase weight after sudden weight loss, weakness of the heart and anemia.
- It contains a lot of calcium, phosphorus, iron, thiamine, riboflavin, niacin and vitamin C.
- It promotes metabolism and prevents sugar from getting converted into fat. Hence it is good to cure obesity as it is also very low in calories.
- Cough, cold, asthma, bronchitis, influenza, fever, sinusitis can be controlled without any side effects using ash gourd.
- It is also good for bleeding gums. Regular gargling with the juice of ash gourd helps to protect teeth and gums.
- Severe asthma can be controlled by regular se of ash gourd. It is also useful to control thyroid related problems. It also helps to fight mouth cancer.
- Being alkaline in nature, it acts as a good antacid. It maintains the chemistry between the alkaline and the internal acids thus maintaining the PH of the body. Hence it keeps the digestive function in place and reduces acidity, constipation and tones the digestive system.
- The unripe ash gourd is cooked as a vegetable. The ripe fruit is largely used for making sweet meat. The leaves and buds are also cooked.

BOTTLE GOURD (*Lagenaria siceraria*)

Various medicinal uses of the leaves, fruits and seeds have been recorded from various countries for example as a pectoral, as an anthelmintic, a purgative and even

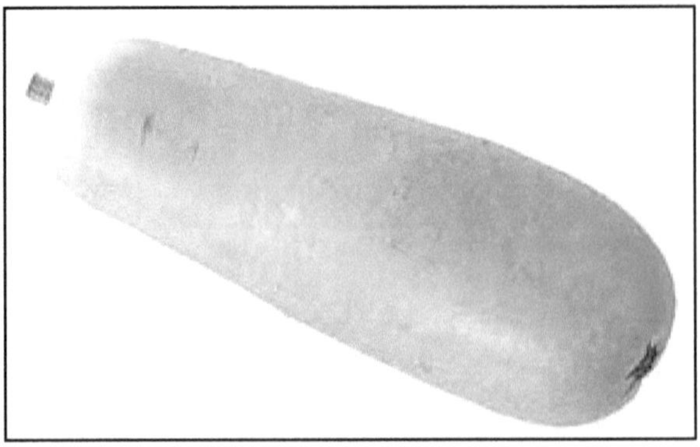

as a headache remedy. The pulp is good for overcoming constipation, cough and night blindness and as antidote against certain poisons. A decoction made from the leaf is very good medicine for curing jaundice. It is good for people suffering from biliousness and indigestion. Its juice is used in the treatment of stomach acidity, indigestion, ulcers, epilepsy and other nervous diseases. According to Ayurveda Bottle gourd (lauki) can help lose weight. Cooked bottle gourd is a diuretic and good antidote to urinary disorders. Its juice is prescribed for stomach acidity, indigestion and ulcers. It is rich in dietary fibre with very low content of fat and cholesterol. Approximately 96 per cent of bottle gourd is water. Apart from the iron content it is also rich in vitamin B and vitamin C, it also contains sodium, potassium, essential minerals and trace elements. It is a suitable vegetable for light, low-cal diets as well as for children, people with digestive problems, diabetics and convalescents. It is a suitable vegetable for light, low-cal diets as well as for children, people with digestive problems, diabetics and convalescents. The vegetable is very effective in treating constipation and other digestive disorders.

Bottle Gourd Nutrition

- ✰ Bottle gourd is rich in dietary fibre with very low content of fat and cholesterol. Approximately 96 per cent of bottle gourd is water and is therefore a great thirst quencher. Apart from the iron content it is also rich in vitaminB and vitamin C. it also contains sodium, potassium, essential minerals and trace elements.
- ✰ High in sodium and potassium, bottle gourd is an excellent vegetable for hypertension patients. It is a suitable vegetable for light, low-cal diets as well as for children, people with digestive problems, diabetics and convalescents.
- ✰ High in water content, calabash gourd is cooling, calming, diuretic and easy to digest when cooked. The vegetable is very effective in treating constipation and other digestive disorders. It helps in treating urinary disorders, since it serves as an alkaline mixture for treating burning sensation in urinary passage, caused by high acidity of urine.
- ✰ Bottle gourd is recommended by ayurvedic physicians for balancing the liver function, when the liver is inflamed and cannot efficiently process food for maximum nutrition and assimilation.
- ✰ Drinking a glass of bottle gourd juice once in the early morning helps in treating grey hair. Mixing the juice with sesame oil provides an effective medicine for insomnia. Massage the scalp with this preparation every night.

- ☆ Bottle gourd juice is very helpful in treating insanity, epilepsy, stomach acidity, indigestion, ulcers and other nervous diseases. The bitter variety of calabash gourd is considered as a cardiac tonic, an antidote for poisoning and a tonic for alleviating bronchitis, cough, asthma, and biligenic affections.
- ☆ A glass of bottle gourd juice, mixed with a little salt, helps in preventing excessive loss of sodium from the body. It quenches thirst, prevents fatigue and keeps one refreshed during summers. The juice extracted from the leaves of bottle gourd is beneficial in jaundice.

Bottle Gourd Calories

- ☆ 100 gm of bottle gourd contains 15 calories.0.1gm fat in 100gm edible portion. It is comprised of 96 per cent of water. Low in Saturated Fat, and Cholesterol, High in Dietary Fibre, Vitamin C, Riboflavin, Zinc, Thiamin, Iron, Magnesium and Manganese.
- ☆ The cooked vegetable is cooling, diuretic, sedative and antibilious and gives a feeling of relaxation after eating it.
- ☆ It is valuable for in urinary disorders. A glass of fresh juice prepared by grating the whole fruit, mixed with teaspoonful of lime juice if taken daily, is best treatment for burning sensation in the urinary passage due to high acidity of urine. As it serves as an alkaline mixture.
- ☆ Its juice is also a valuable medicine for excessive thirst due to severe diarrhea, diabetes and excessive use of fatty foods. A glassful of plain juice with a pinch of salt if taken daily will overcome this condition. Its daily use (juice) during summer prevents excessive loss of sodium, quenches thirst thus helps prevent fatigue.
- ☆ Mixture of fresh juice along with sesame oil acts as an effective medicine for insomnia. It should be massaged over scalp every night.
- ☆ Bottle gourd juice is used in the treatment of insanity, epilepsy, stomach acidity, indigestion, ulcers and other nervous diseases. It helps constipation.

Famous 'YOGI' Swami Ram Dev Ji Maharaj, who is a household name today in his daily sermons, advises use of its juice along with **Pranayama** (meditation), for the treatment of various diseases viz obesity, increased cholesterol, hypertension and other heart ailments. Ayurveda says lauki can help lose weight. Cooked lauki is a diuretic and good antidote to urinary disorders. The juice of the fruit is prescribed for stomach acidity, indigestion and ulcers.

Dr Charu Dua, chief nutritionist at the Pushpanjali Crosslay Hospital, said, It's of vital importance that people check the lauki before drinking its juice. If bitter, it carries a highly toxic compound called **Tetracyclic Triterpenoid Cucurbitacins** that can cause serious side-effects and even death. However, as a vegetable it is healthy because it contains a lot of water. This makes it very low in calories.

BITTER GOURD (*Morordica charantia*)

Bitter melon is the immature pod vegetable, popular in many Asian countries. This widely grown as edible pod is, in fact, among the most bitter of all culinary vegetables. Bitter gourd has excellent medicinal virtues. The fresh juice and the extract of the unripe fruit has been found in various studies to have a blood sugar lowering effect. **It** is a herb that regulates blood sugar levels and keeps body functions operating normally. It is anti-pyretic tonic, appetizing, stomachic, anti-bilious and laxative. The fruit is said to wormicidal and a cure for stomach disorder. It is beneficial for diabetic patients and persons suffering from arthritis, rhematism and asthama. Its leaves are known to act as galactogogs, while the roots are astringent. Its juice from its leaves is used against lysmenorrhea, as well as for external eruptions, burns or boils. A powder prepared from the plants is good for treating ulcers. The seeds of better gourd contain "**Gourdin**"- which activates the inactive insulin present in blood and rejuvenates the pancrease, thus being beneficial to patients suffering with diabetes. Seeds also contain vaccine that trigger symptoms of **favism** in susceptible individuals. It contains **Gurmarin**, a polypeptide considered to be similar to bovine insulin, which has been shown in experimental studies to achieve a positive sugar regulating effect by suppressing the neural response to sweet stimuli. Bitter gourd contains a unique phyto-constituent that has been confirmed to have a **hypoglycemic** effect called **charantin**. There is also another insulin like compound (momordica), a polypeptide (Protein) which reduces blood sugar levels, in much the same way as insulin, when injected into insulin- dependent diabetics.

An anti diabetes drug gourdin extracted from the seeds of better gourd, showed lowering of blood pressure(Pushpa Khanna, 2007). It also helps in purifying the blood. It contains four very promising bioactive compounds. These compounds activate a pectin called AMPK, which is well known for regulating fuel metabolism and enabling glucose uptake, processes which are impaired in diabetes. Drinking, 50 – 60ml of juice has also shown positive results in clinical trials.

Bitter gourd contains a **lectin** that has insulin like activity. The insulin like bioactivity of this **lectin** is due to its linking together 2 insulin receptors. This lectin lowers blood glucose concentrations by acting on peripheral tissues and also, similar to insulin's effects in brain, suppresses the appetite. This lectin is likely to a major contributor to the hypoglycemic effect that develop after eating bitter gourd and it is why it is recommended as a way of managing adult onset diabetes. According to Kaiydey Nighuntu- it is bitter and Vata Karak. It promotes digestive functions, is purgative, bitter and cold. It is Aphrodiastic, and light pittakarak. It cures cough, anaemia, infection of wounds, burning sensation, dermatoses and is antipyretic. Whileas, according to Bhava Prakash- it is cool, purifies excretory system, purgative, slightly bitter and does not cause vata. It cures fever, blood disorders, cough, jaundice, diabetes, and is anthelmintic. Its fragrant volatile oil contains **carotene**, **glucosides**, **saponin** and **mamoridisine** alkaloid.

Bitter Gourd is Beneficial

- It is used as a folk medicine for diabetes, several research proved that it contains a hypoglycaemic insulin like principle, designed as **plant insulin** which has been found highly beneficial in lowering the blood and urine sugar levels.
- It is highly beneficial in the treatment of blood disorders like blood boils, scabies, itching, psoriasis, ringworm and other fungal diseases.
- The juice of fresh leaves is valuable in piles, for the treatment of alcoholism and is effective medicine in early stages of cholera and other types of diarrhea during summer.
- Regular consumption of bitter gourd juice has been proven to improve energy and stamina level. Even sleeping patterns have been shown to be improved\ stabilized.
- The high beta-carotene and other properties in bitter gourd make it one of the finest vegetable-fruit that help alleviate eye problems and improving eyesight.
- Bitter gourd juice may be beneficial in the treatment of a hangover for its intoxication properties. It also helps cleanse and repair and nourish liver problems due to alcohol consumption.
- It juice can help build your immune system and increase your body's resistance against infection.
- It hastens healing, when the paste of its roots are applied over the piles.
- Regular consumption of its juice has also shown to improve psoriasis condition and other fungal infections like ring worm and athletes feet.
- Toxemia – It cleanses the blood from toxins. If two teaspoonful of its juice are sipped daily, it helps cleanse the liver. Also helps in ridding jaundice for the same reason.
- Precaution should be taken by the pregnant women in taking too much of its juice, as it may stimulate the uterus that may lead to preterm labour.

Health Benefits

- Very low in calories provides only 17 cal per 100g. The pods are rich in phytonutrients like dietary fibre, minerals, vitamins and anti-oxidants.
- Bitter melon notably contains phyto-nutrient, polypeptide-P; a plant insulin known to lower blood sugar levels. In addition it also contain hypoglycemic agent called charantin. *Charantin* increases glucose uptake and glycogen synthesis in the cells of liver, muscle and adipose tissue. Together, these compounds are thought to be responsible for reduction of blood sugar levels in the treatment of type-2 diabetes.

- Fresh pods are an excellent source of folates; contains about 72 mcg/100g (Provides 18 per cent of RDA). Folate helps reduce incidence of neural tube defects in pregnant mothers when taken during early pregnancy.
- Fresh bitter melon is an excellent source of vitamin-C (100g of raw pod provides about 140 per cent of RDI). Vitamin-C, one of the powerful natural antioxidant, helps body scavenge deleterious free radicals one of the reasons for cancers development.
- The vegetable also an excellent source of health benefiting flavonoids such as b-carotene, a- carotene, lutein, zeaxanthins. It also contains good amount of vitamin A. These compounds help act as protective scavengers against oxygen-derived free radicals and reactive oxygen species (ROS) that play a role in aging, cancers and various disease processes.
- Bitter melon stimulates digestion and peristalsis which can be helpful in relieving indigestion and constipation problems.
- The vegetable is also good source of Niacin (vitamin B-3), Pantothenic acid (vit. B-5), Pyridoxine (vit. B-6) and minerals such as iron, zinc, potassium, manganese and magnesium.
- Early laboratory tests suggest that compounds in bitter melon might be effective for treating HIV infection.

Caution

Bitter gourd may contain alkaloid substances like quinine and morodicine, resins and saponic glycosides which may be intolerable by some people. The bitterness and toxicity may be reduced somewhat by parboiling or soaking in salt water for up to 10minutes. Toxicity symptoms may include excessive salivation, facial redness, dimness of vision, stomach pain, nausea, vomiting, diarrhea, muscular weakness.)

PUMPKIN (*Cucurbita* spp.)

Pumpkin is rich in vital anti-oxidants and vitamins. This humble vegetable is very low in calories yet good source of vitamin A, flavonoid poly-phenolic antioxidants like leutin, xanthins and carotenes. In structure, the pulp feature golden-yellow to orange colour depending up on the poly-phenolic pigments in it. The fruit has hollow center, with numerous small, off-white coloured seeds interspersed in the net like structure. Pumpkin seeds are great source of protein, minerals, vitamins and **omega-3 fatty acids.** It offers vitamins, minerals, fibre and all important antioxidants that our body needs to stay healthy and young. Pumpkin seed oil (PSO) has high Tocopherol (TC) and unsaturated fatty acids (UFA) making it well suited for improving human nutrition. PSO has been implicated in preventing prostate growth, retarding hypertension, mitigating hypercholesterolemia and arthritis, improved bladder compliance, alleviating diabetes by promoting hypoglycemic activity and reduced gastric, breast, lung and colourectal cancer. The high unsaturated and tocoperol content could potentially improve human nutrition. In a study, Kyung-Soo Hahm, Yoonk yung Park and colleagues extracted proteins from pumpkin rinds to see if the proteins inhibit the growth of microbes, including *Candida albicans* (*C. albicans*). That fungus causes vaginal yeast infection, diaper rash infection in infants and other health problems. One protein had powerful effects in inhibiting the growth of *C. albicans*, in cell culture experiments, with no obvious toxic effects. The study suggests that the pumpkin protein could be developed into a natural medicine for fighting yeast infections in humans. The protein also blocked the growth of several fungi that attack important plant crops and could be useful as an agricultural fungicide, the research found. The study has been published in the current issue of ACS journal of Agricultural and Food Chemistry- a bi-weekly publication (Tribune, 29th oct.2009). Pumpkin makes for a great moisturiser and exfoliating masks depending upon the ingredients you use. The enzymes in it give you a younger look and natural **alpha-hydroxy acids** in pumpkin helps to remove cells to reveal younger, fresher skin. Infact, a lot of face products in the market use pumpkin as their main ingredient. Pumpkin nurishes tired or dry skin instantly giving you a healthy glow any time your skin needs a much- needed boost! Pumpkin peels can also help when you are having one of those dreaded break outs and acne problems. The nutrients can help soothe and reduce inflammation naturally, helping skin recover from acne –related damage (Anonymous, 2011). Pumpkin contains a compound (Alpha Linolenic Acid or ALA)that the body converts into a similar type of beneficial omega fatty acid. Pumpkin seeds have lots of vitamin E, a powerful antioxidant that helps slow down skin ageing, and those omega-3 fatty acids that keep your skin smooth and your hair shiny. Pumpkin seeds are packed with zinc. Studies also have shown them to reduce the appearance of acne and other skin problems.

Health and Nutrition Benefits of Eating Pumpkin

- **Rich in fibre:** Pumpkin is fairly high in fibre. One cup of cooked, mashed pumpkin contains 3 grams of fibre, or roughly 11 per cent of your recommended daily intake of fibre. Pumpkin seeds also contain some fibre, about 1.1 g in 28 gm of seeds. You can consume boiled, roasted or baked pumpkin. You can also use pumpkin as a major ingredient in soup, breads and pies. Pumpkin is a great home remedy for a constipated pet because of pumpkin's fibre content and great taste.
- **Excellent source of vitamins A, K and C:** Vitamin A is a key nutrient for keeping our eyes healthy and our vision good. Moreover, it also helps promote bone growth, keeps the immune system strong and maintains a vigorous reproductive system. Half a cup of canned pumpkin has 953 mg of vitamin A and only 42 calories and that vitamin A comes in the form of beta-carotene. Yet another reason why pumpkin is good for you is because it contains a lot of Vitamin K, about 40 per cent of the recommended daily dose of it! Vitamin K is the overlooked bone builder and heart protector. In fact, one serving of pumpkin contains almost 20 per cent of your

recommended daily dose of Vitamin C which is needed for the growth and repair of tissues in all parts of your body.

- **Loaded with iron, potassium and magnesium:** One small serving of pumpkin contains 250 milligrams of potassium and the same small serving of pumpkin also contains a good amount of iron. One cup has a little less than 10 per cent of your recommended daily allowance of iron. Moreover, unlike other sources of iron, pumpkin is fat-free.
- Our body needs magnesium for maintaining normal muscle function and for boosting our immune systems, among other things. Pumpkin seeds provide us with enough magnesium to perform these functions well.
- Pumpkin is very rich in carotenoids, which is known for keeping the immune system of an individual strong and healthy.
- Beta-carotene, found in pumpkin, is a powerful antioxidant as well as an anti-inflammatory agent. It helps prevent build up of cholesterol on the arterial walls, thus reducing chances of strokes.
- Being rich in alpha-carotene, pumpkin is believed to slow the process of aging and also prevent cataract formation.
- Pumpkins have been known to reduce the risk of macular degeneration, a serious eye problem than usually results in blindness.
- The high amount of fibre, present in a pumpkin, is good for the bowel health of an individual. Being loaded with potassium, pumpkin is associated with lowering the risk of hypertension.
- The presence of zinc in pumpkins boosts the immune system and also improves the bone density.
- It is one of the vegetables which is very low calories; provides just 26 cal per 100 g and contains no saturated fats or cholesterol; but is rich a source of dietary fibre, anti-oxidants, minerals, vitamins. Recommended by dieticians in cholesterol controlling and weight reduction programs.
- Pumpkin is a storehouse of many anti-oxidant vitamins such as vitamin-A, vitamin-C and vitamin-E.
- It is one of the vegetable in the cucurbitaceae family with highest levels of **vitamin-A**, providing about **246 per cent of RDA**. Vitamin A is a powerful natural anti-oxidant and is required by body for maintaining the integrity of skin and mucus membranes. It is also an essential vitamin for vision. Research studies suggest that natural foods rich in vitamin A helps body protect against lung and oral cavity cancers.
- It is also an excellent source of many natural poly-phenolic flavonoid compounds such as α and ß carotenes, cryptoxanthin, lutein and zeaxanthin. Carotenes convert into vitamin A inside the body.
- **Zea-xanthin** is a natural anti-oxidant which has UV (ultra-violet) rays filtering actions in the macula lutea in reitina of the eyes. Thus, it helps protect from "age related macular disease" (ARMD) in the elderly.

- Rich in B-complex group of vitamins like folates, niacin, vitamin B-6 (pyridoxine), thiamin and pantothenic acid.
- It is also rich source of minerals like copper, calcium, potassium and phosphorus.

Health and Nutrition Benefits of Pumpkin Seeds

- **Pumpkin seeds** are good source of dietary fibre and mono-unsaturated fatty acids which are good for heart health. In addition, they are very good in protein, minerals and many health benefiting vitamins. For example 100 g of pumpkin seeds provide 559 cal, 30 g of protein, 110 per cent RDA of iron, 4987 mg of niacin (31 per cent RDA), selenium (17 per cent of RDA), zinc (71 per cent) etc. but no cholesterol.
- The seeds are an excellent source of health promoting amino acid tryptophan. Tryptophan is converted to GABA in the brain.
- They promote overall prostate health, apart from alleviating the problem of difficult urination that is associated with an enlarged prostate.
- They comprise of L-tryptophan, a compound that has been found to be effective against depression.
- They are believed to serve as a natural protector against osteoporosis.
- They have been known to reduce inflammation, without causing the side effects of anti-inflammatory drugs.
- Studies have revealed that they help prevent calcium oxalate kidney stone formation.
- Being rich in phytosterols, they have been associated with reducing the levels of LDL cholesterol.
- Pumpkin seeds are high in zinc, which is good for the prostate and building the immune system.
- They also contain fatty acids that kill parasites.
- Raw pumpkin seeds contain essential fatty acids and beneficial proteins.
- For maximum nutritional benefits, seeds should be eaten raw.
- Roasted seeds contain damaged fat that can lead to plaque in the arteries.

SQUASH (*Cucurbita morschata*)

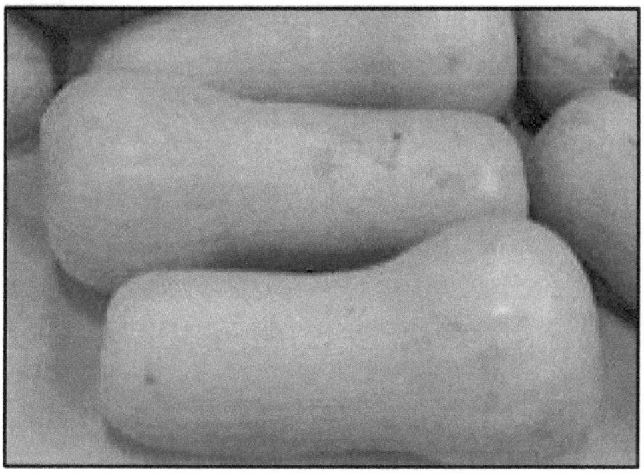

Squash Butternut is the most popular among winter squash varieties. Oftentimes, the squash is recognized as large pear shaped golden-yellow pumpkin fruit, which is put for sale in the markets. Butternuts are annual long trailing vines. They usually cultivated in warmer climates of South and Central American regions for their edible fruits, flowers, as well as seeds.

The butternut plant is monoecious as in pumpkins, and feature different male and female flowers that require honeybees for effective pollination. Butternut, in-fact, is the most common among winter squash fruits.

Externally, the squash is better described as large sized fruit featuring thick neck attached to a pear shaped bottom and has smooth, ribbed skin. However, the fruit varies widely in its shape and size; with individual fruit may weigh up to 15 kg. Interiorly, its flesh is yellow to orange in colour. Cross-section of lower bulb part feature central hollow cavity containing mesh-like mucilaginous fibres interspersed with large, flat, elliptical seeds similar to pepita (pumpkin seeds). The fruit's unique golden yellow colour comes from yellow-orange pigments in their skin and pulp.

Butternut squash seeds are used as nutritious snack food since they contain 35-40 per cent oil and 30 per cent protein. In Argentina, the fruit is also used to feed livestock.

Health Benefits Squash (Butternut)

- ✰ Butternut squash contains many vital poly-phenolic anti-oxidants and vitamins. Similar to other cucurbitaceae members, it is very low in calories; provides just 45 cal per 100 g. It contains no saturated fats or cholesterol; but is rich source of dietary fibre and phyto-nutrients. Squash is one of the common vegetable that is often recommended by dieticians in the cholesterol controlling and weight reduction programs.
- ✰ It has more **vitamin A** than that in pumpkin. At 10630 IU per 100 g, it is perhaps the single vegetable source in the cucurbitaceae family with highest levels of **vitamin-A**, providing about **354 per cent of RDA**. Vitamin A is a powerful natural anti-oxidant and is required by body for maintaining the integrity of skin and mucus membranes. It is also an essential vitamin for vision. Research studies suggest that natural foods rich in vitamin A helps body protect against lung and oral cavity cancers.
- ✰ Furthermore, butternut squash has plentiful of natural poly-phenolic flavonoid compounds like α and ß-carotenes, cryptoxanthin-ß, and lutein. These compounds convert to vitamin A inside the body and deliver same protective functions of vitamin A on the body.
- ✰ It is rich in B-complex group of vitamins like folates, riboflavin, niacin, vitamin B-6 (pyridoxine), thiamin, and pantothenic acid.
- ✰ It has similar mineral profile as pumpkin, containing adequate levels of minerals like iron, zinc, copper, calcium, potassium, and phosphorus.
- ✰ **Butternut squash seeds** are good source of dietary fibre and mono-unsaturated fatty acids that are good for heart health. In addition, they are very good in protein, minerals, and numerous health benefiting vitamins. The seeds are excellent source of health promoting amino acid **tryptophan**. Tryptophan converts to health benefiting GABA neuro-chemical in the brain.

SUMMER SQUASH

Summer squash are believed to be originated in Central America and Mexico. Different cultivars of summer squash are grown throughout the United States during the warm, frost free season. Almost all the members of the squash family feature tender skin and flesh, small edible seeds and high moisture content.

Like other members of the summer squash group, the zucchini plant has the bush habit rather than the vine spread of the winter squashes. Its fruits are ready for harvesting about 40-50 days after seed implantation.

Health Benefits of Summer Squash

One of the very low calories vegetable that is used during weight reduction and cholesterol control programs. Zucchinis provide only 17 calories per 100 g. Contains no saturated fats or cholesterol. Its peel is good source of dietary fibre that helps reduce constipation and offers some protection against colon cancers.

- Courgette is relatively moderate source of folates, consists of 24 mcg or 6 per cent of RDA per 100 g. Folates are important in cell division and DNA synthesis. When taken adequately before pregnancy, it can help prevent neural tube defects in the fetus.
- It is a very good source of potassium, an important intra-cellular electrolyte. Potassium is a heart friendly electrolyte; helps reduce blood pressure and heart rates by countering effects of sodium.
- Fresh fruits are rich in vitamin A; provide about 200 IU per 100 g.
- Furthermore, zucchinis, especially golden skin variety are rich in flavonoid poly-phenolic antioxidants such as carotenes, lutein and zeaxanthin. These compounds help scavenge harmful oxygen-derived free radicals and reactive oxygen species (ROS) from the body that play a role in aging and various disease process.
- Fresh fruit is good source of anti-oxidant vitamin-C. Provide about 17.9 mcg or 30 per cent of RDA per 100g.
- In addition, they are also good in B-complex group of vitamins like thiamin, pyridoxine, riboflavin and minerals like iron, manganese, phosphorus, zinc and potassium. Potassium in an important component of cell and body fluids, helps controlling heart rate and blood pressure.
- Zucchini and other summer squash varieties contain vitamins A and C.
- They also contain potassium and calcium.
- The flavour of zucchini is best when it is less than six inches long.
- Zucchinis can grow as large as baseball bats but have little flavour when they reach this size.

MUSKMELON (*Cucurmis melo*)

In general, melons feature round or oblong shape, measure 4.5- 6.5 inches in diameter and weigh 450 – 850 gm. Internally, flesh range from orange-yellow to salmon colour, has soft consistency and juicy texture with a sweet, musky aroma that emanates best in the ripe fruits. At the center, there is a hollow cavity filled with small off-white colour seeds encased in a web of mucilaginous netting.

It is mainly a dessert fruit which is very wholesome and nutritious. Muskmelon (cantaloupe) are a good source of potassium, vitamin A and folate. The potassium is helpful in preventing kidney stones and cantaloupes in general are a useful laxative. It is low in calories and very nutritious. The hardy green cover protects the rich red juicy flesh inside which has the properties of a powerful anti-oxidant. Consuming about a coup full of musk melon juice provides you healthy antioxidants that will make you immune to fight against the infections. The potassium present in the fruits helps maintain the blood pressure, regulates heart beat, prevents muscle cramping, maintain electrolyte balance in ectracellular fluid, helps reduce water retension, providing sufficient oxygen to the brain so that it can function normally and also helps in preventring stroke. Musk melons have good esdative properties and help those suffering from insomnia. It acts as a mild anti-depressant. Its juice is helpful in curing lack of appetite, acidity, ulcer, weight loss and urinary tract infection, reduces the heat in the body and is also helpful in reducing heat related disorders.

Beneficial to be consumed during the following conditions:

- ☆ Lack of appetite
- ☆ Weight loss
- ☆ Urinary tract infections
- ☆ Constipation, acidity and ulcers

Health Benefits of Muskmelon

- ☆ Wonderfully delicious with rich flavour, muskmelons are very low in calories (100 g fruit has just 34 calories) and fats; but rich in numerous health promoting poly-phenolic plant derived substances, vitamins and minerals that are required for optimum health.
- ☆ The fruit is an excellent source of Vitamin A, (provides about 112 per cent of recommended daily levels) *one of the highest among fruits.* Vitamin A is a

powerful antioxidant and is essential for vision. It is also required for maintaining healthy mucus membranes and skin. Consumption of natural fruits rich in vitamin A is known to help to protect from lung and oral cavity cancers.

- It is also rich in antioxidant flavonoids such as ***beta-carotene***, *lutein*, *zeaxanthin* and *cryptoxanthin*. These antioxidants have the ability to help protect cells and other structures in the body from oxygen free radicals and hence; offer protection against colon, prostate, breast, endometrial, lung, and pancreatic cancers.
- **Zeaxanthin,** an important dietary carotenoid, selectively absorbed into the retinal macula lutea in the eye where it is thought to provide antioxidant and protective light-filtering functions; thus it protects eyes from "Age related macular degeneration" (ARMD) disease in the elderly.
- It is a good source of potassium. Potassium in an important component of cell and body fluids and helps control heart rate and blood pressure; thus offers protection against stroke and coronary heart diseases.
- The fruit also contain good amounts of B-complex vitamins, such as niacin, pantothenic acid and **vitamin C** and minerals like manganese. Consumption of foods rich in vitamin C helps body develop resistance against infectious agents and scavenge harmful oxygen free radicals. Manganese is used by the body as a co-factor for the antioxidant enzyme superoxide dismutase. Commercially, muskmelons are being used to extract an enzyme known as *superoxide dismutase* (SOD) which is essential for maintaining strong antioxidant defenses in the human body.

Other Properties

- It reduces heat in the body to a great extent.
- Relieves tiredness
- Increases appetite
- Increases kapha more
- Reduces vata and pitta
- It is very good to digest
- Sweet in taste
- Unctuous in quality
- Cold in potency
- Diuretic and aphrodisiac
- Laxative and strengthening
- It does not contain cholesterol.
- It counteracts the effect of cigrette smoke on your lungs by restoring the vitamin A level in the body.
- Rich in vitamin A and betacarotene, which are important vision nutrients.

- It helps maintain healthy skin because of vitamin A presence.
- Rich in folic acid which is essential for pregnant women.
- It helps to create healthy fetuses and can even prevent cervical cancer and osteoporosis.

WATERMELON (*Citrullus lanatus* Thunb)

Watermelon has everything you need to beat the scorching summer heat. Wonderfully delicious, Coolest thirst-quenching melons are the great source of much needed water (92 per cent) and electrolytes to beat the summer temperatures. It is not only cooling, but is also a natural diuretic. It is very alkaline in nature and has a soothing effect on the stomach and prevents biliousness. It is the thirst-quencher and is beneficial for those suffering from kidney problems and urinary tract infections. Pumpkin has folic acid, vitamin A, and vitamin C. Bitter gourd has polypeptide P(an insulin – like compound that can replace insulin).

They are indeed the best thirst quenchers, but are rich in anti-oxidants as well. Juicy watermelons are indeed packed with some of the best antioxidants in nature. Research has shown that they are good for cardiovascular diseases, colon cancer and diabetes too. They are rich in Vitamin A, B and C. What's more, they are low in calories and very nutritious. It is an ideal health drink because it is 92 per cent water, and does not contain any fat or cholesterol. It also fills you up fast because of its water content. So have it along with your barbeque or grills as watermelons will prevent you from overeating.

A team of US researchers led by Dr. Bhimangouda Bhimu Patil, Director of Texas A and M's Fruit and Vegetable Improvement, Center in College Station says that watermelon has ingredients that deliver Vigra-like effects to the body's blood vessels and may even increase libido. The study further says that the list of its very important healthful benefits grow longer with each study. Many fruits and vegetables contain nourshing ingredients known as phyto-nutrients, naturally occurring compounds that are bio-active, or able to react with the human body to trigger healthy reactions. In watermelon, these include lycopene –which is typically found in tomatoes; beta-carotene; and the rising star among its phyto-nutrients- **citrulline-** an amino acid, whose beneficial functions include the ability to relax the blood vessels, much like **Viagra** and other impotence drugs do. Scientists say that when watermelon is consumed, citrulline is converted to Arginine through certain enzymes. Arginine is an amino acid that helps the heart and circulation system and maintains a good immune system (Chidanand, 2008). Juice from the rejected water melons is a potential source of biofuel as it can be fermented into ethanol says a study.

Watermelon is the richest edible natural source of **L-citrulline**, which is closely related to **L-arginine,** the amino acid required for the formation of nitric acid oxide essential to the regulation of vascular tone and healthy blood pressure. Once in the body, the L-citrulline is converted in to L-arginine. Simply consuming L-arginine as a dietary supplement is not an option for many hypertensive adults, because it can cause nausea, gastrointestinal tract discomfort and diarrhoea. In addition to the vascular benefits of citrulline, watermelon provides vitamin A, B6, C fibre, potassium and lycopene, a powerful antioxidant. Watermelon may help to reduce serum glucose levels. It is both watermelon and its extract is the best source for L-citrulline. Indivisiduals with increased blood pressure and arterial stiffness-especially those who are older and those with chronic diseases such as type 2 diabetes-would benefit from L-citrulline in either the synthetic or natural (watermelon) form The optimum appears to be 4-5 grames a day (Figueroa and Arjmandi, 2010). It contains about 6 per cent sugar by weight, the rest being mostly water. As with many other fruits, it is a source of vitamin C. It is not significant source of other vitamins or minerals unless one eats several kilograms per day. It contains a significant amount of **citrulline** and after consumption of several kilograms an elevated concentration is measured in the blood plasma, this could be mistaken for **citrullinaemia** or other urea cycle disorders. It is a traditional food plant in Africa, this fruit has potential to improve nutrition, boost food security, foster rural development and support sustainable land care. It is also packed with a giant dose of glutathione, which helps boost our immune system. They are also a key source of lycopene - the cancer fighting oxidant. Other nutrients found in watermelon are vitamin C and Potassium.

Watermelon juice is the best medicine for those suffering from blood pressure because the potassium and magnesium controls the blood pressure in the body. One of the greatest quality in this fruit is it contains 92 per cent of alkaline water, which is not there in any fruit. Adding watermelon to your weight loss diet can help you lose weight. Watermelon is a water-heavy fruit that is extra filling. Watermelon has 50 calories per cup. Many people are concerned with watermelon's high glycemic index ranking of 72, but watermelon has a glycemic load score of 4 out of 40. Watermelon will not dramatically effect your blood sugar because it has very few carbohydrates per serving.

This food is very low in Saturated Fat, Cholesterol and Sodium. It is also a good source of Potassium, and a very good source of Vitamin A and Vitamin C, but large portion of the calories in this food come from sugars.

They are indeed the best thirst quenchers, no doubt, but did you know that they are rich in anti-oxidants as well? Juicy watermelons are indeed packed with some of the best antioxidants in nature. Research has shown that they are good for cardiovascular diseases, colon cancer and diabetes too. They are rich in Vitamin A, B and C. What's more, they are low in calories and very nutritious. It is an ideal health drink because it is 92 per cent water, and does not contain any fat or cholesterol.

Health Benefits

- ✰ Rich in electrolytes and water content, melons are nature's gift to beat tropical summer thirst.

- Watermelons are very low in calories (just 30 cal per 100 g) and fats yet very rich source of numerous health promoting phyto-nutrients and anti-oxidants that are essential for optimum health.
- Watermelon is an excellent source of Vitamin-A, which is a powerful natural anti-oxidant. It is essential for vision and immunity. Vitamin-A is also required for maintaining healthy mucus membranes and skin. Consumption of natural fruits rich in vitamin-A is known to protect from lung and oral cavity cancers.
- It is also rich in anti-oxidant flavonoids like *lycopene*, beta-carotene, *lutein, zeaxanthin and cryptoxanthin.* These antioxidants are found to be protective against colon, prostate, breast, endometrial, lung, and pancreatic cancers. Phyto-chemicals present in watermelon like **Lycopene** (4532 mcg/100 g) and carotenoids have the ability to help protect cells and other structures in the body from oxygen free radicals. Studies have also shown that *lycopene* protects skin damage from UV rays and from prostate cancer.
- Watermelon fruit is a good source of potassium; Potassium is an important component of cell and body fluids that helps controlling heart rate and blood pressure; thus offers protection against stroke and coronary heart diseases.
- It also contains good amount of vitamin-B6 (pyridoxine), thiamin (vitamin B-1), vitamin-C and manganese. Consumption of foods rich in vitamin-C helps body develop resistance against infectious agents and scavenge harmful oxygen free radicals. Manganese is used by the body as a co-factor for the antioxidant enzyme *superoxide dismutase*.
- It is 92 per cent water by weight. And it is also mildly diuretic.
- Watermelon with red flesh is a significant source of **lycopene** and a very concentrated source of the carotenoid.
- Choosing to regularly eat lycopene-rich fruits, such as watermelon, *and* drink green tea may greatly reduce a man's risk of developing prostate cancer, suggests research published the *Asia Pacific Journal of Clinical Nutrition*
- Watermelon is an excellent source of vitamin C and a very good source of vitamin A, notably through its concentration of beta-carotene. Pink watermelon is also a source of the potent carotenoid antioxidant, lycopene. These powerful antioxidants travel through the body neutralizing free radicals.
- Watermelon is rich in the B vitamins necessary for energy production. Our food ranking system also qualified watermelon as a very good source of vitamin B6 and a good source of vitamin B1, magnesium, and potassium.
- This sweet, crunchy, cooling fruit is exceptionally high in **citrulline**, an amino acid our bodies use to make another amino acid, arginine, which is used in the urea cycle to remove ammonia from the body, and by the cells lining our blood vessels to make nitric oxide. Nitric oxide not only relaxes

blood vessels, lowering high blood pressure, it is the compound whose production is enhanced by Viagra to prevent erectile dysfunction. Arginine has been shown to improve insulin sensitivity in obese type 2 diabetic patients with insulin resistance. Am J Physiol Endocrinol Metab. 2006 Nov;291(5)

THYME (*Thymus vulgarius*)

Thyme is one of the sweet smelling herb, popular among culinary plants which has been used traditionally for common winter ailments. Thyme herb is packed with numerous health benefiting phyto-nutrients (plant derived compounds), minerals and vitamins that are essential for wellbeing. It has been in use extensively for the high **thymol**, **phenol** and **carvol** content in the thyme oil. The flowering thyme tops contain an essential oil consisting primarily of thymol and carvacrol, along with tannins, bitter compounds Saponins and organic acids. Thymes best use medicinally is as an antiseptic, but it also has expectorant, anti spasmodic, and deodorant properties. It aids in digestion and as such is excellent when combined with fatty meats that often causes gastrointestinal problems. It is also a popular culinary herb. The essential oil of common thyme (*Thymus vulgaris*) is made up of 20 -55 per cent **thymol.** Thymol, an antiseptic is the main active ingredient in Listerine mouth wash. Before the advent of modern antiseptics, it was used to medicate bandages. It has also been shown to be effective against the fungus that commonly infects toe nails. Medicinally thyme is used for respiratorym infections in the form of a Tincture, tisane, salve, syrup or by steam inhalation. Because it is antiseptic, thyme boiled in water and cooled is very effective against inflammation of the throat when gargled 3 times a day. The inflammation will normally disappear in 2 – 5 days. Other infections and wounds can be dripped with thyme that has been boiled in water and cooled.

Thyme is a good tonic for stomach and nerves. Its tea relieves flatulence, promotes appetite, strengthens digestion, loosens phlegm and increases prespiration. Infusions have a calming effect, relax muscle spasms and alleviate exhaustion. Extracts and infusions are also used to treat bronchitis, laryngitis and coughs. Thyme vinager was used for centuries to relieve headaches. Essential oil of thyme makes a good antiseptic mouthwash, and is also used externally for warts and for relaxing bath (Michael Shron, 2009).

Health Benefits

- ☆ Thyme contains many active principles that are found to have disease preventing and health promoting properties.

- Thyme herb contains **thymol,** one of the important essential oils, which scientifically have been found to have antiseptic, anti-fungal characteristics. The other volatile oils in thyme include *carvacolo, borneol* and *geraniol.*
- Thyme contains many flavonoid **Phenolic antioxidants** like *zeaxanthin, lutein, pigenin, naringenin, luteolin,* and *thymonin*. Fresh thyme herb has one of the highest antioxidant levels among herbs, a total ORAC (Oxygen Radical Absorbance Capacity) value of 27426-umol TE/100 g.
- Thyme is packed with minerals and vitamins that are essential for optimum health. Its leaves are one of the richest sources of **potassium, iron, calcium, manganese, magnesium,** and **selenium**. Potassium is an important component of cell and body fluids that helps controlling heart rate and blood pressure. Manganese is used by the body as a co-factor for the antioxidant enzyme, *superoxide dismutase.* Iron is required for red blood cell formation.
- The herb is also a rich source of many important vitamins such as B-complex vitamins, **beta-carotene,** vitamin A, vitamin K, vitamin E, vitamin C and folic acid.
- Thyme provides 0.35 mg of **vitamin B-6** or pyridoxine; furnishing about 27 per cent of daily recommended intake. Pyridoxine keeps up GABA (beneficial neurotransmitter in the brain) levels in the brain, which has stress buster function.
- Vitamin C helps body develop resistance against infectious agents and scavenge harmful, pro-inflammatory free radicals.
- Vitamin A is a fat soluble vitamin and antioxidant that is required maintaining healthy mucus membranes and skin and is also essential for vision. Consumption of natural foods rich in flavonoids like vitamin A and beta-carotene helps protect from lung and oral cavity cancers

Caution

Too much thyme can overstimulate the thyroid and cause poisoning symptoms.

CORIANDER (*Coriandrum sativum*)

It has been used as a folk medicine for the relief of anxiety and insomnia in Iranian folk medicine. Experiments on mice support its use as an anxiolytic. Coriander seeds are used in traditional Indian medicine as a diuretic by boiling equal amounts of coriander seeds and cumin seeds, then cooling and consuming the resulting liquid. In holistic and some traditional medicine, it is used as a carminative and for general digestive aid. Additionally, coriander juice (mixed with either turmeric powder or mint juice) is used by some as a treatment for **acne**, applied to the face like toner. It is stimulant, aromatic and carminative. The powdered fruit, fluid extracted and oil are chiefly used for medicinally as flavouring to disguise the taste of active purgatives and correct their griping tendencies. It is an ingredient of the following compound preparations of the pharmacopceia confection, syrup and tincture of senna and tincture of syrup of Rhubrab and enters also into compounds with angelica gentian, jalap, quassia and lavender. As a corrigent it is considered superior to other aromatics. If used too freely the seeds become nacrotic. Coriander water was formerly much esteemed as a carminative for windy colic. In general it is anthelmintic and appetite stimulant. It cures tridosas, swelling and oedema. It relieves pain and also thirst. It is a natural purgative and improves digestion. It excites liver and is rakta pitta shamak. It is beneficial for heart and cures cough, urine infections and related disorders. It is anti-pyretic and promotes physical strength. It reduces sperm count (Acharya Balkrishna, 2008). Coriander seeds, used as spice, are round to oval in shape, yellowish brown in colour with vertical ridges and have flavour that is aromatic, sweet and citrus, but also slightly peppery. This popular culinary herb is quite similar to dill in the sense that both its leaves and seeds can be used as a seasoning condiment.

Health Benefits

- ☆ Adding 1 or 2 teaspoons of corriander juice to fresh butter milk is ideal for treating nausea and dysentery.
- ☆ It is beneficial for women especially those suffering from heavy menstrual flow. Six grams of corriander seeds should be boiled in 500ml water and after adding sugar, it can be consumed while warm.
- ☆ It is known to lower blood sugar by stimulating the secretion of insulin. It has also been called as an anti-diabetic plant.
- ☆ It helps the heart by lowering bad cholesterol and raising good cholesterol.
- ☆ The herb is also great for the skin. Its juice mixed with turmeric powder can help treat pimples and black heads (Anonymous, 2012).
- ☆ Coriander herb contains no cholesterol; but is rich in anti-oxidants and dietary fibre which help reduce LDL or "bad cholesterol" while increasing HDL or "good cholesterol" levels.
- ☆ The leaves and seeds contain many essential volatile oils such as *borneol, linalool, cineole, cymene, terpineol, dipentene, phellandrene, pinene* and *terpinolene.*
- ☆ The leaves of the coriander plant have anti-septic and carminative properties.
- ☆ The leaves and stem tips are also rich in numerous anti-oxidant polyphenolic flavonoids such as *quercetin, kaempferol, rhamnetin* and *epigenin*.
- ☆ The herb is a good source of minerals like potassium, calcium, manganese, iron, and magnesium
- ☆ It is also rich in many vital vitamins including folic-acid, riboflavin, niacin, vitamin-A, beta-carotene, vitamin-C that are essential for optimum health. Cilantro leaves provides 30 per cent of daily recommended levels of vitamin-C.

- ☆ It provides vitamin-A, which is an important fat soluble vitamin and anti-oxidant, is also required for maintaining healthy mucus membranes and skin and is also essential for vision. Consumption of natural foods rich in vitamin-A and flavonoids helps body protect from lung and oral cavity cancers.
- ☆ Coriander is one of the richest *herbal sources for* ***vitamin K.*** Vitamin-K has potential role in bone mass building by promoting osteotrophic activity in the bones. It also has established role in the treatment of Alzheimer's disease patients by limiting neuronal damage in their brain.
- ☆ The coriander seeds oil have found application in many traditional medicines as analgesic, aphrodisiac, anti-spasmodic, deodorant, digestive, carminative, fungicidal, lipolytic (weight loss), stimulant and stomachic.
- ☆ The herb contains many phytochemical compounds; phenolic flavonoid antioxidants like quercitin and essential oils have found application in many traditional medicines as analgesic, aphrodisiac, antispasmodic, carminative, depurative, deodorant, digestive, carminative, fungicidal, lipolytic, stimulant and stomachic.
- ☆ Coriander seeds contain many compounds that are known to have anti-oxidant, disease preventing, and health promoting properties.
- ☆ The seeds are excellent source of minerals like iron, copper, calcium, potassium, manganese, zinc and magnesium
- ☆ Unlike other dry spice seeds that are lack in vitamin C, coriander seeds contain ample amount of this anti-oxidant vitamin.
- ☆ Furthermore, the seeds indeed are storehouse for many vital B-complex vitamins like thiamin, riboflavin, and niacin.
- ☆ The characteristic aromatic flavour of coriander seeds comes from the many fatty acids and essential volatile oils. Some important fatty acids in the dried seeds include *petroselinic acid, linoleic acid (omega-6), oleic acid,* and *palmitic acid.* In addition, the seeds contain essential oils such as *linalool (68 per cent), a-pinene (10 per cent), geraniol, camphene, terpine* etc. Together, these active principles are responsible for digestive, carminative, and anti-flatulent properties of the seeds.
- ☆ In addition, dietary fibres bind to bile salts (produced from cholesterol) and decrease their re-absorption in colon, thus help lower serum LDL cholesterol levels. Together with flavonoid anti-oxidants, fibre composition of coriander helps protect the colon mucus membrane from cancers.

CLOVE (*Syzygium aromaticum* L.)

This tiny little spice has many curative properties. Traditionally cloves are used as a table spice and mixed with chillies, cinnamon, turmeric and other spices in the preparation of curry powder. It is hot and pungent. It is beneficial for eyes and promotes digestive functions. It enhances taste and cures Kapha, Pitta and blood disorders. It also quenches thirst and cures vomiting and flatulence caused by indigestion or accumulation of gas. It is beneficial in curing breathing problems, caught and tuberculosis. Its oil promotes digestive power, cures cough and vomiting in case of pregnant women (Acharya Balkrishna, 2008). It a volatile oil contains euginol, emitil and caryophyllene. Both cinnamon and cloves are used extensively in Indian cooking and have been found to improve the function of insulin and to lower glucose, total cholesterol, LDL and triglycerides in people with type 2 diabetes. Cloves boost your energy level. They also have one of the best aromatherapy scents that improve your sexual bevahour.

Clove oil is used in the manufacture of perfumes, soaps, bath salts and as a flavouring agent in medicine and dentistry.

Cloves help stimulate sluggish circulation and thereby promote digestion and metabolism.

In Chinese medicine, it is used for vomiting, indigestion and other related problems. If you are feeling stressed then, boil some water with basil leaves, mint leaves and clove. Then you can use this water for black tea. Add a little bit of honey to it. This will help you ease tension.

- ☆ **Tooth ache**- The use of a clove in toothache decreases pain. It also helps to decrease infection due to its antiseptic properties. Clove oil, applied to a cavity in a decayed tooth, also relieves tooth ache.
- ☆ **Digestive disorders**- Cloves promote enzymatic flow and boost digestive funtioning. They are used in various forms of gastric irritability and dyspepsia. Licking the powder of fried cloves mixed with honey is effective in controlling vomiting. The anesthetic action of clove helps in dealing with stomah pain and stops vomiting.
- ☆ **Coughs**- Chewing a clove with a crystal of common salt eases expectoration, relieves the irritaton in the throat. Chewing a burn clove is also effective medicine for cough.

Clove has Lots of Healing Properties

They are also used to flavour paan. Clove oil is used in the manufacture of perfumes, soaps, bath salts and as a flavouring agent in medicine and dentistry. Cloves help stimulate sluggish circulation and thereby promote digestion and metabolism. In Chinese medicine, it is used for vomiting, indigestion and other related problems. If you are feeling stressed then, boil some water with basil leaves, mint leaves and clove. Then you can use this water for black tea. Add little bit of honey to it. This will help you ease tension.

- ☆ **Toothache** -The use of a clove in toothache decreases pain. It also helps to decrease infection due to its antiseptic properties. Clove oil, applied to a cavity in a decayed tooth, also relieves toothache.
- ☆ **Digestive disorders** -Cloves promote enzymatic flow and boost digestive functioning. They are used in various forms of gastric irritability and dyspepsia. Licking the powder of fried cloves mixed with honey is effective

in controlling vomiting. The anesthetic action of clove helps in dealing with stomach pain and stops vomiting.

- ✩ **Coughs** -Chewing a clove with a crystal of common salt eases expectoration, relieves the irritation in the throat. Chewing a burnt clove is also an effective medicine for coughs.

MINT (*Mintha spicata*)

It is one of the oldest and most popular herbs that is grown around the world. There are many different varieties of mint, each having its own subtle flavour and aroma. This herb is used in a range of dishes from stuffing to fruit salads. Mint is an essential ingredient in many Indian and Middle Eastern cuisine and is popularly mixed with natural plain yogurt to make a 'raita' or brewed with tea to make the famous Indian 'Pudina Chai'. In Thai cooking, it is added to soups and to some highly-spiced curries. Mint grown in Asia is much more strongly flavoured than most European mints, with a sweet, cool aftertaste. Peppermint has been one of the popular herbs known since antiquity for its distinctive aroma and medicinal value. The herb has a characteristic refreshing cool breeze sensation when eaten on taste buds, palate and throat, and on nasal olfaction glands when inhaled. This unique quality of mint is due **menthol,** an essential oil in it. Mint sooths the digestive tract, provides relief from stomach aches and is known to have cancer-prevention properties.

Mint as Minta Spacata is a plant that has been long used in diverse cultures, such as India, Middle East and Europe. Mint has a sweet flavour, with a cooling after-sensation. Both, fresh and dried mint are used in preparing a large number of recipes, including curries, soups, chutneys, salads, juices, and ice creams.

The herb contains **menthol** and is popular as both a flavouring agent and a tea. It is also well known for its carminative effects – that is, it helps to relieve flatulence and indigestion. Since menthol is also an appetite stimultant, digestive and sedative, peppermint can be used to prevent or alleviate cramps, insomnia and vomiting. It is also claimed to be an aphrodisiac. In addition, peppermint has antiviral properties; its tannins have been found to suppress the activity of flu virus and inhibit Herpes simplex. Pepermint oil is used to relieve the symptoms of irritable bowel syndrome (IBS). Mint contains a host of medicinal properties. To relieve headache, migraine, cold and stomache upsets, chew on some fresh mint leaves. Also regular consumption of mint helps in flushing out toxins from the body.

Recent study has shown that spearmint tea may be used as a treatment of fevers, headaches, digestive diseases and for mild hirsutism in women. Its anti -androgenic properties reduce the level of free testosterone in the blood, while leaving total testosterone and DHEA unaffected. A major compound of oil is **R-carvone**; pure R-carvone is sufficient to produce a smell which people identify as a spearmint smell.

The fresh flowering herb on distillation yields 0.25 to 0.50 per cent of volatile oil, known as spearmint oil. The characteristic constituent of the oil is **1-carvone**. The herb is considered stimulant, carminative, diuretic, restorative, stimulant, anti-emetic, anti-spasmodic and stomachic (Duke and Ayensu, 1985). A sweetened infusion of the herb is given as a remedy for infantile troubles, vomiting in pregnancy and hysteria. The leaves are used in fevers and bronchitis. The oil is a counter irritant. The leaves should be harvested when the plant is just coming it to flower and can be dried for later use. The stems are macerated and used as a poultice on bruises. The essential oil in the leaves is antiseptic, though it is tonic in large doses. Both the essential oil and the stems are used in the folk remedies for cancer. A poultice prepared from the leaves is said to remedy tumours (Duke and Ayensu, 1985).

Peppermint has been one of the popular herbs known since antiquity for its distinctive aroma and medicinal value. The herb has a characteristic refreshing cool breeze sensation when eaten on taste buds, palate and throat, and on nasal olfaction glands when inhaled. This unique quality of mint is due **menthol,** an essential oil in it. Peppermint is not just an an ingredient in flavoured chewing gums. The herb is a hybrid of two plants, water mint and spearmint and is beneficial both cosmetically as well as medically.

When extracted as an pil, it is one of the many essential oils and is used extensively in perfumes and colognes. Peppermint oil is both rejuvenating and refreshing. It not only strengthens the immune system but also eases mental exhaustion. The oil can be used to give you a healthy scalp and healthy hair. Add a few drops of peppermint oil to your daily shampoo and it can help reduce that dandruff. Another benefit is it allows taming frizzy hair and brings about a shine to it. Peppermint oil can also bring about hair growth. The oil usage is not just a boon to the hair but also the skin and lips. Not only does it help control oil secretion of the skin but also clears it and brightens tired skin. The oil prevents acne and helps open pores. Peppermint oil helps heal cracked lips, and this is the reason why this oil is an active ingredient in chapsticks and lip glosses. The oil has a soothing effect (because of menthol that's present in it), and so can be used on wind-chapped or sun-burnt lip as it will give a cooling and refreshing effect on the skin. Please note that pepperment oil is really stong and when used, it should be diluted with a carrier oil, else it can cause side effects.

Because of its active compounds contained, mint has sedative, disinfectant and cicatrizing properties. It can be successfully used in gastro-intestinal disorders; it helps the liver and calms indigestion. It contains menthol, menthone, menthofuran, a-pinene, limonene, cardinene, acetic aldehide, isovaleriana, vitamin C and antibiotic substances.

Health Benefits of Peppermint

- ☆ Mint contains numerous chemical compounds that are known to have anti-oxidant, disease preventing and health promoting properties.
- ☆ It contains no cholesterol; but is rich in anti-oxidants and dietary fibre which helps to control blood cholesterol and blood pressure levels.

- ☆ It contains many essential volatile oils like *menthol, menthone, menthol acetate* that act on cold-sensitive receptors in the skin, mouth and throat, the property which is responsible for the well known cooling sensation that it provokes when inhaled, eaten, or applied to the skin.
- ☆ The essential oil, **menthol,** also has analgesic (pain-killer), local anesthetic and counter-irritant properties.
- ☆ Essential oils in the peppermint relax smooth muscles in the intestinal wall and sphincters by blocking calcium channels at cellular level. This property of mint has been applied as an anti-spasmodic agent in the treatment of "irritable bowel syndrome" or IBS and other colic pain disorders.
- ☆ Peppermint-herb is an excellent source of minerals like potassium, calcium, iron, manganese and magnesium. Potassium in an important component of cell and body fluids that helps control heart rate and blood pressure. Manganese and copper are important co-factors for antioxidant enzyme *superoxide-dismutase.*
- ☆ It is also rich in many antioxidant vitamins including vitamin A, beta-carotene, vitamin-C and vitamin E. The leaves of mint also contain many important B-complex vitamins like folates, riboflavin and pyridoxine (vitamin B-6); and the herb is also an excellent source of vitamin-K.
- ☆ It is soothing the digestive tract and if you are having stomach ache then it can be of great help.
- ☆ Drinking herbal mint tea reduces irritated bowel syndromes, cleanses the stomach and also clear up skin disorders such as acne.
- ☆ Mint acts as a cooling sensation to the skin and helps in dealing with skin irritations.
- ☆ Mint helps in eliminating toxins from the body.
- ☆ Crushed mint leaves helps in whitening teeth and combat bad breath.
- ☆ Mint is a very good cleanser for the blood.

Medicinal Uses

- ☆ As mentioned above, the essential oils in the peppermint act on cold-sensitive receptors in the skin, mouth and throat, the property which is responsible for the well known cooling sensation that it provokes when inhaled, eaten, or applied to the skin. This property of mint can be applicable in the preparation of cough/cold reliving remedies like syrups, lozenges and nose inhaler.
- ☆ Peppermint oil has analgesic, local anesthetic and counter-irritant properties and has been used in the preparation of topical muscle relaxants and analgesics.
- ☆ It is also being used in oral hygiene products and bad-breath remedies like mouthwash, toothpaste, mouth and tongue-spray, and more generally as a food flavour agent; *e. g.* in chewing-gum, candy.

☆ The essential oils in the peppermint can relax smooth muscles in the intestinal wall and sphincters by blocking calcium channels in them. This property of mint has been applied in treating irritable bowel syndrome or IBS and as an anti-spasmodic agent.

Caution

Individuals with *gastro-esophageal reflex disease* (GRD) are advised to limit peppermint in their diet since compounds in mint leaves relaxes smooth muscles in the esophageal wall and sphincters by blocking calcium channels in them which can aggravate their reflux condition.

FENNEL (*Foeniculum vulgare*) Saunf

Its succulent enlarged bulb imparts special "anise like" sweet flavour to the recipes. Bulb fennel is cultivated as vegetable for its beautiful, squatted stems in many regions of the country. Infusions of seeds and roots relieve flatulence, strengthen digestion, help suppress appetite and as a result aid weight loss. Fennel is also effective in treating colics and ulcers. The seeds and leaves are used to flavour dishes and the strems are used as a vegetable. Fennel contains an estrogen like substance (estirol) that turns out libido. Fennel contains the antioxidant flavonoid quercetin. This herb is anticarcinogenic and can be useful for cancer patients undergoing chemotherapy or radiation. Fennel can be useful for indigestion and spasms of the digestive tract. It also helps expel phlegm from the lungs. Armed with phyto-nutrients and antioxidants, cancer cells have nothing but to accept defeat when the spice is fennel. Anethole- a major constituent of fennel resists and restricts the adhesive and invasive activities of cancer cells. It suppresses the enzymatic regulated activities behind cancer cell multiplication. A tomato -fennel soup with garlic or fresh salads with fennel bulbs make for an ideal entree prior to an elaborate course meal. Roasted fennel with parmesan can be another star pick

Health Benefits of Fennel Bulb

☆ Fennel bulb is a versatile vegetable, used since ancient times for its nutritional and medicinal properties. This winter season has some noteworthy essential oils, flavonoid anti-oxidants, minerals, and vitamins that have known health benefits.

☆ Bulb fennel is one of very low calorie vegetables. 100 g bulb provides just 31 calories. Further, it contains generous amounts of fibre (3.1 g/100 g or 8 per cent of RDI), very little fat and zero cholesterol.

- Fresh bulbs give sweet anise-like flavour. Much of it is due to high concentration of aromatic essential oils like *anethole, estragole, and fenchone (fenchyl acetate)*. Anethole has been found to have anti-fungal and anti-bacterial properties.
- The bulbs have moderate amounts of minerals and vitamins that are essential for optimum health. Their juicy fronds indeed contain several vital vitamins such as pantothenic acid, pyridoxine (vitamin B-6), folic acid, niacin, riboflavin, and thiamin in small but healthy proportions. 100 g fresh bulbs provide 27 µg of folates. Folic acid is essential for DNA synthesis and cell division. Their adequate levels in the diet during pregnancy can help prevent neural tube defects in the newborn babies.
- In addition, fennel bulb contain average amount of water-soluble vitamin, vitamin-C. 100 g of fresh bulbs provide 12 mg or 20 per cent of vitamin C. Vitamin C helps body develop resistance against infectious agents and scavenge harmful, pro-inflammatory free radicals. Further, it has small amounts of vitamin A.
- The bulbs have very good levels of heart-friendly electrolyte potassium. 100 g provides 414 mg or 9 per cent of daily-recommended levels. It is an important electrolyte inside the cell. Potassium helps reduce blood pressure and rate of heartbeats by countering effects of sodium. Fennel also contains small amounts of minerals such as copper, iron, calcium, magnesium, manganese, zinc, and selenium.

LOTUS (*Nelumbo nucifera*)

The entire plant is used in medicine. The sacred water lotus has been used in the Orient as a medicinal herb for well-over 1500 years. The leaf juice is used in the treatment of diarrhea and is decocted with liquorice (*Glycyrrhiza spp*) for the treatment of sun-stroke. A decoction of the flowers is used in the treatment of premature ejaculation. The flowers are recommended as a cardiac tonic. A decoction of the floral receptacle is used in the treatment of abdominal cramps, bloody discharges etc. The flower stalk is used in treating bleeding gastric ulcers, excessive menstruation, post partum hermorrhage. The stamens are astringent and used in treating urinary frequency, premature ejaculation, hemolysis, epistasis and uterine bleeding. A decoction of the fruit is used in the treatment of agitation fever, heart complaints etc. The seed is used in the treatment of poor digestion, enteritis, chronic diarrhea, insomnia, palpitation etc. The plumule and radicle are used to treat thirst in high febrile disease, hypertension, insomania and restlessness. The root starch is used in the treatment of diarrhea, dysentery etc, a paste is applied to ring worm and other skin aliments. It is also taken internally in the treatment of hermorrhages, excessive menstruation and nose bleeds. The roots are harvested in autum or winter and dried for later use. The root nodes are used in the treatment of nasal bleeding, hermoptysic, hermaturia and functional bleeding of uterus. The plant has a folk history in the treatment of cancer, modern research has isolated certain compounds from the plant that shows anti-cancer activities. The leaves, which have anti-pyretic and refrigerant properties, are used against symptoms of summer-heat, such as head ache, respiratory congestion, chronic thirst and dark scanty urine. The peduncle relieves stomachaches, clams restless fetus and controls leucorrhea.

Health Benefits of Lotus Root

- ☆ Lotus root is one of the moderate calorie root vegetables, composed of several health benefiting phyto-nutrients, minerals, and vitamins.
- ☆ Lotus rhizome is very good source of dietary fibres.
- ☆ Lotus root is one of the excellent sources of vitamin C
- ☆ It contains moderate levels of some of valuable B-complex group of vitamins such as pyridoxine, folates, niacin, riboflavin, pantothenic acid, and thiamin.
- ☆ The root provides some important minerals like copper, iron, zinc, magnesium, and manganese.
- ☆ The crunchy, sweet yet delicate flavour of root lotus is because of its optimum electrolyte balance.It has agreeable ratio of sodium to potassium at the value 1:4. While sodium gives the sweet taste to the root, potassium acts to counter negative effects of sodium by regulating heart rate and blood pressure.

Caution

Lotus root may harbour parasites like *Fasciolopsis buski*, which is a trematode that commonly infests in aquatic plants like lotus, water caltrop (*Trapa natans*), Chinese water-chestnut (*Eleocharis dulcis*) etc. The symptoms may include stomach pain, vomiting, diarrhea, fever, and intestinal obstruction. Thorough washing and cooking in the steam destroys the larvae.

ROSEMARY (*Rosmarinus officinalis*)

A beautiful, fragrant ever green herb, it has a variety of herbal uses. An infusion of the leaves and flowering tops is a sedative that can be used to relieve flatulence and headaches, promote perspiration, increase bile flow and stimulate menstruation. Rosemary, especially its young leaves, contain **carnosic acid**, a powerful antioxiodant. Recent research identified rosemary as one of the first plants that can improve memory by inhibiting **acetylcholinesterase** (a brain chemical which switches off the connection between nerve cells). Rosemary leaves are used as a culinary seasoning and to make essential oils for aromatherapy. They are also used externally mainly in champoos and other hair preparations. Apart from culinary and medicinal purpose rosemary shoots, flowers and leaves are used in ceremonies such as weddings and festivals for decorating banquet halls as incense to ward off bad influences.

Rich in nutrients including iron, calcium, vitamins and antioxidants, rosemary in various forms —rosemary extract, rosemary oil and rosemarinic acid —is effective for many skin types and is found in skin care cleansers, soaps, face masks, toners and creams. This herb is used widely in aromatherapy as its stimulant properties refresh, rejuvenate and help you feel energised. Here's how this fragrant herb can benefit your skin (Anonymous,2013).

Health Benefits

- **Prevents ageing of skin:** Since it contains naturally powerful antioxidants, rosemary oil strengthens the capillaries and helps to slow the effects of ageing on skin. It also tightens sagging and loose skin to help it look firmer and more elastic. Additionally, rosemary stimulates biological activity and cell growth to reduce fine lines and wrinkles.
- **Treats acne:** With its powerful disinfectant and antibiotic properties, the herb can help to treat acne and oily skin conditions. When massaged into the face, rosemary oil can help lightens dark spots and blemishes on the skin resulting in an improved skin complexion.
- **Prevents infection:** The anti-inflammatory properties of rosemary extract help to reduce swelling and puffiness of the skin. It also helps to heal burns and soothe the skin. This herb's medicinal qualities makes it a powerful cure for chronic skin conditions including dermatitis, eczema and psoriasis.
- Rosemary leaves contain certain phyto-chemical compounds that are known to have disease preventing and health promoting properties.
- The herb parts especially flower tops contain phenolic anti-oxidant ***rosmarinic acid*** as well as numerous health benefiting volatile essential oils such as *cineol, camphene, borneol, bornyl acetate, α-pinene* etc. These compounds are known to have rubefacient (counter irritant), anti-inflammatory, anti-allergic, anti-fungal and anti-septic properties.
- Rosemary leaves provide just 131 calories per 100 g and contains no cholesterol.
- The herb is exceptionally rich in many B-complex group of vitamin, such as **folic acid**, pantothenic acid, **pyridoxine**, and riboflavin.
- It is one of the herbs contain high levels of folates; providing about 109 mcg per 100 g (about 27 per cent of RDA). Folates are important in DNA synthesis and when given during peri-conception period can help prevent neural tube defects in the newborn babies.
- Rosemary herb contains very good amounts of **vitamin A**. Vitamin A is known to have antioxidant properties and is essential for vision. It is also required for maintaining healthy mucus membranes and skin. Consumption of natural foods rich in vitamin A is known to help body protect from lung and oral cavity cancers.

- Fresh rosemary leaves are good source of antioxidant vitamin; vitamin-C containing about 22 mg per 100 g, about 37 per cent of RDA. It is required for the collagen synthesis in the body. Collagen is the main structural protein in the body required for maintaining the integrity of blood vessels, skin, organs, and bones.
- Rosemary herb parts, whether fresh or dried, are rich source of minerals like potassium, calcium, iron, manganese, copper, and magnesium. Potassium is an important component of cell and body fluids, which helps control heart rate and blood pressure. Manganese is used by the body as a co-factor for the antioxidant enzyme, *superoxide dismutase.*
- This herb is an excellent source of **iron**, contains 6.65 mg/100 g of fresh leaves.
- It increases circulation, can help relieve excess muscle tension.
- Can help control dandruff and boost shine.
- Several compounds (essential oils) in rosemary help alleviate cold symptoms and loosen chest congestion.
- Rosemary has strong effects on the adrenal and other glands of the body.
- It has anti-parasitical, anti-fungal and anti- bacterial properties.
- It increases the ability to memorize key facts and concepts and increases test performances in students.
- It helps to boost recall and mental clarity by working directly on the central nervous system.
- Fatigue and depression brought on by weariness or feeling " run down" are relieved by the stimulating and energizing effects of rosemary essential oil.

Medicinal Uses

- Rosmarinic acid, a natural polyphenolic antioxidant found in rosemary, has been found to have anti-bacterial, anti-inflammatory, and anti-oxidant functions. Sage, peppermint, oregano, thyme also contain appreciable levels of rosmarinic acid.
- Rosemary oil, distilled from the flowering tops, contains volatile essential oil such as camphene, cineol, borneol, bornyl acetate and other esters. These compounds are known to have tonic, astringent, diaphoretic, and stimulant properties.
- Its herbal oil is also being used externally as a rubefacient to soothe painful ailments in gout, rheumatism and neuralgic conditions.
- Rosemary herb extractions when applied over scalp have stimulating function on the hair-bulbs and help preventing premature baldness. It forms an effectual remedy for the prevention of scurf and dandruff.
- Rosemary tea is a good remedy for removing nervous headache, colds, and depression.

Caution

Rosemary herb might cause abortion in pregnant woman when eaten in large amounts. In some cases, rosemary oil products may cause allergic skin reactions. In toxic doses, rosemary has been found to cause kidney dysfunction, and might exacerbate existing neurological conditions like epilepsy, neuroses etc.

CURRY LEAF [*Murraya koenigii* (L.) Sprengel]

Curry leaf is found almost throughout India up to an altitude of 1500 mtrs. It is much cultivated for its aromatic leaves. The leaves are highly valued for their distinctive flavour, making them very popular in the cuisines of South and West India and Sri Lanka. They are also used as a herbal tonic in ayurveda. It is therefore, a good idea to include curry leaves in your diet. They can be used both raw or cooked. They can be ground and also used as a paste in curries. Incorporating curry leaves into your daily diet can help you lose weight. The leaves flush out fats and toxins, reducing fat deposits that are stored in the body as well as reducing bad cholesterol levels. If you are over weight, incorporate eight to ten curry leaves into your diet daily. Chop them finely and mix them into a drink or sprinkle them over a meal.

Curry leaf is a good source of vitamin A, calcium, and folic acid. Its richness in vitamin A and anti-oxidant may help explain its use in preventing early development of cataract. Being a fairly good source of folic acid, the leaves can also help in absorption of iron, other proposed benefits include boost in circulation and anti – inflammation. It is also anti-diabetic, anti-oxidant, anti-microbial, hepatoprotective, hypo-cholestrolemic and delays premature greying. With its anti-inflammatory benefits, it is used in treating bruises and skin eruptions. It can also be used as a sedative and a hair tonic. Its midly properties aid digestion too. The studies have shown it to have anti-fungal activities, which explains why it is used to fight bad breath and gum diseases. Some practioners of herbal medicine advise consuming a few leaves in the morning, while others recommend therapeutic doses as juice. However, it should not exceed more than 15grams (Life style. In. msn. com). *Curry leaf is an essential ingredient in Indian cooking especially in South India. However, it's customary for most of us to simply remove and throw the leaf from our food and not consume it.* Curry leaf has many medicinal properties. It stimulates digestive enzymes and helps break down food more easily. Curry leaves contain 2.5 per cent oil, alpha-selinene, beta-bisabolene, beta-cadinene, beta-caryophyllene, beta-elemene, beta-gurjenene, beta-phellandrene, beta-thujene and beta-transocimene. These ingredients give curry leaves it typical aroma. The leaf is used in South India as a natural flavouring agent in various curries. Volatile oil is used as a fixative for soap perfume. The leaves, bark and root of the plant are used in the indigenous medicine as a tonic, stimulant, carminative and stomachic.

Health Benefits

- ☆ Cury leaf has many medicinal properties: It stimulates digestive enzymes and helps break down food more easily (have a glass of butter milk mixed with a little hing (asafoetida) with a few curry leaves thrown after meal for good digestion).
- ☆ A good remedy for nausea and indigestion: Extract juice of curry leaves, squeeze a lime and add a pinch of sugar.
- ☆ Chew a few leaves every day to lose weight.
- ☆ Curry leaves are also known to improve eye sight, so make sure you do not throw away the leaves while eating. It is also believed to prevent cataract.
- ☆ Cury leaves are also good for hair growth and colour: If you do not like its raw taste, you can buy the curry leaf powder widely available in the market and have it with dosa or hot rice.
- ☆ Alternatively, you can also add a few curry leaves to your hair oil and boil it for a few minutes. Apply this to hairi tonic which will keep your hair healthy.
- ☆ Using curry leaves with food strengthens the functions of stomach and promotes its action. It is good for indigestion problem (Reshmi, 2012).
- ☆ It also helps to control Diarrhea, and upset stomach problems.
- ☆ It also helps prevent premature hair greying.
- ☆ Mix lime juice, honey or sugar crystals to the juice of curry leaf and drink it. This prevents nausea and dry vomiting.
- ☆ **For Hair:** Heat oil and add some curry leaves. When the curry leaves turns dark black remove the oil from heat and let it cool down. Store in a glass container. Apply this oil on your head and leave it overnight. This helps prevent premature aging of grey hair and also nourishes the hair roots. This oil also acts as a very good stimulant and helps in hair growth.
- ☆ Mix curry leaf with finely chopped ginger. Mix it with rice and take it early in the morning instead of water which will reduce dizziness, stomach disorders and constipation.
- ☆ Grind Curry leaf, black pepper corns, dried ginger, little cumin seeds and salt to a fine powder. Mix it with rice, make two small balls and eat it.

SAGE (*Salvia officinalis*)

Traditionally used to improve mental acuity, new research now suggests that sage may restore mental functions and improve memory, Sage tea is astringent, sedative and expels gas; it clears the respiratory tract, makes a good gargle for soe throats and helps overcome colds. Sage is useful for night sweats as it reduces sweating. It also reduces milk flow in nursing mothers prior to weaning, prevents the formation of kidney stones by dissolving resides of uric acid and regularizes menstruation. An infusion of sage can be applied to the scalp to reduce dandruff.

The sage herb is greatly beneficial for its astringent, stimulant, nervine, diuretic, expectorant, memory boosting, anxiolytic, anti-inflammatory, anti microbial, antioxidant effects. It will benefit the swellings, the cuts, sprain, mouth ulcers, sore throat, dyspepsia, abdominal troubles, menopausal symptoms and joint pain.

Health Benefits

- ✰ Sage herb parts have many notable plant derived chemical compounds, essential oils, minerals, vitamins that are known to have disease preventing and health promoting properties.
- ✰ The primary biologically active component of common sage appears to be its essential oil which contains mainly ketone α- and β-thujone. In addition, sage leaf contains numerous other substances including cineol, borneol, tannic acid; bitter substances with cornsole and cornsolic acid; fumaric, chlorogenic, caffeic and nicotinic acids; nicotinamide; flavones; flavone glycosides; and estrogenic substances. These compounds are known to have counter-irritent, rubefacient, anti-inflammatory, anti-allergic, anti-fungal and anti-septic properties.
- ✰ **Thujone** is GABA and Serotonin (5-HT3) receptor antagonist. It enhances concentration, attention span and quickens the senses; hence sage infusion has long been recognised as "thinker's tea". Its effects help deal with grief and depression.
- ✰ Thee lobe sage (S. triloba) has flavone called **salvigenin**. Research studies found that vascular relaxant effect of salvigenin may provide benefits in the cardiovascular diseases.
- ✰ This herb is exceptionally very rich source of many B-complex groups of vitamins, such as **folic acid**, thiamin, **pyridoxine** and riboflavin with many vitamins several times more than recommended levels.
- ✰ The herb contains very good amounts of **vitamin A** and beta-carotene levels. 100 g dry ground herb provides 5900 IU; about 196 per cent of RDA. Vitamin A is a powerful natural antioxidant and is essential for vision. It is also required for maintaining healthy mucus membranes and skin. Consumption of natural foods rich in vitamin A known to helps body protect from lung and oral cavity cancers.
- ✰ Fresh sage leaves are good source of antioxidant vitamin; vitamin-C. Vitamin C helps in the synthesis of structural proteins like collagen. Adequate levels in the body help maintain integrity of blood vessels, skin, organs, and bones. Regular consumption of foods rich in vitamin-C helps body protect from **scurvy**; develop resistance against infectious agents (boosts immunity) and scavenge harmful, pro-inflammatory free radicals from the body.
- ✰ Sage herb parts, whether fresh or dried, are rich source of minerals like potassium, zinc, calcium, iron, manganese, copper, and magnesium.

Potassium is an important component of cell and body fluids which helps control heart rate and blood pressure. Manganese is used by the body as a co-factor for the antioxidant enzyme *superoxide dismutase.*

- ☆ The fresh leaves or the tender branches are used as tooth brush as margossatree is used very commonly in rural India, as herbal tooth brush cum paste. It strengthens gums and cleans the oral cavity and teeth. Herbal tooth powders contain this herb.
- ☆ Dipping a cloth in its hot tea works like a compress. It can be applied to forehead to relieve tension, headache. A compress applied to abdominal areas can help in cramps, flatulence and general intestinal and digestive discomfort. Sprains also respond to this compress. This hot compress can be used in itching and swelling after insect bites.
- ☆ For purpose of formentatrion or poultice, the bags of its powder are used.
- ☆ The oil is used in preparations for joints pain cream or liniments.

Medicinal Uses

- ☆ The essential oil obtained from sage has been found to have acetylcholinesterase (Ach) enzyme inhibition activities. This help rise Ach levels in the brain. Ach improves concentration and may play a role in the treatment methods for memory loss associated with the disease like Alzheimer's.
- ☆ Sage oil, distilled from the flowering tops, contains volatile essential oil such as camphene, cineol, borneol, bornyl acetate and other esters. These compounds are known to have tonic, astringent, diaphoretic and stimulant properties.
- ☆ Sage herb oil is also being used externally as a rubefacient to soothen painful ailments like muscle stiffness, rheumatism and neuralgic conditions.
- ☆ Used as blended massage oil or in the aromatic therapy sage oil helps with nervousness, anxiety, headaches, stress and fatigue.

Caution

Sage herb when used in large amounts causes nervous irritation, convulsions and death. Hence its use is prohibited in known epileptic conditions. The herb should not be used in pregnancy as chemical compounds like **thujone** in it may cause uterine stimulation resulting in abortion.

WATER CRESS (*Nasturtium officinalis*)

It is rich in vitamins and minerals, especially vitamin A and C, zinc and iron. The plant is also rich in potassium, calcium, phosphorus and iron, with good quantities of iodine, sodium and magnesium.

Although the healing powers of water cress have long been known, recent scientific studies have shown that it can inhibit the growth of some cancerous tumours. New research has revealed resh watercress contains high levels of PEITC **(phenethyl**

isothiocyanate), which neutralizes a dangerous carcinogen in tobacco called NNK and one study indicated that the consumption of 40g of fresh watercress three times a day for three days will help to protect smokers from lung cancer. For more lasting protection, this process can be repeated once a month.

A tea prepared from the leaves can strengthen digestion, increase urination and cleanse the respiratory system by releasing phlegm and mucus. Watercress is also recommended for the treatment of catarrh, anaemia, weak digestion and gout.

Health Benefits

- This rich flavoured green leafy vegetable is store house of many phytonutrients that have health promotional and disease prevention properties.
- One of the very low calorie green leafy vegetables (11 kcal per 100 g raw leaves) and very low in fats; recommended in cholesterol controlling and weight reduction programs.
- Cress leaves and stems contain gluconasturtiin, a glucosinolate compound that gives peppery flavour. Research studies suggest that the hydrolysis product of gluconasturtiin, 2-phenethyl isothiocyanate (PEITC), is believed to be cancer preventing by inhibition of phase I enzymes (mono-oxygenases and cytochrome P450s).
- Fresh cress has more concentration of ascorbic acid (vitamin C) than some of fruits and vegetables. 100 g of leaves provide 47 mg or 72 per cent of RDA of vitamin C. As an anti-oxidant, vitamin C helps to quench free radicals and reactive oxygen species (ROS) through its reduction potential properties. Lab studies suggests that regular consumption of foods rich in vitamin C helps maintain normal connective tissue, prevent iron deficiency, and also helps body develop resistance against infectious agents by boosting immunity.
- It is one of the excellent vegetable sources for **vitamin-K**; 100 g provides over 200 per cent of daily recommended intake. Vitamin K has potential role bone health by promoting osteotrophic (bone formation and strengthening) activity. Adequate vitamin-K levels in the diet helps limiting neuronal damage in the brain; thus, has established role in the treatment of patients suffering from **Alzheimer's disease.**
- Cress is also excellent source of **vitamin-A** and flavonoids anti-oxidants like **ß carotene**, lutein and zeaxanthin.
- It is also rich in B-complex group of vitamins such as riboflavin, niacin, vitamin B-6 (pyridoxine), thiamin and pantothenic acid that are essential for optimum cellular metabolic functions.

- ☆ It is also rich source of minerals like copper, calcium, potassium, magnesium, manganese and phosphorus. Potassium in an important component of cell and body fluids that helps controlling heart rate and blood pressure by countering effects of sodium. Manganese is used by the body as a co-factor for the antioxidant enzyme *superoxide dismutase.* Calcium is required as bone/teeth mineral and in the regulation of heart and skeletal muscle activity.

Regular inclusion of cress in the diet is found to prevent osteoporosis, anaemia, vitamin A deficiency and believed to protect from cardiovascular diseases and colon and prostate cancers.

Caution

Excessive or prolonged use of watercress may cause kidney problems and it should not be used in pregnancy. Since, nowadays, wild water cress grows mainly in polluted water, it may contain various pollutants as well as the deadly liver fluke. Therefore, wild watercress is unsafe for gathering and only water cress grown commercially is filtered, shallow, gravelled beds should be consumed.

DANDELION (*Taraxacum officinale*)

The whole plant is highly nutritious and can be used as a tonic, diuretic and mild laxative. When prepared as an infusion, it can be used to stimulate bile formation and is also said to relieve the symptoms of jaundice and gall stones. It is useful for relief of water retension and to cleanse the body of poisons. The fresh, young leaves which are best used before the flower forms, can be added to salads or used for juice.

Dandelion with Golden Yellow Flower

It is a natural diuretic that increases urine production by promoting the excretion of salts and water from the kidney. Dandelion may be used for a wide range of conditions requiring mild diuretic treatment, such as poor digestion, liver disorder and high blood pressure. It is a source of potassium, a nutrientl often lost through the use of other natural and synthetic diuretics. The roots are stout, fusiform and fleshy, dark brown externally and white pulp inside somewhat yam-like. It contains bitter milky latex; more concentrated than in stems and leaves. Roots are generally dug when the plant turns second year of life. Generally, roots are harvested in summer for medicinal purposes or autumn for drying and grinding for coffee.

Dandelion Greens

- ☆ Dandelion is beneficial to digestion and is an antiviral that may be useful in the treatment of AIDS and herpes

- ☆ It may also be useful in treating jaundice, cirrhosis, edema due to high blood pressure, gout, eczema and acne
- ☆ Dandelion is also used to treat and prevent breast and lung tumors and premenstrual bloating
- ☆ Dandelion greens are high in vitamin A in the form of antioxidant carotenoid and vitamin C
- ☆ They also contain calcium and potassium
- ☆ Dandelion root contains inulin, which lowers blood sugar in diabetics

Health Benefits

- ☆ Fresh dandelion greens, flower tops, and roots contain valuable constituents that are known to have anti-oxidant, disease preventing, and health promoting properties.
- ☆ Fresh leaves are very low in calories; providing just 45 calories per 100 g. It is also good source of dietary fibre (provide about 9 per cent of RDA per 100 g). In addition, its latex is a good laxative. These active principles in the herb help reduce weight and control cholesterol levels in the blood.
- ☆ Dandelion root as well as other plant parts contains bitter crystalline compounds **Taraxacin**, and an acrid resin, **Taraxacerin**. Further, the root also contains *inulin* (not insulin) and *levulin*. Together, these compounds are responsible for various therapeutic properties of the herb.
- ☆ Fresh dandelion herb provides one of the highest source of vitamin-A among culinary herbs. Vitamin A is an important fat-soluble vitamin and anti-oxidant, required for maintaining healthy mucus membranes and skin and vision.
- ☆ Its leaves are packed with numerous health benefiting flavonoids such as *carotene-β*, *carotene-α*, *lutein*, *crypto-xanthin* and *zea-xanthn*. Consumption of natural foods rich in vitamin-A and flavonoids (carotenes) helps body protect from lung and oral cavity cancers. Zeaxanthin has photo-filtering functions and protects retina from UV rays.
- ☆ The herb is good source of minerals like **potassium**, calcium, manganese, iron, and magnesium. Potassium is an important component of cell and body fluids, which helps regulate heart rate and blood pressure. Iron is essential for red blood cell production. Manganese is used by the body as a co-factor for the antioxidant enzyme, *superoxide dismutase*.
- ☆ It is also rich in many vital vitamins including folic acid, riboflavin, pyridoxine, niacin, vitamin -E and vitamin-C that are essential for optimum health. Vitamin-C is a powerful natural antioxidant. Dandelion greens provide 58 per cent of daily-recommended levels of vitamin-C.
- ☆ Dandelion is probably the *richest herbal sources of* **vitamin K; provides about 650 per cent of DRI.** Vitamin-K has potential role in bone mass building by promoting osteotrophic activity in the bones. It also has established role in

the treatment of Alzheimer's disease patients by limiting neuronal damage in the brain.

- ☆ Dandelion has traditionally been used internally for gall bladder and urinary disorders, gall stones, jaundice, cirrhosis, dyspepsia with constipation, edema associated with high blood pressure and heart diseases, chronic joint and skin complaint, gout, eczema and acne, soothing effect on bee stings, sores etc.
- ☆ Skin care products to rejuvenate
- ☆ Laxative **properties.**
- ☆ Source of vitamin A and C, potassium, calcium, lethicin, iron, magnesium, niacin, phosphorus.
- ☆ Cirrhosis- the herb increases bile production and cleanses the blood stream.
- ☆ It is a natural diuretic
- ☆ Arthritis - relieves the stiffness in the joints.
- ☆ It also preovides rest and sleep in those with fever.
- ☆ Bile production by liver and urinary out put from the kidney is increased with the use of the herb.
- ☆ The leaves are particularly strong, being equivalent to frusemide, a drug used to treat hypertension.
- ☆ It is powerful remedy, not only for hypertension but also for cardiac oedema, hepetogenic dropsy and water retension, due to stasis or congestion in the blood vessels serving the liver.
- ☆ The diuretic effect of dandelion is helpful in the treatment of a number of other conditions, particularly chronic disorders like rheumatism, gout and eczema.
- ☆ When the stomach is irritated and where active treatment would be injurious the decoction or extract of dandelion administration 3 or 4 times a day will often prove a valuable remedy.
- ☆ It has a good effect in increasing the appetite and promoting digestion.

Medicinal Uses

Almost all the parts of dandelion herb found place in various traditional as well in modern medicine.

- ☆ The principle compounds in the herb have laxative and diuretic functions.
- ☆ The plant parts have been used as herbal remedy for liver and gall bladder complaints.
- ☆ The herb is also a good tonic, appetite stimulant and is a good remedy for dyspeptic complaints.
- ☆ The inside surface of the flower stems used as a smoothening agent for burns and stings (for example in stinging nettle allergy).

Caution

Although dandelion herb contains some bitter principles, it can be safely used in healthy individuals without any reservations. However, in patients on potassium sparing diuretic therapy, it may aggravate potassium toxicity. Dandelion herb can also induce allergic contact dermatitis in some sensitive individuals.

SWEET CORN (*Zea mays* var. *Saccharata*)

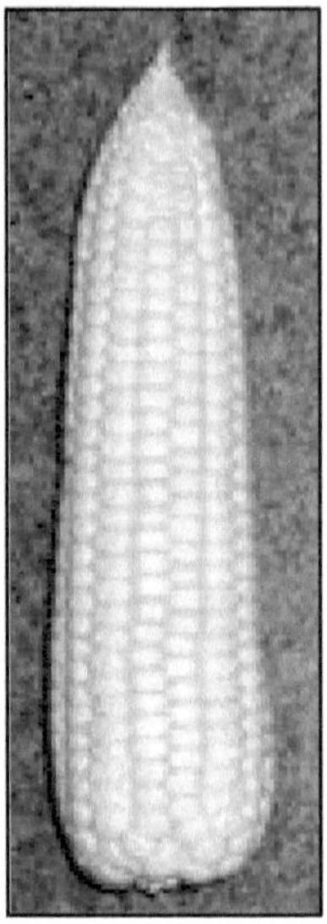

Sweet corn is a special maize variety in which its sweet kernels eaten as vegetable. In contrast to traditional field corn, sugar corn variety is harvested when the ears just reach milk stage and used fresh as the sugars in the kernels convert quickly to starch. Baby corns are very young, miniature ears harvested when they are very small. The kernels are in incipient stage and its core is sweet and tender enough to eat raw. Baby corns measure about 3-5 inches in length and weigh about 20-50 g.

It provides a good source of vitamin A, B1, B2, niacin and minerals such as iron, copper, phosphorus, magnesium, zinc, copper, iron and selenium. It has small amount of potassium. Corn silk, the fine tassel on the top of the corn cob, can be made into an infusion which soothes the urinary passages and act as a diuretic. This can be very benefical in cases of kidney stones and cystitis, but to be effective several cups of the infusion should be drunk each day. Rich in vitamin B1, which is used in the metabolism of carbohydrates, vitamin B5, which helps with physiological functions, folate that helps generating new cells and vitamin C, which helps fight diseases.

Corn has vitamin B (Thiamin, Vitamin B6, Niacin, Riboflavin, and Folate). It has traces of vitamin A and vitamin E. Being rich in folate, corn helps the generation of new cells, especially important before and during pregnancy. Those suffering from anemia have shown positive effects after consuming corns. Regular consumption of corn, in moderate quantities, has been associated with better cardiovascular health

Corn is high in fibre, which helps you fight digestive problems like constipation. Fibre also helps lower cholesterol levels, reduces the risk of colon cancer and is also useful in helping to lower blood sugar levels in diabetics. Folic acid present in corn is known to prevent neural- tube birth defects. Expert recommend corn for people who suffer from anaemia. Therefore, corn should be a must in your diet.

Health Benefits

- ☆ At 86 calories per 100 g, sugar corn kernels are moderately high in calories on comparison to other vegetables. However, fresh kernels have much lower in calories than field corn and other cereals like wheat, rice etc. Their calorie mainly comes from more simple carbohydrates like glucose, sucrose than complex sugars like amylose and amylopectin as in cereals.

- ☆ Sweet corn is gluten free cereal and may be used safely much like rice, quinoa etc, in celiac disease individuals.
- ☆ Corn features high quality phyto-nutrition profile comprising of dietary fibre, vitamins, and antioxidants in addition to moderate proportions of minerals. It is one of the finest source dietary fibres; 100 g kernels provide 2 g or 5 per cent of daily-requirement of dietary fibre. Together with slow digesting complex carbohydrates, moderate amounts of fibre in the food regulates gradual rise in blood sugar levels.
- ☆ Yellow variety corn has significant levels of phenolic flavonoid pigment antioxidants such as ß-carotenes, and lutein, xanthins and cryptoxanthin pigments along with vitamin A. 100 g fresh kernels provide 187 IU or 6 per cent of daily-requirement of vitamin A. Altogether, these compounds are required for maintaining healthy mucus membranes, skin and vision. Consumption of natural foods rich in flavonoids helps to protect from lung and oral cavity cancers.
- ☆ Corn is a good source of phenolic flavonoid antioxidant, **ferulic acid**. Several research studies suggest that ferulic acid plays vital role in preventing cancers, aging, and inflammation in humans.
- ☆ It also contains good levels of some of valuable B-complex group of vitamins such as thiamin, niacin, pantothenic acid, folates, riboflavin, and pyridoxine. Many of these vitamins functions as co-factors to enzymes in metabolism.
- ☆ It contains healthy amounts of some important minerals like zinc, magnesium, copper, iron, and manganese.
- ☆ Baby corn is a good source of folate and vitamin B. It also contains potassium, vitamin B6, riboflavin, vitamin C and fibre.
- ☆ Good for people on a diet as it contains a good amount of nutrients, fibre and is low in calories.
- ☆ Corn is an excellent source of dietary fibre and potassium and contains vitamin C, beta-carotene and niacin (one of the B group vitamins which assist in the functioning of the digestive system, skin, and nerves. It is also important for the conversion of food to energy.)
- ☆ Being rich in folate, corn helps the generation of new cells, especially important before and during pregnancy. Those suffering from anemia have shown positive effects after consuming corns. The Pantothenic acid present in corns helps with the physiological functions of the body.
- ☆ Owing to the presence of thiamin, corns have been said to help in the metabolism of carbohydrates. Corn has been found to be helpful in treating kidney problems, including renal dysfunction.
- ☆ Regular consumption of corn, in moderate quantities, has been associated with better cardiovascular health.
- ☆ The beta-cryptoxanthin in corn makes it good for the health of the lungs and may even help prevent lung cancer.

SWISS CHARD (*Beta vulgaris-cicla* Group)

Chard plant features distinctly large dark green leaves with well-developed edible stalks. Generally, chard leaves are harvested at various stages of maturity. While whole plant with tender young leaves harvested for salad preparation; individual matured large sized leaves with slightly tougher stems picked up for sautéing and cooking.

Health Benefits

- ☆ Swiss chard, like spinach, is the store-house of many phytonutrients that have health promotional and disease prevention properties.
- ☆ Chard is very low in calories (19 kcal per 100 g fresh, raw leaves) and fats, recommended in cholesterol controlling and weight reduction programs.
- ☆ Chard leaves are an excellent source of anti-oxidant vitamin, **vitamin-C.** Its fresh leaves provide about 33 per cent of recommended levels per 100 g. As an anti-oxidant, vitamin C helps to quench free radicals and reactive oxygen species (ROS) through its reduction potential properties. Research studies suggests that regular consumption of foods rich in vitamin C helps maintain normal connective tissue, prevent iron deficiency, and also helps body develop resistance against infectious agents by boosting immunity.
- ☆ Chard is one of the excellent vegetable sources for **vitamin-K**; 100 g provides about 700 per cent of recommended intake. Vitamin K has potential role bone health by promoting osteotrophic (bone formation and strengthening) activity. Adequate vitamin-K levels in the diet helps limiting neuronal damage in the brain; thus, has established role in the treatment of patients suffering from **Alzheimer's disease.**
- ☆ It is also rich source of omega-3 fatty acids; **vitamin-A** and flavonoids anti-oxidants like **ß carotene**, α-carotene, lutein and zeaxanthin.
- ☆ It is also rich in B-complex group of vitamins such as folates, niacin, vitamin B-6 (pyridoxine), thiamin and pantothenic acid that are essential for optimum cellular metabolic functions.
- ☆ It is also rich source of minerals like copper, calcium, sodium, potassium, iron, manganese and phosphorus. Potassium is an important component of cell and body fluids that helps controlling heart rate and blood pressure by countering effects of sodium. Manganese is used by the body as a co-factor for the antioxidant enzyme, *superoxide dismutase*. Iron is required for cellular oxidation and red blood cell formation.
- ☆ Regular inclusion of swiss chard in the diet is found to prevent osteoporosis, iron deficiency anemia, vitamin A deficiency and believed to protect from cardiovascular diseases and colon and prostate cancers.

Caution

Because of its high vitamin K content, patients taking anti-coagulants such as warfarin are encouraged to avoid this food since it increases the vitamin K concentration in the blood, which is what the drugs are often attempting to lower. This effectively raises the effective dose of the drug and causes toxicity. Chard contains oxalic acid, a naturally occurring substance found in some vegetables which may crystallise as oxalate stones in the urinary tract in some people. It is, therefore, advisable to avoid eating chard in people with known oxalate urinary tract stones. Adequate intake of water is therefore advised to maintain normal urine output.

ALOE VERA [*Aloe vera* (L.) Burn]

Aloe vera is a clear thin gelatinous material that comes from inside the aloe vera leaves. Aloe vera products are available in various forms like capsules, gel, juice and those that are applied directly to the skin. it has cooling effect and bitter in taste, It contains aloin that is responsible for its purgative action so it is known to relieve constipation. It regulates the peristaltic movements of intestines and promotes digestion, the liver and spleen function are stimulated by the use of this herb. Research work carried out over many years points conclusively to a toxic colon being the cause of a very wide range of illness, in establishing a regime of regular daily dose of aloe vera juice it may not only be found that the effects are gradual gentle and with no irritant or harmful side effects, but also blood circulation will be improved due to aloe Vera's ability to detoxify. Also being a natural healer, any internal ulcers or lesions will be soothed and healing will be enhanced. Aloe vera leave has some ingredients like Vitamin, Minerals, Amino acid, Polysaccharides, Enzymes, Plant steroids, Saponins, Lignin, Anthraquinones, Salicylic acid which are necessary for human bodies.

Aloe Vera (barbadensis miller) contains over 20 minerals, all of which are essential to the human body. The human body requires 22 amino acids for good health — eight of which are called "essential" because the body cannot fabricate them. **Aloe Vera** contains all of these eight essential amino acids, and 11 of the 14 "secondary" amino acids. Aloe Vera has Vitamins A, B1, B2, B6, B12, C and E. Vitamins cannot be manufactured within the body, and some cannot be stored by the body, so it is necessary for the diet to sustain a continuous supply. Therefore to get a steady supply of these minerals, amino acids and vitamins - one should drink between two to four ounces of Aloe Vera Gel twice daily.

The most popular among all beauty foods, this magic herb contains adequate amounts of amino acids (20 of them), calcium, sodium, magnesium, enzymes, vitamins, nitrogen and more. Capsules of aloe vora help among other things in blood circulation which in turn boosts the appearance of skin. The aloe vera juice contains many important vitamins and minerals, as well as β-carotene, enzymes, amino acids and complex carbohydrate mucopolysaccharide. These are responsible for its many healing actions, which include the soothing of inflammation of the digestive tract and the relief of constipation, flatulence and symptoms of irritable bowel syndrome (IBS). The gelatinous juice of the fresh leaves can be used externally. It is reputed to heal wounds and can also be rubbed on the skin to alleviate sunburns, wrinkles, skin irritations and minor cuts. Infusions are good for bathing wounds and eyes. The juice has a somewhat repellent taste so, when taken orally, it is usually blended with fruit juice to make it more palatable.

Aloe vera juice is excellent if you want younger-looking skin says Deepshikha and adds, "Have 30 ml aloe vera juice diluted with 100 ml water early morning on empty stomach. And after 20-30 minutes you can have your regular breakfast."Aloe vera works as Anti-septic, Antibacterial, and Anti infla-mmatory. It cures Eczema, Diabetes, Arthritis and Prevent infections. It also improves human immune system and digestive system.

But do you know the health benefits of drinking aloe vera juice? Completely safe and versatile, aloe vera juice stems from the fact that it naturally contains many different nutrients: vitamins, minerals, amino acids and other trace elements.

- ☆ **Helps digestion:** Drinking aloe vera juice naturally allows the body to cleanse the digestive system. It encourages the bowels to move and helps with elimination if a person is constipated. And if you have diarrhea, it will help slow it down.
- ☆ **Increases energy levels**: Our diets include many substances which can cause fatigue and exhaustion. Taken regularly, aloe vera juice ensures a greater feeling of well-being, allowing energy levels to increase and also helps maintain a healthy body weight.
- ☆ **Builds immunity**: It is especially great for those who have chronic immune disorders like polysaccharides or fibromyalgia since the polysaccharides in aloe vera juice stimulate macrophages, the white blood cells that fight viruses.
- ☆ **Detoxifies:** Aloe vera juice is a great natural aid to detox. With our stressful lives, the pollution around us and the junk foods we eat, we all need to cleanse our systems from time to time. Drinking aloe vera juice provides a fantastically rich cocktail of vitamins, minerals and trace elements to help our bodies deal with these stresses and strains every day.
- ☆ **Reduces inflammation**: It improves joint flexibility and helps in the regeneration of body cells. It strengthens joint muscles, which therefore reduces pain and inflammation in weakened or aged joints.

ENDIVE (*Cichorium endiva*)

Endive, commonly popular as **escarole,** is a green leafy vegetables with a hint of bitter flavour. However, this well known salad plant is much more than just a leafy green. Escarole is packed with numerous health benefiting plant nutrients such as vitamin C, vitamin A etc.

Health Benefits

- ✰ Endive is a very low calorie leafy vegetable. 100 g fresh leaves provide just 17 calories but contribute about 8 per cent of RDA of fibre.
- ✰ Current research studies suggest that high inulin and fibre content in escarole helps reduce high glucose and LDL cholesterol levels in diabetes and obese patients.
- ✰ Endive is enriched with good amount Vitamin A and beta-carotenes. Both of these compounds are known to have antioxidant properties. Carotenes convert to vitamin A in the body. Also, vitamin A is required for maintaining healthy mucus membranes and skin. In addition, it is also essential vitamin for vision. Consumption of natural vegetables/greens rich in vitamin A helps to protect from lung and oral cavity cancers.
- ✰ It contains good amounts of many essential B-complex groups of vitamins such as folic acid, pantothenic acid (vitamin B5), pyridoxine (vitamin B6) and thiamin (vitamin B1), niacin (B3). These vitamins are essential in the sense that body requires them from external sources to replenish and required for fat, protein and carbohydrates metabolism.
- ✰ Escarole also good source of minerals like manganese, copper, iron, and potassium. Manganese is used as a co-factor for the antioxidant enzyme *superoxide dismutase.* Potassium is an important intracellular electrolyte helps counter the hypertension effects of sodium.

Caution

Although this greeny leaf vegetable contains high concentrations of bitter glycosides and inulin, no known side effects so far notified when eaten in moderate.

RHUBARB (*Rheum rhabarbarium*)

Rhubarb is easy to grow and lives for many years (10-15 years) once established. The plant is usually propagated by dividing the old rhizomes (roots). Well grown plant feature broad heart shaped, dark green leaves with 12 to 18 inches long leaf

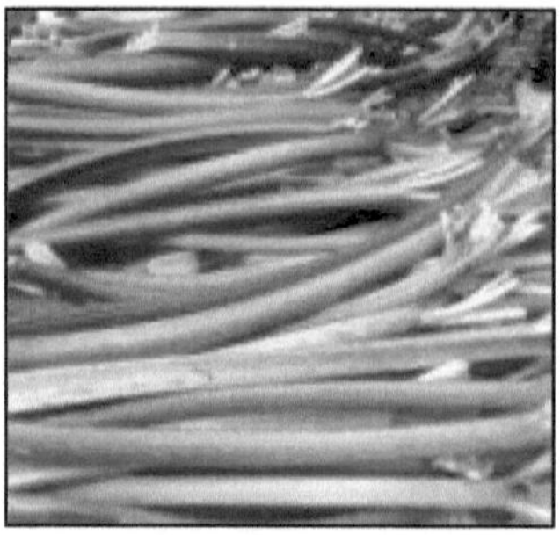

Fresh Rhubarbe Stalks

Rhubarbe Plant

petioles. It is these stalks, which are used, and their top greens discarded, as they are unfit for human consumption. Usually its stalks can be harvested from second year onwards after planting when the foliage spread and stalks reached sufficient girth of about one to two inches thick.

It is used mainly for its roots and rhizomes. It is an appetite stimultant and astringent, It is effective for both constipation and diuarrhoea, depending on the amounts used. Larger amounts can promote diarrhoea while tiny amounts will have a constipative effect.

Health Benefits of Rhubarb

- ☆ Rhubarb is one of the least calories vegetable. 100 g fresh petioles provide just 21 calories. Nonetheless, it contains some vital phyto-nutrients such as dietary fibre, poly-phenolic anti-oxidants, minerals, and vitamins. Further, its petioles contain no saturated fats or cholesterol.
- ☆ The stalks are rich in several B-complex vitamins such as folates, riboflavin, niacin, vitamin B-6 (pyridoxine), thiamin, and pantothenic acid.
- ☆ Red colour stalks contain more vitamin-A than green varities. Furtehr, the stalks also contain small amounts of poly-phenolic flavonoid compounds like ß-carotene, zeaxanthin, and lutein. These compounds convert to vitamin A inside the body and deliver same protective effects of vitamin A on the body. Vitamin A is a powerful natural anti-oxidant and is required by body for maintaining the integrity of skin and mucus membranes. It is also an essential vitamin for vision. Research studies suggest that natural foods rich in vitamin A helps body protect against lung and oral cavity cancers.
- ☆ Like in other greens like kale, spinach; rhubarb stalks also provide good amounts of **vitamin-K**; 100 g of fresh stalks provide 29.3 µg or about 24 per cent of daily recommended intake. Vitamin K has potential role in bone health by promoting osteotrophic (bone formation and strengthening) activity. Adequate vitamin-K levels in the diet helps limiting neuronal damage in the brain; thus, has established role in the treatment of **Alzheimer's disease.**
- ☆ Its stalks also contain healthy levels of minerals like iron, copper, calcium, potassium, and phosphorus. However, most of them may not absorb into

the body as they are chelated by oxalic acid into insoluble complexes and excreted out.

Caution

Top green part of rhubarb leaf (blade) contains unusually high amounts of oxalic acid, a naturally occurring substance found in some vegetables. 100 g of leaves contain about 0.59 - 0.72 mg of oxalates. Lowest published lethal dose (LDLo) of oxalate in humans is 600 mg/kg. Oxalate can cause severe symptoms even at much lower concentrations than this on the human body. Symptoms may include burning in the eyes, mouth, and throat; skin edema, difficulty breathing. In severe cases it can result in kidney failure, convulsions, coma, and death. The leaves are high in oxalic acid, which can bind calcium and promote calcium crystals in people susceptle to kidney stones.

ARUGULA (*Eruca sativa*)

Arugula, also known as salad rocket, is a nutritious leafy green vegetable of Mediterranean origin. It belongs to the *brassicaceae* family like mustard greens, cauliflower, kale, etc. and has scientific name **Eruca sativa.** Rocket-salad is a low growing annual herb features dandelion like succulent, elongated, lobular leaves with green veins. Young plant features plain light green colour leaves which appear somewhat identical to that of spinach. Young plant has mildly sweet, less peppery leaves.

Health Benefits

- Like other greens, arugula is one of very low calorie vegetable. 100 g of fresh leaves provides just 25 calories. Nonetheless it has many vital phytochemicals, anti-oxidants, vitamins, and minerals that can immensely benefit health.
- Rocket salad is rich source of certain phytochemicals such as *indoles, thiocyanates, sulforaphane* and *iso-thiocyanates*. Together they have been found to counter carcinogenic effects of estrogen and thus help benefit against prostate, breast, cervical, colon, ovarian cancers by virtue of their cancer cell growth inhibition, cytotoxic effects on cancer cells.
- In addition, di-indolyl-methane (DIM), a lipid soluble metabolite of indole has immune modulator, anti-bacterial and anti-viral properties (by potentiating Interferon-Gamma receptors and production). DIM has currently been found application in the treatment of recurring respiratory papillomatosis caused by the Human Papilloma Virus (HPV) and is in Phase III clinical trials for cervical dysplasia.

- Rocket is very good source of folates. 100 g of fresh greens contain 97 mcg or 24 per cent of folic acid. When given around conception period it helps prevent neural tube defects in the newborns.
- Like kale, salad rocket is an excellent source of vitamin A. 100 g fresh leaves contain 1424 mcg of beta-carotene and 2373 IU of vitamin A. Beta-carotenes converts into vitamin A in the body. Studies found that vitamin A and flavonoid compounds in in green leafy vegetables help protect from skin, lung and oral cavity cancers.
- This vegetable also rich in B-complex group of vitamins such as thiamin, riboflavin, niacin, vitamin B-6 (pyridoxine), and pantothenic acid those are essential for optimum cellular enzymatic and metabolic functions.
- Fresh rocket leaves contain good levels of vitamin C. Vitamin C is a powerful, natural anti-oxidant. Foods rich in vitamin C helps body protect from scurvy disease; develop resistance against infectious agents (boosts immunity) and scavenge harmful, pro-inflammatory free radicals from the body.
- Salad rocket is one of the excellent vegetable sources for vitamin-K; 100 g provides about 90 per cent of recommended intake. Vitamin K has potential role bone health by promoting osteotrophic (bone formation and strengthening) activity. Adequate vitamin-K levels in the diet helps limiting neuronal damage in the brain; thus, has established role in the treatment of patients suffering from *Alzheimer's disease.*
- Arugula is good in minerals especially copper and iron. In addition it has small amounts of some other essential minerals and electrolytes such as calcium, iron, potassium, manganese and phosphorus.

YAMS (*Dioscorea* spp.)

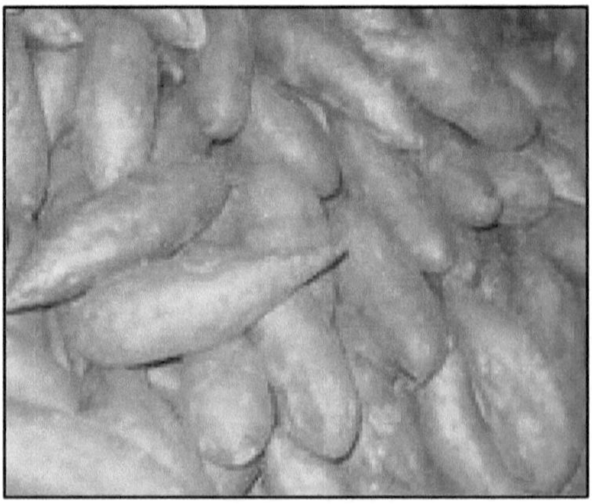

Yams Tubers

The plant is a perennial vine cultivated for its large, edible, underground tuber, which can grow up to 120 pounds in weight and up to 2 meters in length. Yam is similar in appearance to sweet potato; but not at all related to it. Important differences: yams are monocotyledons, larger in size, features thick, rough, dark brown to pink skin depending up on cultivar type whereas sweet potatoes (*Ipomoea batatas*) are dicotyledonous, relatively smaller in size and possess very thin peel. Unlike sweet potatoes which can be eaten raw, yams should not be eaten raw since they contain many naturally occurring plant toxins including dioscorin, diosgenin and tri-terpenes. They must be peeled and cooked in order to remove these bitter proteins.

Health Benefits

- Yam is a good source of energy; 100 g provides 118 calories. It mainly composed of complex carbohydrates and soluble dietary fibre. Together, they raise blood sugar levels rather very slowly than simple sugars and therefore recommended as low glycemic index healthy food. In addition, dietary fibre helps reduce constipation, decrease bad or "LDL" cholesterol levels by binding to it in the intestines and prevent colon cancer risks by preventing toxic compounds in the food from adhering to colon mucosa.
- The tuber is excellent source of B-complex group of vitamins. Provides adequate daily requirements of pyridoxine (viamin B6), thiamin (vitamin B1), riboflavin, folic acid, pantothenic acid and niacin. These vitamins mediate various metabolic functions in the body.
- Fresh root also contains good amounts of anti-oxidant vitamin, vitamin-C. Provides about 29 per cent of recommended levels per 100 g. Vitamin C has important roles in anti-aging, immune function, wound healing, bone growth.
- Also contains good amount of vitamin-A and beta-carotene levels. Carotenes convert to vitamin A in the body. These compounds are strong antioxidants. Vitamin A has many functions like maintaining healthy mucus membranes and skin, night vision, growth and protection from lung and oral cavity cancers.
- This tuber is indeed one of the vegetable rich sources of minerals like copper, calcium, potassium, iron, manganese and phosphorus. 100 g provides about 816 mg of Potassium. Potassium is an important component of cell and body fluids which helps controlling heart rate and blood pressure by countering hypertensive effects of sodium. Copper is required in the production of red blood cells. Manganese is used by the body as a co-factor for the antioxidant enzyme *superoxide dismutase*. Iron is required for red blood cell formation.

Medicinal Uses

The mucilaginous tuber milk contains allantoin, a cell-proliferant that speeds the healing process when applied externally to ulcers, boils and abscesses. Its decoction is also used to stimulate appetite and to relieve bronchial irritation, cough etc.

Caution

Yams of African species must be cooked to be safely eaten, because various natural toxin substances such as dioscorine can cause illness if consumed.

BASIL (*Ocimum basilicum*)

The king of herbs, **basil herb** is one of the oldest and popular herbal plant rich in many notable health benefiting phyto-nutrients. This highly prized plant is revered as "holy herb" in India. It is also known as Tulsi, is known to possess antioxidant properties and boosts metabolism. It is great for skin because it contains vitamin C, calcium, phosphorus and carotein. The leaves have appetite-stimulating properties and are generally used as a culinary flavouring. Infusions of leaves can be used to relieve flatulence, fermentation, stomach cramps and constipation and the plant is also said to relieve nausea. To relieve stomach ache or an upset stomach, chew fresh basil leaves.

Basil with Pink Flowers and Leaves

Health Benefits of Basil Herb

- Basil leaves contain many chemical compounds that are known to have disease preventing and health promoting properties.
- Basil herb contains many *polyphenolic flavonoids* like **orientin** and **vicenin.**
- Basil leaves contains many health benefiting essential oils such as *eugenol, citronellol, linalool, citral, limonene* and *terpineol.* These compounds are known to have anti-inflammatory and anti-bacterial properties.
- The herbs parts are very low in calories and contain no cholesterol, but are very rich source of many essential nutrients, minerals, and vitamins that are essential for optimum health.
- Basil herb contains exceptionally high levels of *beta-carotene,* ***vitamin A,*** *cryptoxanthin, lutein* and *zea-xanthin.* These compounds help act as protective scavengers against oxygen-derived free radicals and reactive oxygen species (ROS) that play a role in aging and various disease process.
- **Zeaxanthin**, a yellow flavonoid carotenoid compound, is selectively absorbed into the retinal macula lutea where it found to filter harmful UV rays from reaching retina. Herbs, fruits, and vegetables that are rich in this compound help to protect from *age related macular disease* (AMRD), especially in the elderly.
- Vitamin A is known to have antioxidant properties and is essential for vision. It is also required for maintaining healthy mucus membranes and skin. Consumption of natural foods rich in vitamin-A has been found to help body protect from lung and oral cavity cancers.
- Vitamin K in basil is essential for many coagulant factors in the blood and plays vital role in the bone strengthening function by helping mineralization process in the bones.
- Basil herb contains good amount of minerals like potassium, manganese, copper, and magnesium. Potassium is an important component of cell and

body fluids, which helps control heart rate and blood pressure. Manganese is used by the body as a co-factor for the antioxidant enzyme, *superoxide dismutase.*

- ☆ Basil leaves are an excellent source of **iron**, contains 3.17 mg/100 g of fresh leaves (about 26 per cent of RDA). Iron, being a component of hemoglobin inside the red blood cells, determines the oxygen carrying capacity of the blood.

Medicinal Uses

- ☆ Basil leaves contains many health benefiting essential oils such as eugenol, citronellol, linalool, citral, limonene and terpineol. These compounds are known to have anti-inflammatory and anti-bacterial properties.
- ☆ An important essential oil, **eugenol** has been found to have anti-inflammatory function by acting against the enzyme *cycloxygenase* (COX), which mediates inflammatory cascade in the body. This enzyme-inhibiting effect of the eugenol in basil makes it an important remedy for symptomatic relief in individuals with inflammatory health problems like rheumatoid arthritis, osteoarthritis, and inflammatory bowel conditions.
- ☆ Oil of basil herb has also been found to have anti-infective functions by inhibiting many pathogenic bacteria like *Staphylococcus, Enterococci, shigella* and *Pseudomonas.*
- ☆ Basil tea (basil water-brewed) helps relieve nausea and is thought to have mild anti-septic functions.

SAFFRON (*Crocus sativus*)

It is both an expensive aromatic spice and a medicinal herb with several benefits. Saffron is also believed to strengthen the appetite, soothe the alimentary canal, increase bile flow, clear liver stagnancy, help menopausal difficulties and relieve phlegm. In small doses, saffron has been used to treat coughs, bloated stomach, colic and insimnia and it is sometimes used in herb liqueurs as an appetite stimultant. A natural carotenoid dicarboxylic acid called 'Crocetin' is the primary cancer fighting element that saffron contains. It not only inhibits the progression of the disease but also decreases the size of the tumour by half, guaranting a complete goodbye to cancer. Though it is the most expansive spice in the world for it is derived from around 250,000 flower stigmas (saffron crocus) that make just about half a kilo, a few saffron threads come loaded with benefits you won`t regret paying. Saffron threads can be used in various ways.

Health Benefits

- ☆ Saffron contains many chemical compounds known to have anti-oxidant, disease preventing and health promoting properties.

- ☆ The flower stigma are composed of many essential volatile oils but the most important being **safranal**, which gives saffron its distinct hay-like flavour. Other volatile oils in saffron are cineole, phenethenol, pinene, borneol, geraniol, limonene, p-cymene, linalool, terpinen-4-oil, etc.
- ☆ It has many non-volatile active components; the most important of them is **α-crocin**, a carotenoid compound, which gives the stigmas their characteristic golden yellow colour. It also contains other carotenoids including zeaxanthin, lycopene, α- and β-carotenes. These are important antioxidants that helps protect body from oxidant-induced stress, cancers, infections and acts as immune modulators.
- ☆ The active components in saffron have many therapeutic applications in many traditional medicines as antiseptic, anti-depressant, anti-oxidant, digestive, anti-convulsant.
- ☆ Is a good source of minerals like copper, potassium, calcium, manganese, iron, selenium, zinc and magnesium.
- ☆ It is also rich in many vital vitamins including vitamin A, folic acid, riboflavin, niacin, **vitamin-C** that are essential for optimum health.

Medicinal Uses

- ☆ The active components present in saffron have many therapeutic applications in many traditional medicines since long time ago as anti-spasmodic, carminative, diaphoretic.
- ☆ **Safranal**, a volatile oil found has antioxidant, cytotoxicity towards cancer cells, anticonvulsant and antidepressant properties.
- ☆ **Alfa-crocin**, a carotenoid compound, which gives the spice its characteristic golden yellow colour, has anti-oxidant, anti-depressant, and anti-cancer properties.

Caution

Saffron contains a poison that can damage the kidney and nerves and 10g can be a fatal dose for humans. It should therefore, be used only in small amounts. High doses of saffron can cat as uterine stimulant and in severe cases can cause miscarriage. Therefore, pregnant woman may be advised to avoid it in their dishes.

CARAWAY (*Carum carvi*)

This biennial herbaceous plant blooms once in every two years to produce creamy flowers in umbels. It grows to about 30cm in height bearing small feathery leaves. The seeds which appear like cumin are crescent shaped and dark brown with up to five vertical ribs. The seeds are harvested during early hours of the day to avoid spilling on the ground. The cut plants are then staked till they dry and then the seeds are threshed. Stimulate the appetite, relieve flatulence and improve digestion. It can also promote the onset of menstruation and alleviate utrine cramps. To prepare an infusion, use one teaspoon of crushed seeds boiled in half a cup of water.

Health Benefits of Caraway Seeds

- In addition to its use as medicinal values, caraway indeed has many health benefiting nutrients, minerals, vitamins and anti-oxidants.
- Seds are rich source of **dietary fibre**. They increase bulk of the food and helps prevent constipation by speeding up movement of food through the gut.
- Fibre also binds to toxins in the food and helps protect the colon mucus membrane from cancers. In addition, dietary fibres bind to bile salts (produced from cholesterol) and decrease their re-absorption in colon, thus help lower serum LDL cholesterol levels.
- Caraway has many essential oils and the principle volatile compounds are *carvone, limonene, carveol, pinen, cumuninic aldehyde, furfurol,* and *thujone.* These active principles in the caraway seeds are known to have antioxidant, digestive, carminative and anti-flatulent properties.
- It has many health benefiting flavonoid anti-oxidnats such as *lutein, carotene, crypto-xanthin* and zeaxanthin. These compounds act as an powerful anti-oxidants by removing harmful free radicals from the body thus protect from cancers, infection, aging and degenerative neurological diseases.
- Caraway spice is an excellent source of minerals like **iron, copper, calcium, potassium, manganese, selenium, zinc** and **magnesium**.
- The seeds indeed are storehouse for many vital vitamins. Vitamin A, vitamin E, vitamin C as well as many B-complex vitamins like thiamin, pyridoxine, riboflavin and niacin particularly are concentrated in the caraway seeds.

Medicinal Uuses

- Caraway water is sometimes used in treating flatulence and indigestion in traditional medicines, especially used to relive infantile colic.
- It is also used in pharmaceuticals as flavouring agent in mouth-wash and gargle preparations.
- Caraway extraction is used as rubefacient (to soothe muscle sores), clear the cold, as a remedy in bronchitis and ***irritable bowel syndrome*** in many traditional medicines.

BLACK PEPPER (*Piper nigrum* L.)

Popular black pepper often referred as "king of spice" is a popular spice known to the world since ancient times. The peppercorn plant is native to tropical evergreen rain forest of South Indian state, **Kerala,** from where it spread to rest of the world. The

Pepper fruit, also known as *peppercorn*, is actually a berry obtained from this plant. And the peppercorns are collected as whole berries or as ground black pepper. Peppers have been in use since ancient times for its anti-inflammatory, carminative, anti-flatulent properties. Improves digestion and promotes intestinal health, prevents gas formation, has diuretic(promotes urination) properties, has impressive antioxidant and antibacterial effect. It stimulates liver and increases salivation. Cures bronchitis and bronchial asthama. It is a pain reliever, anthelmintic, purifies urine and is beneficial in erectile dysfunction, amenorrhoea, skin disorders, fever and dermatoses (Acharya Balkrishna, 2008). It is also good for eyes. The important alkaloids that cause bitterness in black pepper are **piperin, piperidin, piperetine** and **chavicine.** Technically, the pepper fruit is a drupe, measuring about 5 mm in diameter, containing single large seed at its center.

Black pepper has more healthy properties than most people know about. Black pepper is a spice that has the potential to make food more beneficial when used in various recipes and as a table spice. In Kerala, many start their day with a cup of black coffee with a pinch of black pepper powder. The aroma of freshly ground black pepper is difficult to ignore.

Research studies have shown that black pepper has quite a handful of healthy properties.

It has the ability to enhance the function of the digestive tract. Black pepper is a great way to combat this. More than just a spice, it has been shown to improve digestion and stimulate the secretion from the taste buds. This taste bud stimulation tells the stomach that it's supposed to increase its own digestive juices, namely hydrochloric acid. This, in turn breaks down protein in the stomach, improving the process of digestion. Black pepper is known to have a great amount of antioxidant properties. It also has benefits against bacterial growth, particularly in the intestinal tract.

The research pinpoints piperine – the pungent-tasting substance that gives black pepper its characteristic taste, concluding that piperine also can block the formation of new fat cells. Soo-Jong Um, Ji-Cheon Jeong and colleagues describe previous studies indicating that piperine reduces fat levels in the blood stream and has other beneficial health effects.

Black pepper and the black pepper plant, they noted, have been used for centuries in traditional Eastern medicine to treat gastrointestinal distress, pain, inflammation and other disorders.

Their laboratory studies and computer models found that piperine interferes with the activity of genes that control the formation of new fat cells. In doing so, piperine may also set off a metabolic chain reaction that helps keep fat in check in other ways. The group suggests that the finding may lead to wider use of piperine or black-pepper extracts in fighting obesity and related diseases.

So, just add freshly ground black pepper powder to your dish and see the difference.

Health Benefits

- ☆ The chemical compounds present are known to have disease preventing and health promoting properties.
- ☆ Peppercorns contain many health benefiting essential oils such as **piperine,** an amine alkaloid, which gives strong spicy pungent character to the pepper. It also contains numerous monoterpenes hydrocarbons such as *sabinene, pinene, terpenene, limonene, mercene* etc that gives aromatic property to the pepper.
- ☆ The above-mentioned active principles in the pepper may increase the motility of the gastro-intestinal tract as well as increase the digestion power by increasing gastro-intestinal enzyme secretions. It has also been found that piperine can increase absorption of selenium, B-complex vitamins, beta-carotene, as well as other nutrients in the food.
- ☆ Black peppercorns contain good amount of minerals like potassium, calcium, zinc, manganese, iron, and magnesium.
- ☆ They are also excellent source of many vital B-complex groups of vitamins such as Pyridoxine, riboflavin, thiamin and niacin.
- ☆ Peppercorns are rich source of many anti-oxidant vitamins such as **vitamin-C** and **vitamin-A.**

Medicinal Uses

- ☆ Peppers have been used therapeutically in dentistry as an antiseptic for tooth-decay and gum swellings.
- ☆ Peppercorns are also being used in traditional medicines in treating flatulence and indigestion in traditional medicine.

AMARANTH (*Amaranthus* spp.)

Amaranth's great nutritional qualities are it's high protein, particularly in the amino acid, **Lysine,** the building blocks for the synthesis of protein, which is low in the cereal grains. In fact, Amaranth has the highest lysine content of all the grains with Quinoa coming in a close second. About 3.5 oz of amaranth seeds provide 15 per cent of the recommended daily allowance of calcium, 76 per cent of the iron, and over 25 per cent of the folic acid recommended in diets today. **Flour made from amaranth seeds would have been a critical food source in diets that lack protein rich foods. To make your whole wheat bread a complete protein, substitute about 25 per cent of your wheat flour with Amaranth flour. Amaranth, by itself, has a really nice amino acid blend**. Just 150 grams of the grain is all that's required to supply an adult with 100 per cent of the daily requirement of protein. Amaranth is one of the highest grains in fibre content. This makes Amaranth an effective agent against cancer and heart disease. Amaranth is also the only grain that contains significant amounts of phytosterols which scientists are just now learning play a major part in the prevention of all kinds of diseases. Amaranth is also rich in many vitamins and minerals. The crimson flowers have an astrigent property and infusion of these can be effective in the treatment of diarrhoea and dysentery. Amaranth has a high nutritional value because of the high levels of essential micronutrients like carotene, vitamin C, iron and calcium. It is especially rich in lysine, and essential amino acid that is lacking in diets based on cereals and tubers. The protein found in young plants can be important for people without access to meat or other sources of protein. Amaranth seed contains more protein than other grains such as wheat, maize, rice, sorghum or beans. It also contains high levels of minerals especially iron, phosphorous and magnesium more than what is found in animal products like milk and meat. It also has high levels of vitamin A, Bs, and E. The protein value of grain amaranth is highlighted when amaranth flour is mixed with other cereal grain flours on a ratio of 1:1 or 1:2. It is the flowers and leaves which are of value for medicinal purposes.

Amaranth Plant with Dark Green Leaves

The fat content in amaranth seed is high (7 – 8 per cent) double other cereals. An utmost important constituent squalene is present with 4 – 6 per cent, more than four fold concentration compared with olive oil. Grain amaranth is highly recommended for infants because of its protein digestibility, absorption and retention by the baby's body system. These seeds are one of the most nutritious foods grown. Not only are they richer in protein than the major cereals, but the amino acid balance of their protein comes closer to nutritional perfection for the human diet than that in normal cereal grains.

Amaranth has been found to be having medicinal values, which can reduce or combat common diseases such as diabetes, hypertension, liver disease, hemorrhage, TB, HIV/AIDS, wound healing, skin disease among others. Amaranth seeds and biomass are rich in soluble and insoluble diet fibres important in prevention of coronary heart diseases of the colon. The compounds in amaranth can enhance human growth and development, improve general health, and strengthen immune responses to combat diseases. The food is low in saturated fat and sugar but very low in cholesterol and sodium. It is also a good source of iron, manganese, magnesium and phosphorus.

Part Used and Uses

- **Roots, leaves, flowers, fruits and seeds**: Antidote for snake-bites.
- **Root**: Antidote for scorpion-stings; Muscle spasticity; Dermatitis.
- **Root decoction**: Haematemesis; Menorrhagia; Leucorrhoea; Heals boils and sores.
- **Leaf**: Dysuria; Gonorrhoea; Urolithiasis; Hepatitis; Haemorrhoids; Stops epistaxis; Antidote for spider toxin.

CUMIN (*Cuminum cyminum* L.)

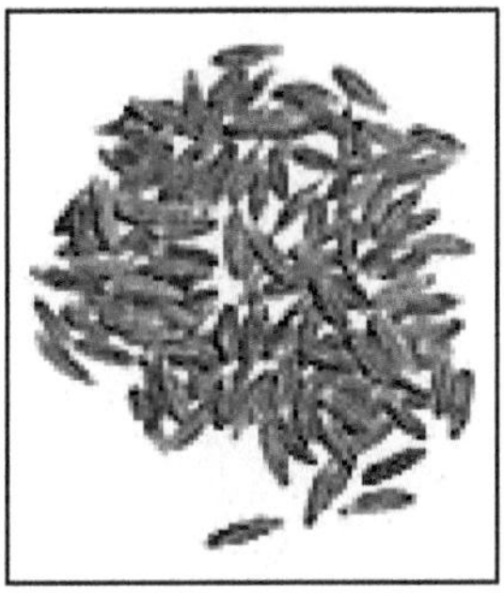

A portent herb with anti-oxidant characteristics, cumin seeds contain a compound called **Thymoquinone** that checks proliferation of cells responsible for prostate cancer. It aids digestion and probably that is why we like chewing a handfull of cumin seeds at the end of every meal. All the types of cumin seeds are dry, hot, carminative and pungent. They promote potency and digestive functions, cure pitta, promote intellect, purify uterus. They are anti-pyretic, digestive, enhance taste, cures cough and are beneficial for eyes. They promote longevity, relieve flatulence, cure vomiting and diarrhea (Acharya Balkrishna, 2008). The seeds and their essential oils are also used medicinally as an aid to stimulate gastric juices, increase appetite and relieve flatulence. Cumin is also said to increase milk secretion in nursing mothers. It bears small, gray-yellow coloured, oblong shaped seeds with vertical ridges on its outer surface and single centrally placed seed that closely resemble caraway seeds in appearance.

Distinctive flavour and strong, warm aroma of cumin's is due to its essential oils. The main constituent and important aroma compound is **cuminaldehyde** (4-isopropylbenzaldehyde).

Health Benefits of Cumin Seeds

- **Common cold**: Antiseptic properties of cumin can help fight flu, by boosting your immune system. A cup of water boiled with cumin seeds, ginger, basil leaves and honey, can give great relief.
- **Anemia**: Cumin seeds contain a good amount of iron. Iron is an essential element for the formation of hemoglobin in the blood required for transport of oxygen. Anemia [lowered levels of hemoglobin in the blood] has always been a concern in **women**, children and adolescents. It's a good idea to include jeera in everyday preparations like parathas, curries, cookies, soups, rice, lentil preparations.
- **Digestion**: It's a common practice in India to find people popping in Jeera Goli after meals, but did you know it works well for digestive disorders? Cumin helps control stomach pain, indigestion, diarrhoea, nausea and morning sickness.

- ☆ Jeera is also found to be effective in stimulating menstrual cycle in women. Cumin can be used in the treatment of piles due to its fibre content, anti-fungal, laxative and carminative properties. So, make it a point to add a dash of jeera powder in your diet (Deepika Sahu, 2012).
- ☆ Seeds contain numerous phyto-chemicals that are known to have antioxidant, carminative and anti-flatulent properties. The seeds are an excellent source of dietary fibre.
- ☆ Its seeds contain many health benefiting essential oils such as *cuminaldehyde* (4-isopropylbenzaldehyde), pyrazines, 2-methoxy-3-sec-butylpyrazine, 2-ethoxy-3-isopropylpyrazine, and 2-methoxy-3-methylpyrazine.
- ☆ The active principles in the cumin may increase the motility of the gastro-intestinal tract as well as increase the digestion power by increasing gastro-intestinal enzyme secretions.
- ☆ This spice is an excellent source of minerals like iron, copper, calcium, potassium, manganese, selenium, zinc and magnesium.
- ☆ It also contains very good amounts of B-complex vitamins such as thiamin, vitamin B-6, niacin, riboflavin, and other vital anti-oxidant vitamins like vitamin E, vitamin A and vitamin C.
- ☆ The seeds are also rich source of many flavonoid phenolic anti-oxidants such as carotenes, zeaxanthin, and lutein.

Medicinal Uses

Its seeds are used to prepare decoction, which is sometimes used in treating flatulence and indigestion in traditional medicine.

The seeds are used in traditional medicines to stave off common cold.

CINNAMON (*Cinnamomum zeylanicum*)

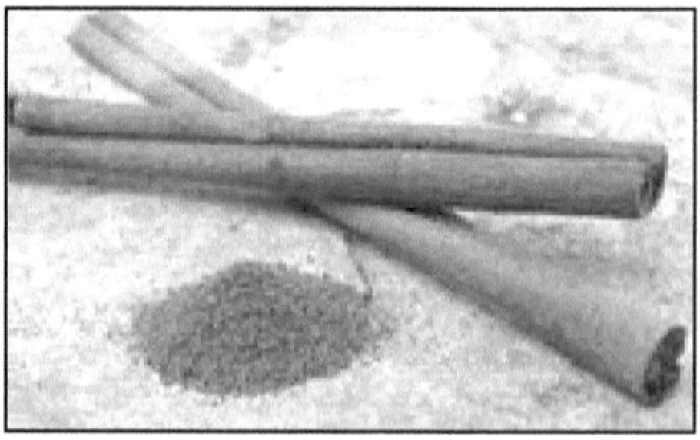

Cinnamona "Quills" with Powder

It is sedative. In folk medicine it has been used in various conditions such as insomnia, menstrual cramps, flatulence and nausea. Cinnamon stabilizes blood sugar levels. A teaspoon of cinnamon a day has recently been reported to prevent or delay the onset of non- insulin dependent diabetes which develops in older age and can also reduce cholesterol. It is a natural food preservative and a source of iron and calcium. It is useful in reducing tumour growth and blocks the formation of new vessels in the human body. It is very low in saturated fat, cholesterol and sodium, but is a good source of vitamin K, iron, dietary fibre, calcium and manganese. Cinnamon is used in winter season as it is warm in nature. It cures common cold, cough and congestion of lungs, digestion troubles, toothaches, bad breath and diarrhea. It is used in the treatment of Type 11 diabetes and insulin resistance and reduces the risk of colon cancer.

It is the brown bark of the cinnamon tree which is available in its dried tubular form known as a quill or as ground powder. Traditionally, the inner bark is bruised with a brass rod, peeled and long incision are made in the bark. Its bark is then rolled by hand and allowed to dry.

It is the bark of the tree from where aromatic essential oil (makes up 0.5 per cent to 1 per cent of its composition) is extracted. Usually, the oil is processed by roughly pounding the bark, macerating it in seawater, and then quickly distilling the whole. The oil features golden-yellow colour, with the characteristic odor of cinnamon and a very hot aromatic taste. Some of the effective ways of including cinnamon in your diet are.

Start your day with a cup of cinnamon tea (in leaf or sachet)

Make your breakfast meal a super healthy one; just add this wonder spice to your morning oatmeal and you are going well.

A fruity delight comparising chopped apples, a few walnuts and your magic potion cinnamon

Honey and cinnamon in your glass of milk before going to bed; no cancer nightmares assured.

Health Benefits

- ✰ The active principles in the cinnamon spice are known to have anti-oxidant, anti-diabetic, anti-septic, local anesthetic, anti-inflammatory, rubefacient (warming and soothing), carminative and anti-flatulent properties.
- ✰ Cinnamon has highest anti-oxidant strength of all the food sources in nature.
- ✰ The spice contains many health benefiting essential oils such as **eugenol**, a phenylpropanoids class of chemical compound, which gives pleasant, sweet aromatic fragrances. **Eugenol** has got local anesthetic and antiseptic properties, hence; useful in dental and gum treatment procedures.
- ✰ Other important essential oils in cinnamon include *ethyl cinnamate, linalool, cinnamaldehyde, beta-caryophyllene*, and *methyl chavicol.*
- ✰ *Cinnamaldehyde* in cinnamon-sticks has been found to have anti-clotting action, prevents clogging of platelets in the blood vessels, and thus helps prevent stroke and coronary artery disease.
- ✰ The active principles in this spice may increase the motility of the intestinal tract as well as help increase the digestion power by increasing gastro-intestinal secretions.
- ✰ This spicy bark is an excellent source of minerals like potassium, calcium, manganese, iron, zinc, agnesium and very good amounts of vitamin A, niacin, pantothenic acid, and pyridoxine.
- ✰ The spice is also very good source of flavonoid phenolic anti-oxidants such as *carotenes, zeaxanthin, lutein* and *cryptoxanthins*.
- ✰ A half teaspoon of cinnamon powder every day keeps cancer risk away.

- It may help to lower cholesterol and reduce inflammation as it contains a myriad of compounds that have anti-oxidantal properties.
- Circulatory stimulant effects of cinnamon have been reported in several books on medicinal plants and Ayurveda. It helps in reduction of total and bad cholesterol (LDL). It also helps improve insulin resistance, thereby making it useful in diabetes management.

Medicinal Uses

- The essential oil, eugenol, has been in therapeutic use in dentistry as a local anesthetic and antiseptic for teeth and gum.
- Eugenol also has been found to reduce blood sugar levels in diabetics, but further detailed studies required to establish its benefits.
- The extraction from the sticks (decoction) is sometimes used in treating flatulence and indigestion in traditional medicine.
- The spice is used in traditional medicines to stave off common cold and oxidant stress conditions.
- It is also used as natural food preservative.
- Along with the improvement in blood sugar, studies have shown improvement in triglycerides, blood pressure and LDL cholesterol.
- Uncooked cinnamon spice can cause choking and respiratory distress. Excessive use of cinnamon stick may cause inflammation of taste buds, gum swelling, and mouth ulcers. Large quantities can cause difficulty breathing, dilate blood vessels, and cause sleepiness, depression, or even convulsions.

CARDAMOM (*Amomum subulatum*)

This is a thermogenic herb that increases metabolism and help burn body fat. Cardamom is consisdered one of the best digestive aids and is believed to soothe the digestive system and help the body process other foods more efficiently. Cardamom is known as a carminative, relieving flatulence, stimulating the stomach and aiding digestion. However, it is mainly used as a cooking spice or for flavouring drinks and medicins. These green wonders increase energy and relieve fatigue and help you rock your love making process.

Both varieties of cardamom feature three sided pods with a thin papery outer cover and small black seeds that are arranged in vertical rows. Elettaria pods are small and light green in colour, while Amomum pods are larger and dark brown. The pods are being used as flavouring base in both food and drink, in cooking recipes and as well as in medicine. Black cardamom (Amomum subulatum) also known as *Nepal cardamom* is relatively big sized pod of same zingiberaceae family. The pod has dark brown rough outer coat, measure about 2-4 cm in length and 1-2 cm in diameter. The pods have camphor like intense flavour commonly used in spicy stews in sub Himalayan plains of India, Pakistan, Nepal as well as in China.

Health Benefits

- ☆ This exotic spice contains many plant derived chemical compounds that are known to have anti-oxidant, disease preventing and health promoting properties.
- ☆

  The spicy pods contain many **essential volatile oils** that include *pinene, sabinene, myrcene, phellandrene, limonene, 1, 8-cineole, terpinene, p-cymene, terpinolene, linalool, linalyl acetate, terpinen-4-oil, a-terpineol, a-terpineol acetate, citronellol, nerol, geraniol, methyl eugenol,* and *trans-nerolidol.*
- ☆ The therapeutic properties of cardamom-oil have found application in many traditional medicines as antiseptic, antispasmodic, carminative, digestive, diuretic, expectorant, stimulant, stomachic and tonic.
- ☆ Cardamom is a good source of minerals like potassium, calcium, and magnesium. Potassium in an important component of cell and body fluids that helps control heart rate and blood pressure. Copper is required in the production of red blood cells.
- ☆ It is also an excellent source of manganese and iron. Manganese is a co-factor for the enzyme superoxide dismutase, which is a very powerful free radical scavenger. Iron is required for red blood cell formation.
- ☆ The pods are rich in many vital vitamins including riboflavin, niacin, vitamin-C that are essential for optimum health.

Medicinal Use

The therapeutic properties of cardamom oil have found application in many traditional medicines as antiseptic and local anesthetic, antioxidant and; health promoting and disease preventing roles.

MUSHROOMS

Mushrooms not only taste good but also a great source of healthy food. Mushrooms of button variety are rich in potassium, phosphorus, copper and iron. They are also a good source of thiamine (vitamin B_1) and riboflavin (vitamin B_2). As mushrooms mature, their caps open and expose their gills. It is best to avoid wide open caps and also to select mushrooms that are firm, not spongy. They are known to be benefical in reducing blood fat levels and to have antibiotic properties; they are also claimed to have anti-tumour actions and to boost the immune system action against disease-producing micro-organisms by increasing white blood cell count. Mushrooms are easily digested and are recommended for any one suffering with digestive problems. Mushrooms are full of proteins, vitamins and minerals, amino acids, anti-biotic and anti-oxidants. Mushrooms have a significant amount of vitamin D. The amount varies according to its type or varieties. Shitake mushrooms are considered to be one of the best sources of vitamin D.

The health benefits of mushroom include the following:

- ☆ The power of mushrooms comes from their ability to enhance the activity of natural killer T cells (NKT).
- ☆ These NKT's attack and remove cells that are damaged or infected by a virus. Mushrooms are associated with decreasing most cancers and significantly reducing the risk of breast cancer in women. They prevent DNA damage, slow cancer or tumor growth, and prevent tumors from acquiring a blood supply.
- ☆ B Vitamins are vital for turning food (carbohydrates) into fuel (glucose), which the body burns to produce energy. They also help the body metabolize fats and protein. Mushrooms contain loads of vitamin B_2 and vitamin B_3.
- ☆ Mushrooms have zero cholesterol, fats and very low carbohydrates. The fibre and certain enzymes in them also help lower **cholesterol** level. The high lean protein content in mushrooms helps burn cholesterol when they are digested.
- ☆ Mushrooms can be an ideal low energy **diet** for diabetics. They have no fats, no cholesterol, very low carbohydrates, high proteins, vitamins and minerals, a lot of water and fibre. Moreover, they contain natural insulin and enzymes which help breaking down of sugar or starch of the food.
- ☆ **Ergothioneine,** a powerful anti-oxidant present in mushrooms is very effective in giving protection from free radicals as well as boosting up immunity. Mushrooms contain natural antibiotics (similar to penicillin, which itself is extracted from mushrooms) which check microbial and other fungal infections.
- ☆ Mushrooms are the only vegetable and the second known source (after cod liver oil) to contain vitamin-D in edible form. They are rich in calcium (good for bones), iron (cures anemia), potassium (good for lowering blood pressure) and selenium. The best source of selenium is animal proteins. So, mushrooms can be the best choice for vegetarians.
- ☆ Mushrooms stimulate the reproduction and activity of immune cells. In Japan they are often used to complement the chemotherapy. inorder to support the immune system.

SOYBEAN (*Vicia faba*)

Fava or broad beans are large, flattened light green pods usually eaten shelled for their delicious beans. Lima beans (Phaseolus lunatus) are large, plump, pale green pods with kidney shaped seeds. Soybeans have been found to be a storehouse of phytochemicals – beneficial antioxidant nutrients such as saponins, phytosterols and phenolic acids - which protect the body from free radical damage, reducing the risk of the degenerative diseases of ageing, such as heart attack, cancer and strokes. Two of the best anticarcinogens in soy are some plant oestrogens, which prevent breast cancer and **genistein** which inhibits the growth of cancer cells. In fact, soya beans have a mild oestrogenic activity due to their **isoflavones,** also called phytoestrogens. Eating soya products can therefore be beneficial for women in HRT and can help offset some post menopausal symptoms. Soybeans have also been found to be beneficial in reducing cholesterol levels and preventing heart attack. A new Candian University study found that three glasses of soya milk and one soya dessert consumed daily lowered the **'bad' LDL colesterol** by 11 per cent and increased the **'good' HDL cholesterol** by 9 per cent in 70 per cent the subjects (Michael Sharon, 2009).

YAM BEAN

It is crisp with a slight sweetness like water chestnuts. There are many nutritional properties of merit in *Pachyrhizus tuberosus;* for example vitamin A, vitamin B complex, vitamin- C, phosphorus can be found. It is most frequently used cooked or fresh. Certain types of soy protein inhibits fat accumulation and reduces inflammation. Soybeans rich in **beta-conglycinins** limit lipid accumulation in fat cells by inhibiting an enzyme called fatty acid synthesis. A specific peptides (digested proteins) to do this mechanism has been identified. This exciting research could lead to the development of nutraceuticals to fight obesity (de Mejia-Food Science and Human Nutrition-AN, Tribune,7th, April.2010)

NUTMEG

It is one of the most popular aphrodisiaces. Research proves that nutmeg has the same effect on mating behavour as vigra.

CASSIA

Its leaves or sap are used to treat fungal infection such as ringworm. They contain a fungicide, **chrysophanic acid.** Because of its antifungal properties, it is a common ingredient in soaps, shampoo's and lotions in Philippines. The effectiveness of this plant against skin diseases, it is also used to treat a wide range of ailments from stomach problems, fever, asthama to snake bite and venereal diseases (syphilis, gonorrhoea). Its leaves contain a chemical called **adenine** which has been documented as an effective platelet aggregating inhibitor (reduces sticky blood and arterial plaque).

SWEDE

Its fibre content stimulates regular bowel movements, while its high levels of potassium provide it with natural diuretic properties. In addition, it is very low in calories.

POINTED GOURD

The fruit is particularly recommended during convalescence. It is easily digested and is diuretic, laxative and cardiatonic. It is effective against bronchitis, high fever and nervousness.

WINTER SQUASH

It provides an outstanding source of alpha and beta-carotenes (provitamin A) and contains good amounts of calcium, phosphorus and potassium. Squash can benefit the stomach, reduce inflammations and improve circulation, and squash seeds are reputed to be effective in destroying worms. Some healers recommend that a handful of seeds should be eaten daily for three weeks in order to eliminate parasitic worms.

SEAWEEDS

Seaweeds or marine algae, is a general descriptive name for sea vegetable. They are extremely rich in **iodine** and in other minerals, such as calcium, iron, and fluorine and has been known for centuries for their ability to promote health. They are especially beneficial to thyroid function, and can also lower cholesterol, alkalize the blood, remove radioactivity residues, help with weight loss and are important in bone mineralization and density. Seaweeds are used to treat conditions such as goiter, water retension (oedema) and swollen lymph glands. They come in several colours- brown, red, green, blue-green and yellow-green- each of which has its own individual properties in addition to the properties to all.

RUE (*Ruta graveolens*)

The aromatic leaves are high in **rutin,** the bioflavonoid that strengthens capillaries and blood vessels. An infusion of the leaves is diuretic, slightly increases blood pressure and has an abortifacient effect (induces abortion). In Yemenite folk medicine, rue is used to treat nervour breakdowns and to stimulate the onset of deylayed menstruation.

Caution

Rue must not be taken during pregnancy.

BATHUA (*Chenopodium album*)

It is also known as allgood or goose foot. It is mainly eaten in the form of saag and in raita. It is rich in vitamin A, calcium, phosphorus and potassium.

The leaves and stems are edible and absolutely delicious, with an earthy, mineral- can be compared rich flavour that to spinach. This herb is one of the most nutritious wild foods you can eat. According to Ayurveda, the whole herb is generally used as medicine. It improves appetite. It is oleaginous, anthelmintic, laxative (at over dose), diuretic, aphordisiac and tonic. It is useful in treatment of biliousness, abdominal pains, eye diseases, throat infections, piles, blood disorders and troubles of heart and spleen. The leaves are antiscorbutic and yield ascaridole, which is used to treat round and hookworms. The leaves are sometimes dried and used for relieving some diseases. The powered leaves are used externally as an antiseptic and the leaf juice is used to relieve burns. A decoction of the leaves and stems, mixed with alcohol is used as a rub to relieve arthritis and rheumatism. The seeds of the herb are used to treat hepatic disorders and spleen enlargement. It contains eight essential amino acids, vitamins, minerals and fibre. It gives instant energy and keeps you active. It's easy to cook and can be cooked like rice in only 10 minutes.

Health Benefits

- ☆ **For kidney stone**: it is very effective and useful in problem of kidney stone. Take tender leaves and branches of leaves and grind them to extract its juice and take 10-15 gm of it daily with or without water. This also reduces the tendency of stone formation.
- ☆ **For inter swelling:** When bathua taken internally it helps to reduce internal swelling.
- ☆ **For external swelling:** For external swelling cook leaves of bathua in steam and apply at place of swelling this help to reduce swelling.
- ☆ **For jaundice:** Take bathua and giloy ras and mix them in equal proportion and take 25- 30 gm of it daily twice a day.
- ☆ **For irregular period:**Take bathua seeds and sonth (dry ginger powder) make powder 15-20n gm boil in 400 gm of water when it reduces to 100 gm drain it and drink twice a day.
- ☆ **For curing infections after delivery:** Take 10 gm of bathua, ajwain, methi and jaggery add dashmol and take for 10-15 days.
- ☆ **For anemia:** Take 25-50 gm of bathua ras (juice extract from leaves after crushing) daily.
- ☆ **For urinary tract infections**: Take 10 gm of bathua leaves ras add 50 ml of water and take with misri (candy sugar) take regularly.
- ☆ **For Blood purification:** When bathua ras taken with 4-5 neem leaves ras it works as blood purifier.

Bathua finds a use in the treatment of some skin conditions, and the oil made from these leaves is used to treat hook worms.

It is rich in vitamin A, B-complex vitamins, vitamin C, calcium, potassium, phosphorus, trace minerals, iron and fibre

Caution

Please also note that you should eat bathua in moderation as consuming it more may result in diarrhea. In Ayurveda, it is said that pregnant women should not eat Bathua as it may result miscarriage.

Dark Green Leafy Vegetables

It is not possible to meet your nutritional needs without having leafy vegetables in your diet. They are available in abundance and are huge source of many vitamins and minerals your body needs to stay healthy, such as vitamins A, C and K, folate, iron and calcium. They are also great sources of fibre. They are also a vision protector and provides four essential minerals, *i.e.* calcium, magnesium, iron and potassium. Therefore, try to have them daily in your diet and darker the better. Research suggests that the nutrients found in dark green vegetables may prevent certain types of cancers and promote heart health. Green, especially leafy vegetables are to be eaten in the raw form, if you want to get the best of their nutritional properties. It is recommended that teenage girls eat 3 cups of dark green vegetables per week or about ½ a cup every day. Dark green vegetables are also high in fat -soluble vitamins such as vitamins A, K, D and E. These vitamins require a little bit of dietary fat in order for the body to absorb them. Therefore, when you eat dark green vegetables, make sure to add a teaspoon of dietary fat, such as butter, olive or canola oil, cheese or salad dressing to make sure your body absorbs all of the vitamins you eat. Examples of dark green leafy vegetables:

- ✰ **Arugula** has a peppery taste and is rich in vitamins A, C and calcium. It can be eaten raw in salads or added to stir-fry, soups and pasta sauces.
- ✰ **Broccoli** has both soft florets and crunchy stalks and is rich in vitamins A, C, K, folate and fibre. Broccoli can be eaten raw or steamed, sautéed or added to a casserole.
- ✰ **Collard Greens** have a mild flavour and are rich in vitamins A, C, K, folate, fibre and calcium. The best way to prepare them is to boil them briefly and then add to a soup or stir-fry. Collard greens can also be eaten as a side dish. Just add your favourite seasoning and enjoy.
- ✰ **Dandelion Greens** have a bitter, tangy flavour and are rich in vitamin A and calcium. They are best when steamed or eaten raw in salad.
- ✰ **Kale** has a slightly bitter, cabbage-like flavour and is rich in vitamins A, C and K. It is tasty when added to soups, stir- fries and sauces.
- ✰ **Mustard Greens** have a peppery or spicy flavour and are rich in vitamin A, C, K, folate and calcium. They are delicious when eaten raw in salads or in stir-fries and soups.

Popular Leafy Vegetables or Pot herbs (vegetables eaten as the cooked leaves)

Agati	Chard	Kale	Spinach beet
Amaranth	Colocasia	Kohlrabi greens	Nasturtium
Artichoke	Dandelion	Leek	Turnip green
Arugula	Endive	Lettuce	Water cress
Beet greens	Escarole	Mustard greens	Water spinach
Brussels sprout tops	Fenugreek	PurslaneRomaine	Yarrow
Collard greens	Garden cress	Sorrel	
Cabbage	Iceberg lettuce	Spinach	

- ☆ **Romaine Lettuce:** Lettuce contains vitamin B6, C, E, folic acid and minerals like manganese and chromium. Romaine lettuce which is more dark green in colour in comparison to the Iceberg variety is said to be more nutritious. It a nutrient rich that is high in vitamins A, C, K and folate. If your hair is not having an optimum growth, it is good idea to consume lettuce as a regular part of your diet. It is best when eaten raw in salads, sand witches or wraps.
- ☆ **Spinach** has a sweet flavour and is rich in vitamins A and K, folate and iron. Spinach tastes great when eaten raw in salads or steamed.
- ☆ **Swiss Chard** tastes similar to spinach and is rich in vitamins A, C, K, potassium and iron. It is best stir-fried or eaten raw in salads.
- ☆ **Celery** contains vitamin A and B, calcium, sodium, and iron. It is said to have negative calories and digesting them burns more calories. They are good for people who want to lose weight.

Health Benefits

Dark green leafy vegetables are, calorie for calorie, perhaps the most concentrated source of nutrition of any food. They are a rich source of minerals (including iron, calcium, potassium and magnesium) and vitamins, including vitamins K, C, E and many of the B vitamins. They also provide a variety of phyto nutrients including beta-carotene, lutein and zeaxanthin, which protect our cells from damage and our eyes from age-related problems, among many other effects. Dark green leaves even contain small amounts of omega -3 fats. Dark greens *viz.* spinach is a good source of iron a nutrient that may help protect against the sleep robber known as restless legs syndrome.

Perhaps the star these nutrients is vitamin K. A cup of most cooked greens provide at least nine times the minimum recommended intake of vitamin K and even a couple of cups of dark salad greens usually provide the minimum all on their own. Recent research has provided evidence that this vitamin may be even more important than we once thought (the current minimum may not be optional) and many people do not get enough of it.

Vitamin K

- ☆ Regulates blood clotting.
- ☆ Vitamin K is a fat –soluble vitamin, so make sure to put dressing on your salad, or cook your greens with oil.
- ☆ Helps protect bones from osteoporosis
- ☆ May help prevent and possibly even reduce atherosclerosis by reducing calcium in arterial plaques.
- ☆ May be key regulator of inflammation and may help protect us from inflammatory diseases including arthritis
- ☆ May help prevent diabetes.

☆ Vitamin K is a fat –soluble vitamin, so make sure to put dressing on your salad, or cook your greens with oil.

Greens have very little carbohydrates in them, and the carbohydrates that are there are packed in layers of fibre, which make them very slow to digest. That is why, in general, greens have very little impact on blood glucose. In some systems greens are even treated as a " freebie" carb-wise (meaning the carbohydrate does not have to be counted at all). However, some greens contain substances called oxalates which may bind some percentage of the calcium in the greens.

Benefits	*Vegetarian Diet*	*Non vegetarian Diet*
Detoxifies	Contains fibre *e. g.* bottle gourd, pumpkin, spinach, cabbage, which flushes toxins out of the body.	*e. g.* eggs, fish and mutton is a poor source of fibre.
Stronger Bones	Calcium extraction is rare amongst vegetarians.	Gorging on meat can lead to protein over load. This can tax our kidneys, interfere with the absorption of calcium and prompt the body to extract existing calcium from the bones
Carb deficiency	Abundant in root and tuber vegetables	A non-vegetarian diet is a poor source of carbohydrates, which can lead to ketosis – a condition where the body starts breaking down fat (instead of cabs) as a source of energy.
Easy digestion	Complex carbohydrates in vegetarian food are digested gradually providing a steady source of glucose.	Where as meats rich in fat and protein are difficult to digest.
Healthy skin	Eating beet root, tomato, pumpkin and bitter gourd can clear of blemishes.	Not possible
Weight management	Eating whole grains, legumes, vegetables, nuts and fruits lowers cholesterol levels, blood pressure and obesity.	Avoiding meat is the simplest way to reduce fat intake.
Easy on the teeth	Our molars are more suited for grinding grains and vegetables than tearing flesh.	Digestion begins with the sliva, which can only digest complex carhohydrates present in plant foods.
Phyto-nutrients	Diabetes, cancer, kidney disease, stroke and bone loss are partially preventable with a good intake of phytonutrients.	As these are present only in vegetarian diet, the non vegetarians are at a loss.

Bibliography

Acharya Balkrishna.(2008). Secrets of Indian Herbs, for good health. Divya Prakashan Patanjali Yogpeeth.Maharishi Dayanand Gram, Delhi- Hardwar Highway, Bahadarabad, Hardwar-249408, Uttarkhand, India.

Anonymous (2009). Termeric can sooth bowls. Tribune, April 22nd 2009, p13.

Anonymous (2009). Vegetable juice may help people lose weight. Tribune, 22nd April.p.12.

Anonymous (2009). Health & Science.Garlic can treat cancer.Times of India. 9th April, 2009.

Anonymous (2009). Health Notes Proteins in garden peas may help fight high blood pressure, urinary diseases.Tribune, Wednes day, 03. 25th 2009 p.11.

Anonymous (2009). Health Notes. Broccoli, cabbage, can help fight skin cancer, Tribune.04.03.09, p.12.

Anonymous (20011). Eating veggies and fruits cuts the risk of certain blood cancers. Times of India, Sep.28th. 20011, p18.

Anonymous (20011). Pumpkin works magic on your skin. Times of India, IndiaTimes.com life Style / beauty/ pumpkin.

Aonymous (2012). Why men should eat berries.Times of India. 18th April 2012.

Anonymous (2012). Sip a cherry juice to sleep betterTimes of India.10.05.2012.

Anonymous (2012). Viagra, move over, pomerganate's here. Daily Mirror.27th August, 2012.

Anonymous(2012). Eating asparagus may prevent hangover(study). Times of India, 27th Dec. 2012.

Anonymous (2013). Eat Cauliflower to ward off cancer.Times of India: Health and Fitness. March, 28th 2013.

Anonymous (2013). Foods that don't let you slim down.Daily Mirrow. Apr 22, 2013.

Anonymous (2013). Eating tree nuts can boost health and cut heart disease risk. Health Notes.The Tribune, April, 24th, 2013.

Anonymous (2013). 7 reasons you Must eat olives.Times of India.Health & Fitness. May 2, 2013.

Anonymous (2013). 5 Fruits for a glowing skin.Times of India.Health & Fitness.May 16, 2013.

Anonymous (2013). Figs are extremely nutritious.Times of India.Health & Fitness. 28th, May 2013.

Anonymous (2013). Garlic: Wonder drug for healthy life.Times of India. Diet.Jun 3rd, 2013.

Anonymous (2013). Want to live longer? Turn vegan.Times of India. Diet. June 8th.2013.

Anonymous (2013). Rosemary for a glowing, soft skin.Times of India.Beauty.Jul 11, 2013.

Anonymous (2013). Want to live longer? Eat Walnuts.Times of India. Diet. Jul 20, 2013.

Anonymous (2013). Kiwi fruit more nutritious than 27 other fruits. Times of India. Health & Fitness. 6th August 2013.

Anonymous (2013). Fruits, Vegetables may lower women's bladder cancer risk. Times of India. IANS 24th August 2013.

Anonymous (2013). Eating broccoli can help prevent osteoarthritis. Times of India. ANI Science, Aug 28th, 2013.

Anonymous (2013). Must-eat foods that improve your mood. Times of India. Health Me Up. 29TH August.2013.

Anonymous (2013). Walnuts may prevent diabetes and heart disease. Tribune Spectrum, Health Capsules.6th October, 2013.

Anonymous (2013). Pomegranates help burn fat, increase blood flow.Times of India. Health & Fitness.12th October 2o13.

Anand Holla (2012). Spers are important, care them. Mumbai Mirror.1 st. August 2012.

Charanjit Parmar (2012). Fruit facts; Fully ripe Bananas have anti cancer properties. Tribune, 25th March.p.6.

Cummings, J.H; Beatty, E.R; Kingman, S.M, Bingham, S.A; Englyst.(1996). Digestion and Physiological properties of resistant starch in human large bowls.Br. J. Nutr. 75: 733-747.

Deepika Sahu (2012). Health benefits of Cumin.Times of India. Dec.30th 2012.

Duggal, R.K (1995). Native spices of alien plant.Daily Excelsior. 19th Nov. 1995, col 3.p.6.

Englyst, H.N; Kingman, S.N;Cummings, J.H.(1992). Classification and measurement of nutritionally importan starch fractions. Eur. J.Lin. Nutr. 46: 533-550.

Gita Angara.(2007). Fighting Cancer? Eat some cabbage. V. Care foundation, 183, Golden Jubilee Block, Tata Memorial Hospital, Parel, Mumbai.4300012, vol.14. No.2 June 2007 p.6.

Gopalkrishnan, T.R (2007). Vegetable Crops. Horticultural Science. Series-4. Publishing Agency. Pitampora, New Dehli.CU Block; L.S.C Market p.66-75, 308.

Ipshita Mitra (2012) Eight Indian Spices that prevent cancer.Times of India. July 14th 2012.

Ipshita Mitra (2013). Top 10 foods that prevent breast cancer. Times of India. May 11th, 2013.

Ishi Khosla (2013). Good Health. The Sunday Tribune Spectrum, (Other Seeds) Chandigarh. 8th September 2013.

Howard Michael (1987). Traditional Folk Remedies (Remedies, 1987) p.120.

Johnson, E.J(2002). The role of carotenoids in human health. Nutr. Chin. Care. 5: 56-65 or lutein content in Sweet Potato leaves. Hort. Sci.41(5):1269-1271.

Kounteya Sinha (2008). Garlic effective in treating arsenic poisoning.Studies.The Times of India. New Delhi. 28th January. P.7.

Levy, M.T (1966). The medicinal formulary or Agradbin of Al- Kindi Madison Wi: Uni. of Wiscosin Press: 410.

Maneka Gandhi. (2005). Meat may develop Alzheimer's disease. Daily Excelsior, May, 15th p. 3 Sunday Magazine.

Manu Vipin (2012). Eat lychee to fight breast cancer. Times of India, August 4th, 2012.

Methew Meghana (2008). Broccoli may slow down effects of ageing says the study. Good Health. Times of India, April 1sth 2008, p.28.

Michael Sharon.(2009). Nutrients A-Z. A user's guid to foods, herbs, vitamins, minerals and supplements. Carlton Book Ltd. 20 Mortimer Street London. W1 T 3.

Meghna Mukherjee.(2013). Beauty benefits of Walnut oil.Times of India. 10thJan.2013.

Moghul Sobiya. N. (2013). 20 best muscle building foods.Times of India. Health Me 6th, Jun, 2013.

Moghul Sobiya. N. (2013). Why you must eat eggplant.Times of India. Health Me Up 11th June.2013.

Nandkarni, K.N (1954). Allium cepa linn and Allium sativum linn in the Indian Materia Media, 3rd ed (part 1) Puranik, M.V.Popular Book Depo. Mumba.

Parmar, Chiranjit.(2012). Two Bananas give 90minutes' body fuel. Fruit Facts. The Sanday Tribune. Spectrum, 2012.

Parmar, Chiranjit.(2013). Litchi is a super fruit. Fruit Facts.Tribune, July, 14th 2013.

Purvaja Sawant (2013). 3 Best foods for your hair.Times of India. Health & Fitness.May 13th, 2013.

Purvaja Sawant, (2013). Beauty benefits of papaya. Times of India.Health Benefits, Oct 11, 2013.

Renita Tisha Pinto (2013). Top 20 fruits for diabetics.Times of India. Health Me Up, 18.03.2012.

Reuters, K.N (2012) Health benefits of curry leaf. Times of India.1st. August, 2012.

Roy, S.K and Chakraborti, A.K(1993). Vegetable's of Temperate climate. Commercial and Dietary importance in Encyclopedia of Food Science.Food Technology and Nutr. Academic Press London. p.4743-4743.

Sinha, Kounteya (2008). Beetroot juice helps lower blood pressure, says the study. Times of India New Delhi, Feb.8th. 2008, p.12.

Sinah, Kunteya.(2013). Good news for 13 million Indian asthmatics. Ginger abundantly found, shown to ease asthma attack.Times of India. Science.May 20th, 2013.

Sobiya N. Moghul.(2013), Cancer: 12 foods to battle cancer.Health Me Up. Times of India, Apr 14, 2013,.

Sobiya N. Moghul.(2013). 10 Healthy reasons to eat apples.Times of India. Health Me Up. Jun 29th, 2013.

Sumitra Nair (2013). The right way to eat fruits.Times of India. Health & Fitness.10th, Jun 2013.

Sunderagan, P. (2005). Turmeric- a potential weapon against Alzheimer's.The Hindu. 25th April, 2005.

Takuya Sakurai.(2008). Biotechnol Biochemic.(Kyorin University, Japan.72(2), 463-476.

Trina Remedios.(2013). About fertility:Foods for fertility.Times of India. Jan.9th.2013.

Trina Remedios (2013). Are you getting enough fruits ?Times of India. Health me up. 2nd October 2012.

Trina Remedios (2013). Health benefits of pineapple.Times of India. Health & Fitness. Health Me Up. Oct 17, 2013,.

Vatsyayan, R (2010). Use Onion, be healthy.The Tribune. December, 29th 2010, p12.

Vatsyayan, R(2011). Ginger: The universal Medicine. The Tribune, Wednesday, August 3rd, p.14.

Willcox, J.K; Catignani, R.L and Lazarus, S(2003). Tomatoes and Cardio-vascular health. Crit. Rev. Food Science. Nutr. 43: 1-18.

Zeenia Baira (2013). Health benefits of pears.Times of India. Health & Fitness. 8th September. 2013.

Internet

http://timesofindia.indiatimes.com/life-style/beauty/Beauty-benefits-of-walnut oil/articleshow/17887375.cms

http://www.times of india.indiatimes.com/life-style/health-fitnes/diet/benifits-of coriander/articleshow/5912718.cm

http://www.prokerala-com/health/medicininal-herbs/annonasquamosa-sugarapple.php.

http://www.times of india.indiantimes.com/life-style/health-fitness/diet/figs-are-extremely-nutritious/articleshow/7626531.cms

http://timesofindia.indiatimes.com/life-style/health-fitness/health/Health-benefits-of- pomegranate/articleshow/176645

http://timesofindia.indiatimes.com/life-style/health-fitness/diet/About-fertility-Foods-for-fertility/articleshow/17937114.

http://www.nutrition-and-yuy.com/fruit-nutrition.html

http://www.nutrition-and-yuy.com/vegetable-nutrition.html

http://www.nutrition-and-yuy.com/nuts - nutrition.html

http://www.nutrition-and-yuy.com/healthy-herbs.html

http://www.nutrition-and-yuy.com/apple-fruit.html

http://www.nutrition-and-yuy.com/apricot.html

http://www.nutrition-and-yuy.com/banana-fruit.html

http://www.nutrition-and-yuy.com/cherry-fruit.html

http://timesofindia.indiatimes.com/life-style/health-fitness

http://www.nutrition-and-yuy.com/coconut.html

http://www.nutrition-and-yuy.com/dates.html

http://www.livestrong.com/article/87197-nutritional-value-dragon-fruit

http://www.drgranny.com/2011/02/18/health-benefits-of-dragon-fruit/

http://www.nutrition-and-yuy.com/fig-fruit.html

http://www.nutrition-and-yuy.com/guava.html

http://www.nutrition-and-yuy.com/orange-fruit.html

http://www.nutrition-and-yuy.com/grapefruit-html

http://www.nutrition-and-yuy.com/lemon.html

http://www.nutrition-and-yuy.com/papaya-fruit.html

http://www.nutrition-and-yuy.com/grapes.html

http://www.nutrition-and-yuy.com/pears.html

http://www.nutrition-and-yuy.com/peaches.html

http://www.nutrition-and-yuy.com/plums.html

http://www.nutrition-and-yuy.com/lychee.html

http://www.nutrition-and-yuy.com/mango-fruit.html

http://timesofindia.indiatimes.com/life/health-fitness

http://times of india.indiatimes.com/life-style/health-fitness

http://www.nutrition-and-yuy.com/loquat-fruit.html
http://www.nutrition-and-yuy.com/pomegranate.html
http://www.nutrition-and-yuy.com/pineapple.html
http://www.nutrition-and-yuy.com/jackfruit.html
http://www.nutrition-and-yuy.com/strawberries.html
http://www.nutrition-and-yuy.com/raspberry.html
http://www.nutrition-and-yuy.com/gooseberries.html
http://www.nutrition-and-yuy.com/almonds.html
http://timesofindia.indiatimes.com/life-style/health-fitness
http://www.nutrition-and-yuy.com/walnut.html
http://www.nutrition-and-yuy.com/avacados-fruit.html
http://www.nutrition-and-yuy.com/blackberries.html
http://www.nutrition-and-yuy.com/cranberries.html
http://www.nutrition-and-yuy.com/kiwi-fruit.html
http://timesofindia.indiatimes.com/life-style/health-fitness
http://www.liveandfeel.com/nutrition/fruit-vegetable-role.html

Index

L

M

N

O

P

Q

R

S

www.ingramcontent.com/pod-product-compliance
Ingram Content Group UK Ltd.
Pitfield, Milton Keynes, MK11 3LW, UK
UKHW041643031026
15587UKWH00001B/460

9 789351 301660